Springer Undergraduate Mathematics Series

T0222441

For further volumes:
http://www.springer.com/series/3423

R. H. Dyer · D. E. Edmunds

From Real to Complex Analysis

 Springer

R. H. Dyer
Department of Mathematics
University of Sussex
Brighton
UK

D. E. Edmunds
Department of Mathematics
University of Sussex
Brighton
UK

ISSN 1615-2085   ISSN 2197-4144 (electronic)
ISBN 978-3-319-06208-2 ISBN 978-3-319-06209-9 (eBook)
DOI 10.1007/978-3-319-06209-9
Springer Cham Heidelberg New York Dordrecht London

Library of Congress Control Number: 2014936238

Mathematics Subject Classification: 26A42, 54E35, 30-01

Printed on acid-free paper

Springer is part of Springer Science+Business Media (www.springer.com)

Preface

This book evolved from a series of lectures at the University of Sussex and is designed to provide an integrated course in real and complex analysis for undergraduates who have taken first steps in real analysis; the intention is to exhibit something of the interplay between these and other areas of mathematical study. The prerequisites are modest: it would be completely sufficient to have followed preliminary courses in real analysis (involving ε, δ ideas) and algebraic structures. There are many exercises, ranging from the elementary to the quite demanding. To establish notation and terminology, some prerequisites are reviewed in the appendices.

A persistent theme in the text is the search for a primitive. In the case of real analysis, the Riemann integral offers one route in this quest and, with an eye to complex analysis, the improper Riemann integral is an extension consonant with the demands of contour integrals.

Chapter 1 deals with the Riemann theory of integration on the real line using the simple and elegant approach due to Darboux that quickly leads to the basic properties of the integral together with means of evaluation and estimation. It also enables direct, elementary proofs to be given of the results that if f is Riemann-integrable, then (i) the set of its points of continuity is dense in the domain of f, and (ii) $g \circ f$ is Riemann-integrable if g is continuous. A characterization of the class of Riemann-integrable functions, from which these last two assertions follow, is postponed to the next chapter as it is technically more challenging. The Riemann integral is confined to bounded functions defined on closed bounded intervals and requires extension to cope with the demands of later chapters. To allow for some relaxation of these constraints, the improper Riemann integral is introduced. We indicate the limitations of the Riemann integral which led to the development of Lebesgue's integral (which itself would require slight extension for use in the later chapters), of which the former is a special case.

Metric spaces form the theme of Chap. 2; the earlier one provides a wealth of examples of such objects. Detailed coverage is given of the core properties of completeness, compactness, connectedness and simple connectedness: this last property is highlighted. While it has become more common in recent times to present such matters in the context of normed linear spaces, we believe it is important for the student to realize that linear structure is irrelevant to many of the results. Regarding completeness, Cantor's characterization is established as are

Banach's contraction mapping theorem and the Baire category theorem, the last leading to a proof of existence of a continuous, nowhere-differentiable function and also to the fact that the pointwise limit of a sequence of continuous real-valued functions on a complete metric space is continuous on a dense subset of that space. Compactness and connectedness are motivated in a variety of ways, the definitions chosen being intrinsic and applicable in more general contexts. Among the applications of compactness are differentiation under the integral sign, Peano's theorem on the existence of solutions of initial-value problems for certain nonlinear ordinary differential equations, and the characterization of Riemann-integrable functions as functions that are bounded and continuous almost everywhere. With the next chapter in mind, we conclude with the consideration of simply connected spaces. Various forms of homotopy are given especially detailed coverage, strenuous efforts being made to give complete proofs. We show that a metric space is simply connected if and only if it is path-connected and its fundamental group at any (and hence every) point of the space has only one element.

In Chap. 3, we reach our main goal, the theory of complex analysis, surely one of the most wonderful and fertile parts of mathematics. After some basic definitions and results, we deal with power series, branches of the argument and logarithm, continuous logarithms of continuous zero-free functions, the winding number for arbitrary paths in the plane and its invariance under free homotopy, and integrals over contours. Ample justification for the introduction of the winding number is provided by the demands of the proof of the Jordan curve theorem given later (for which the winding number is essential and the index is inadequate as it is undefined for general paths having no smoothness), but in addition we believe that there is a computational and pedagogical advantage in having this concept available. The homology version of Cauchy's theorem is derived by means of the elegant approach of Dixon [6]. Rudin [15] was one of the first to draw attention to the importance of Dixon's contribution and the organisation of complex analysis consequent upon it. Rather than appeal to an interchange of the order of integration, as Rudin does, we follow Dixon's original treatment and use differentiation under the integral sign. This leads to the residue theorem, from which flow such major theoretical results as Rouché's theorem and the open mapping and inverse function theorems; further, at a practical and technical level it is valuable in the evaluation of definite integrals. The penultimate section contains a result of exceptional aesthetic appeal which establishes, for connected open sets $G \subset \mathbf{C}$ (the space of all complex numbers) the equivalence of various statements of an analytic, algebraic and topological character. In particular, it shows that every function analytic on G has a primitive if and only if G is simply connected. In the course of the proof, such famous results as Montel's theorem and the Riemann mapping theorem are obtained. The final section reinforces the links between analysis and topology. Further study of topics introduced earlier, namely continuous logarithms of continuous zero-free functions and the winding number of a path, leads in a very natural way to a proof of the celebrated Jordan curve theorem. For this development of the theory, we acknowledge a major debt to the book [3] by Burckel. A beautiful result due to Borsuk concerning any compact set

$K \subset \mathbf{C}$ emerges in the course of the proof of the Jordan curve theorem: $\mathbf{C}\backslash K$ is connected if and only if every continuous function $f : K \to \mathbf{C}\backslash\{0\}$ has a continuous logarithm.

Our exposition covers aspects of classical analysis due to the efforts of generations of mathematicians. There is no claim to originality save for the selection and presentation of material. We have been greatly influenced by the scholarly and inspirational books by Burckel [3], Remmert [13] and Rudin [15], and hope readers of the present book will go on to consult these more advanced and wider-ranging works.

It is a pleasure to acknowledge our great indebtedness to Dorothee Haroske for her immense help and patience. Finally, we express our appreciation to Joerg Sixt and his staff at the London Office of Springer-Verlag for constant encouragement and advice.

Contents

Chapter 1
The Riemann Integral

In this chapter we give an account of the Riemann integral for real-valued functions defined on intervals of the real line. This integral is of historic interest, has considerable intuitive appeal and possesses great practical value. For economy of presentation we use the approach of Darboux rather than that originally employed by Riemann.

Hidden from immediate view but at the heart of the chapter lies the sense in which integration is the inverse of differentiation. For the class of continuous functions the Riemann integral provides an affirmative answer to the question "Given $f : [a, b] \to$ **R**, where a and b are real and $a < b$, does there exist $F : [a, b] \to$ **R** such that $F' = f$?" With somewhat greater effort, development of the Lebesgue integral would allow us to enlarge this class. However, for the topics covered in this text the answer provided suffices; in particular, it is entirely adequate in the resolution of an analogous question asked in the context of complex analysis, a question which is the focus of our final chapter.

1.1 Basic Definitions and Results

Definition 1.1.1 Let a and b be real numbers, with $a < b$. Any finite set of points $x_0, x_1, \ldots, x_n$ with $a = x_0 < x_1 < \ldots < x_n = b$ is called a **partition** of $[a, b]$ and will often be denoted by P; we put $\Delta x_j = x_j - x_{j-1}$ $(j = 1, \ldots, n)$ and call $w(P) := \max \{\Delta x_j : j = 1, \ldots, n\}$ the **width** of P. The family of all partitions of $[a, b]$ is denoted by $\mathscr{P}[a, b]$, or simply by $\mathscr{P}$ if no ambiguity is possible. Let $\mathscr{B}[a, b]$ (or simply $\mathscr{B}$) be the family of all bounded functions $f : [a, b] \to \mathbf{R}$; given any $f \in \mathscr{B}$ and any $P \in \mathscr{P}$, put

$$M_j = \sup \{f(x) : x_{j-1} \le x \le x_j\}, \ m_j = \inf \{f(x) : x_{j-1} \le x \le x_j\}$$

for $j = 1, \ldots, n$ and call

R. H. Dyer and D. E. Edmunds, *From Real to Complex Analysis*,
Springer Undergraduate Mathematics Series, DOI: 10.1007/978-3-319-06209-9_1,
© Springer International Publishing Switzerland 2014

$$U(P, f) := \sum_{j=1}^{n} M_j \Delta x_j, \quad L(P, f) := \sum_{j=1}^{n} m_j \Delta x_j$$

the **upper** and **lower sums of** f **with respect to** P, respectively.

Note that $U(P, f)$ is the sum of the signed areas of n rectangles, the jth of which has base Δx_j and height M_j; $L(P, f)$ is the same except that the jth rectangle has height m_j. These quantities are familiar to anyone who has tried to estimate the area of the set of points lying between the curve $y = f(x)$ and the lines $x = a, x = b$ and $y = 0$ by drawing the graph of f on squared paper: $U(P, f)$ arises from consistent over-estimation of the area by rectangles above the graph, while $L(P, f)$ comes from a corresponding lower estimation by rectangles below the graph.

Example 1.1.2

(i) Let $f : [a, b] \to \mathbf{R}$ be monotonic increasing and let

$P = \{a = x_0, x_1, \ldots, x_n = b\}$ be a partition of $[a, b]$. Then

$$U(P, f) := \sum_{j=1}^{n} f(x_j)\Delta x_j, \quad L(P, f) := \sum_{j=1}^{n} f(x_{j-1})\Delta x_j.$$

(ii) Let $f : [a, b] \to \mathbf{R}$ be defined by

$$f(x) = \begin{cases} 1, & x \text{ rational,} \\ -1, & x \text{ irrational.} \end{cases}$$

Then given any partition P of $[a, b]$, $M_j = 1$ and $m_j = -1$ $(j = 1, \ldots, n)$ since each interval $[x_{j-1}, x_j]$ contains both rational and irrational points. Hence

$$U(P, f) = b - a, \quad L(P, f) = -(b - a).$$

Now let $f \in \mathscr{B}[a, b]$; that is, let f be a bounded, real-valued function on $[a, b]$. Since f is bounded, there are numbers $m, M \in \mathbf{R}$ such that for all $x \in [a, b]$, $m \le f(x) \le M$. Hence for all $P \in \mathscr{P}[a, b]$,

$$m(b - a) = \sum_{j=1}^{n} m \Delta x_j \le \sum_{j=1}^{n} m_j \Delta x_j \le \sum_{j=1}^{n} M_j \Delta x_j \le \sum_{j=1}^{n} M \Delta x_j = M(b - a);$$

that is,

$$m(b - a) \le L(P, f) \le U(P, f) \le M(b - a).$$

Thus $\{U(P, f) : P \in \mathscr{P}\}$ and $\{L(P, f) : P \in \mathscr{P}\}$ are bounded sets of real numbers; consequently they have a finite infimum and supremum.

Definition 1.1.3 Let $f \in \mathscr{B}[a, b]$. The **upper** and **lower integrals** of f over $[a, b]$ are

$$\overline{\int_a^b} f := \inf\{U(P, f) : P \in \mathscr{P}\}, \quad \underline{\int_a^b} f := \sup\{L(P, f) : P \in \mathscr{P}\},$$

respectively. If these upper and lower integrals are equal, we say that f is **Riemann-integrable over** $[a, b]$ and write

$$\int_a^b f = \overline{\int_a^b} f \left(= \underline{\int_a^b} f\right);$$

$\int_a^b f$, often written $\int_a^b f(x)dx$, is called the **Riemann integral of** f over $[a, b]$. The family of all functions which are Riemann-integrable over $[a, b]$ is denoted by $\mathscr{R}[a, b]$, or simply by $\mathscr{R}$.

Example 1.1.4

(i) Let $c \in \mathbf{R}$ and let $f : [a, b] \to \mathbf{R}$ be defined by $f(x) = c$ for all $x \in [a, b]$. Then for all $P \in \mathscr{P}[a, b]$, $U(P, f) = L(P, f) = c(b - a)$; hence $\overline{\int_a^b} f = \underline{\int_a^b} f = c(b - a)$ and so $f \in \mathscr{R}[a, b]$ with $\int_a^b f = c(b - a)$.

(ii) For the function f of Example 1.1.2 (ii), evidently $\overline{\int_a^b} f = b - a$ and $\underline{\int_a^b} f := -(b - a)$, so that $f \notin \mathscr{R}[a, b]$. However, example (i) above shows that despite this, $|f| \in \mathscr{R}[a, b]$.

We now proceed to investigate the family $\mathscr{R}[a, b]$ and to develop various properties of the integral.

Definition 1.1.5 Given any two partitions P, Q of $[a, b]$, Q is called a **refinement** of P if $P \subset Q$; that is, if every point of P is a point of Q. If $P_1, P_2 \in \mathscr{P}[a, b]$, then $Q := P_1 \cup P_2$ is called the **common refinement** of P_1 and P_2.

Lemma 1.1.6 *Let $f \in \mathscr{B}[a, b]$, let $K \in \mathbf{R}$ be such that $|f(x)| \le K$ whenever $x \in [a, b]$, and let $P \in \mathscr{P}[a, b]$. If $Q \in \mathscr{P}[a, b]$ and Q is a refinement of P with exactly k points in addition to those of P, then*

$$\text{(i)} \quad 0 \le U(P, f) - U(Q, f) \le 2kKw(P)$$

and

$$\text{(ii)} \quad 0 \le L(Q, f) - L(P, f) \le 2kKw(P).$$

Proof It suffices to prove (i), since (ii) follows on the observation that

$$U(P, -f) = -L(P, f)$$

(see Exercise 1.1.10/3).

The proof of (i) when $k = 1$ is almost trivial. Let $P = \{x_0, x_1, \ldots, x_n\}$, let $x_* \in (x_{j-1}, x_j)$ for some $j \in \{1, 2, \ldots, n\}$ and let $Q = P \cup \{x_*\}$. Let

$$M_j^* = \sup \{f(x) : x_{j-1} \leq x \leq x_*\}, M_j^{**} = \sup \{f(x) : x_* \leq x \leq x_j\};$$

evidently $M_j^*, M_j^{**} \leq M_j$. Now

$$U(P, f) - U(Q, f) = (M_j - M_j^*)(x_* - x_{j-1}) + (M_j - M_j^{**})(x_j - x_*),$$

since the other terms of the upper sums cancel. Hence

$$0 \leq U(P, f) - U(Q, f) \leq 2Kw(P).$$

Now suppose (i) is false for some $k \in \mathbf{N}$. Then there is a least $k_0 \in \mathbf{N}$, necessarily greater than 1, and an associated $Q_0 \in \mathscr{P}[a, b]$ with precisely k_0 points in addition to those of P, such that

$$U(P, f) - U(Q_0, f) \notin [0, 2k_0 Kw(P)]. \tag{1.1.1}$$

Delete one point from Q_0 which does not lie in P and let Q_1 be the resulting partition of $[a, b]$. By what has already been proved,

$$0 \leq U(Q_1, f) - U(Q_0, f) \leq 2Kw(Q_1) \leq 2Kw(P).$$

Further, since (i) holds for $k = k_0 - 1$,

$$0 \leq U(P, f) - U(Q_1, f) \leq 2(k_0 - 1)Kw(P).$$

Addition shows that

$$0 \leq U(P, f) - U(Q_0, f) \leq 2k_0 Kw(P),$$

which contradicts (1.1.1) and proves that (i) is true for all k.   □

Lemma 1.1.6 is very useful: it shows that the upper and lower sums are decreasing and increasing respectively on refinement of a partition, and enables the changes in these sums on refinement to be estimated. It plays a key rôle in the proof of the following theorem due to Darboux, a theorem which is a cornerstone of the theory as we shall develop it.

Theorem 1.1.7 *Let $f \in \mathscr{B}[a, b]$ and let (P_n) be a sequence in $\mathscr{P}[a, b]$ such that $\lim_{n \to \infty} w(P_n) = 0$. Then*

$$\lim_{n \to \infty} U(P_n, f) = \overline{\int_a^b} f, \quad \lim_{n \to \infty} L(P_n, f) = \underline{\int_a^b} f.$$

In particular, $f \in \mathscr{R}[a, b]$ if, and only if, $\lim_{n \to \infty} \{U(P_n, f) - L(P_n, f)\} = 0$.

Proof Let $K \in \mathbf{R}$ be such that $|f(x)| < K$ for all $x \in [a, b]$, and let $\varepsilon > 0$. By definition of the upper integral, there exists $Q \in \mathscr{P}[a, b]$ such that

$$U(Q, f) < \overline{\int_a^b} f + \varepsilon/2.$$

Let Q have exactly k points. For each $n \in \mathbf{N}$, $P_n \cup Q$ is a refinement of P_n with at most k additional points; thus by Lemma 1.1.6,

$$\begin{aligned} U(P_n, f) &\leq 2kKw(P_n) + U(P_n \cup Q, f) \\ &\leq 2kKw(P_n) + U(Q, f) \\ &\leq 2kKw(P_n) + \overline{\int_a^b} f + \varepsilon/2. \end{aligned}$$

Now, by hypothesis, there exists $N \in \mathbf{N}$ such that $w(P_n) < \varepsilon/(4kK)$ whenever $n \geq N$. It follows that, for $n \geq N$, we have

$$0 \leq U(P_n, f) - \overline{\int_a^b} f < \varepsilon.$$

Hence $\lim_{n \to \infty} U(P_n, f) = \overline{\int_a^b} f$. Since $\overline{\int_a^b}(-f) = -\underline{\int_a^b} f$, the rest follows directly. □

Corollary 1.1.8 *For all $f \in \mathscr{B}[a, b]$, $\underline{\int_a^b} f \leq \overline{\int_a^b} f$.*

Proof Let (P_n) be a sequence in $\mathscr{P}[a, b]$ such that $w(P_n) \to 0$. By Theorem 1.1.7,

$$\underline{\int_a^b} f = \lim_{n \to \infty} L(P_n, f) \leq \lim_{n \to \infty} U(P_n, f) = \overline{\int_a^b} f.$$

□

The power of Theorem 1.1.7 is considerable. We use it in the next three sections to exhibit large classes of integrable functions, to give a rapid exposition of the basic properties of the integrals defined above, and to provide, at least in principle, a technique for their evaluation. Before engaging in such matters, however, we prove an equivalent version of it.

Theorem 1.1.9 *Let* $f \in \mathcal{B}[a, b]$. *Then*

$$\lim_{w(P)\to 0} U(P, f) = \overline{\int_a^b} f;$$

that is, given any $\varepsilon > 0$, *there exists* $\delta > 0$ *such that* $0 \le U(P, f) - \overline{\int_a^b} f < \varepsilon$ *if* $P \in \mathcal{P}[a, b]$ *and* $w(P) < \delta$. *Moreover*,

$$\lim_{w(P)\to 0} L(P, f) = \underline{\int_a^b} f.$$

Proof To obtain a contradiction suppose that, for some $f \in \mathcal{B}[a, b]$,

$$\lim_{w(P)\to 0} U(P, f) \neq \overline{\int_a^b} f.$$

Then an $\varepsilon > 0$ exists such that, for each $n \in \mathbf{N}$, there is a $P_n \in \mathcal{P}[a, b]$ with the properties

$$\text{(i) } w(P_n) < 1/n, \text{ and (ii) } U(P_n, f) > \overline{\int_a^b} f + \varepsilon.$$

The first property shows that the sequence $(w(P_n))$ converges to zero; the second, in conjunction with Theorem 1.1.7, that

$$\overline{\int_a^b} f = \lim_{n\to\infty} U(P_n, f) \geq \overline{\int_a^b} f + \varepsilon,$$

which is impossible for a positive ε. $\square$

Exercise 1.1.10

1. Let A and B be non-empty, bounded subsets of $\mathbf{R}$ and let $\lambda \in \mathbf{R}$. Put

$$A + B = \{a + b : a \in A, b \in B\} \text{ and } \lambda A = \{\lambda a : a \in A\}.$$

Show that

(i) $\sup(A + B) = \sup A + \sup B$,

(ii) $\inf(A + B) = \inf A + \inf B$,

(iii) $\sup(\lambda A) = \begin{cases} \lambda \sup A & \text{if } \lambda \geq 0, \\ \lambda \inf A & \text{if } \lambda < 0, \end{cases}$

(iv) $\inf(\lambda A) = \begin{cases} \lambda \inf A & \text{if } \lambda \geq 0, \\ \lambda \sup A & \text{if } \lambda < 0. \end{cases}$

2. Let A be a non-empty subset of a set X and let $f : X \to \mathbf{R}$ be bounded. The **oscillation of** f **over** A, $\mathrm{osc}(f; A)$, is defined to be

$$\sup\{|f(x) - f(y)| : x, y \in A\}.$$

Prove that

$$\mathrm{osc}(f; A) = \sup\{f(x) : x \in A\} - \inf\{f(x) : x \in A\}.$$

3. Let $f : [a, b] \to \mathbf{R}$ be bounded and let $P \in \mathscr{P}[a, b]$. Prove that

$$U(P, -f) = -L(P, f) \text{ and } L(P, -f) = -U(P, f).$$

4. Using merely the definition of integrability, show that the function f from $[0, 1]$ to $\mathbf{R}$ defined by $f(t) = t^2$ $(0 \le t \le 1)$ is Riemann-integrable over $[0, 1]$ and that

$$\int_0^1 f = 1/3.$$

[Show that (i) if $P \in \mathscr{P}[0, 1]$ then $U(P, f) \ge 1/3 \ge L(P, f)$; (ii) if P_n is that partition of $[0, 1]$ which divides it into n subintervals of equal length, then

$$U(P_n, f) = (n + 1)(2n + 1)/6n^2 \text{ and } L(P_n, f) = (n - 1)(2n - 1)/6n^2.]$$

5. Suppose $a < b$ and let $f : [a, b] \to \mathbf{R}$ be bounded and such that $f(t) > 0$ for all $t \in [a, b]$. Prove that $\overline{\int_a^b} f > 0$.
   [A subinterval $[c, d]$ of $[a, b]$, with $c < d$, and an $\varepsilon > 0$ exist such that $\sup\{f(t) : \alpha \le t \le \beta\} \ge \varepsilon$ whenever $c \le \alpha < \beta \le d$.]

6. (Riemann's criterion for integrability.) Let $f \in \mathscr{B}[a, b]$. Then $f \in \mathscr{R}[a, b]$ if, and only if, given any $\varepsilon > 0$, there exists $P \in \mathscr{P}[a, b]$ such that

$$U(P, f) - L(P, f) < \varepsilon.$$

7. Let $f : [a, b] \to \mathbf{R}$. Prove that $f \in \mathscr{R}[a, b]$ if, and only if, there exists a real number $A \left(= \int_a^b f\right)$ with the following property: for each $\varepsilon > 0$, there exists $\delta > 0$ such that

$$\left| A - \sum_{j=1}^n f(\xi_j)\Delta x_j \right| < \varepsilon$$

whenever $P = \{x_0, x_1, \ldots, x_n\} \in \mathscr{P}[a, b]$, $w(P) < \delta$ and $\xi_j \in [x_{j-1}, x_j]$ for each $j \in \{1, 2, \ldots, n\}$.

8. Let $f : [0, 1] \to \mathbf{R}$ be defined by $f(x) = \sqrt{x}$ $(0 \le x \le 1)$. Let

$$P_n = \left\{ 0, \left(\frac{1}{n}\right)^2, \left(\frac{2}{n}\right)^2, \ldots, \left(\frac{n-1}{n}\right)^2, 1 \right\}.$$

Calculate $w(P_n)$ and show that $\lim_{n\to\infty} w(P_n) = 0$. Determine $L(f, P_n)$ and $U(f, P_n)$, and show that f is Riemann-integrable over $[0, 1]$ and that

$$\int_0^1 \sqrt{x}\,dx = 2/3.$$

1.2 Classes of Integrable Functions

Clearly, the utility of any theory of integration depends on commonly encountered types of function having an integral under that theory. Continuous real-valued functions defined on closed bounded intervals are of such a type. Beginning with some preliminary discussion of continuity, we show that if $f : [a, b] \to \mathbf{R}$ is continuous then $f \in \mathscr{R}[a, b]$.

Definition 1.2.1 Let I be a non-empty interval in $\mathbf{R}$ and let $f : I \to \mathbf{R}$. The function f is said to be **continuous at** $x_0 \in I$ if, given any $\varepsilon > 0$, there exists $\delta > 0$ such that if $x \in I$ and $|x_0 - x| < \delta$, then $|f(x) - f(x_0)| < \varepsilon$; it is said to be **continuous on** I if it is continuous at each point of I. We say that f is **uniformly continuous on** I if given any $\varepsilon > 0$, there exists $\delta > 0$ such that if $x, y \in I$ and $|x - y| < \delta$, then $|f(x) - f(y)| < \varepsilon$.

Note that the essential difference between continuity and uniform continuity on I is that while for uniform continuity the number δ depends only on ε, in the case of continuity δ depends on ε *and on* x_0: there may be no single δ which will achieve the desired smallness of $|f(x) - f(x_0)|$ *for all* $x_0 \in I$. Although uniform continuity on I evidently implies continuity on I, in general the converse is false. The following examples may help to understand the distinction between these two forms of continuity.

Example 1.2.2

(i) Let $I = [0, 1]$ and suppose that $f(x) = x^2$ for all $x \in I$. Then f is uniformly continuous on I, for if $\varepsilon > 0$ then

$$|f(x) - f(y)| = |(x + y)(x - y)| \le 2|x - y| < \varepsilon$$

if $x, y \in I$ and $|x - y| < \varepsilon/2$. Hence we may take $\delta = \varepsilon/2$.

(ii) Let $I = [0, \infty)$ and again suppose that $f(x) = x^2$ for all $x \in I$. Then f is continuous on I: to see this let $x_0 \in I$ and $\varepsilon > 0$. Given any $\delta > 0$ and any $x \in I$ such that $|x - x_0| < \delta$ it follows that $x + x_0 < 2x_0 + \delta$, and hence

$$|f(x) - f(x_0)| = |(x - x_0)(x + x_0)| < \delta(2x_0 + \delta).$$

The choice of any positive number δ less than $\eta := (x_0^2 + \varepsilon)^{1/2} - x_0$, say $\delta = \eta/2$, now shows that $|f(x) - f(x_0)| < \varepsilon$ if $|x - x_0| < \delta$, and the continuity of f on I is established. Note the dependence of δ on x_0. However, f is not uniformly continuous on I: for if it were, then given any $\varepsilon > 0$, there would exist $\delta > 0$ such that if $x, y \in I$ and $|x - y| < \delta$, then $|x^2 - y^2| < \varepsilon$. But given *any* $\delta > 0$, if we choose $n \in \mathbf{N}$, $x = n$ and $y = n + \frac{1}{2}\delta$, then $|x - y| < \delta$ but $|x^2 - y^2| = |(n + \frac{1}{2}\delta)^2 - n^2| = \delta n + \frac{1}{4}\delta^2$, which can be made arbitrarily large by choosing n large enough.

(iii) Let $I = (0, 1)$ and suppose that $f(x) = 1/x$ for all $x \in I$. This function is continuous on I: for given any $x_0 \in I$ and any $\varepsilon > 0$, we see that if $x \in I$ and $|x - x_0| < \delta < x_0$,

$$|f(x) - f(x_0)| < \delta/\{x_0(x - x_0 + x_0)\} < \delta/\{x_0(x_0 - \delta)\};$$

thus to obtain $|f(x) - f(x_0)| < \varepsilon$ we simply choose $\delta < \varepsilon x_0^2/(1+\varepsilon x_0)$. It is not possible to choose δ independent of x_0; that is, f is not uniformly continuous on I. To see this, it is merely necessary to observe that $\frac{1}{n} - \frac{1}{n+1} \to 0$ as $n \to \infty$, while $\left| f(\frac{1}{n}) - f(\frac{1}{n+1}) \right| = 1$ for all $n \in \mathbf{N}$.

If I is a closed, bounded interval $[a, b]$ the distinction between continuity and uniform continuity on I disappears. To establish this it is convenient to appeal to the famous Bolzano-Weierstrass theorem: every bounded sequence of real numbers has a convergent subsequence. A proof of this theorem is given in Theorem A.4.13 of the Appendix.

Theorem 1.2.3 *Let $a, b \in \mathbf{R}$, with $a < b$. A function $f : [a, b] \to \mathbf{R}$ is continuous on $[a, b]$ if, and only if, it is uniformly continuous on $[a, b]$.*

Proof Suppose first that f is continuous, but not uniformly continuous on $[a, b]$. Then there exists $\varepsilon > 0$ such that given any $n \in \mathbf{N}$, there are points $x_n, y_n \in [a, b]$ with $|x_n - y_n| < 1/n$ and $|f(x_n) - f(y_n)| \geq \varepsilon$. The sequence (x_n) is bounded and so, by the Bolzano-Weierstrass theorem, has a convergent subsequence $(x_{m(n)})$ with $\lim_{n\to\infty} x_{m(n)} = x$, say; clearly $\lim_{n\to\infty} y_{m(n)} = x$. In view of the continuity of f,

$$\lim_{n\to\infty} \left\{ f(x_{m(n)}) - f(y_{m(n)}) \right\} = f(x) - f(x) = 0.$$

But, for all n, $\left| f(x_{m(n)}) - f(y_{m(n)}) \right| \geq \varepsilon$. Thus

$$\lim_{n\to\infty} \left| f(x_{m(n)}) - f(y_{m(n)}) \right| \geq \varepsilon,$$

which gives a contradiction. Hence f is uniformly continuous on $[a, b]$. The converse is obvious. $\qquad\square$

An important result, given next, is an immediate consequence of Theorem 1.2.3. (The reader should also know a direct proof of it, one which bypasses the notion of uniform continuity. See Exercise 1.2.14/1.)

Theorem 1.2.4 *Let $a, b \in \mathbf{R}$, with $a < b$, and let $f : [a, b] \to \mathbf{R}$ be continuous on $[a, b]$. Then f is bounded.*

Proof Since f is uniformly continuous on $[a, b]$, there exists $\delta > 0$ such that $|f(x) - f(y)| < 1$ if $x, y \in [a, b]$ and $|x - y| < \delta$. Choose $P = \{x_0, x_1, \ldots, x_n\} \in \mathscr{P}[a, b]$ such that $w(P) < \delta$. Then

$$\sup_{x \in [a,b]} |f(x)| \leq 1 + \max_{1 \leq i \leq n} |f(x_i)|.$$

The proof is complete.   □

Armed with this equipment we now return to integration.

Theorem 1.2.5 *Let $a, b \in \mathbf{R}$ with $a < b$, and let $f : [a, b] \to \mathbf{R}$. Then:*

(i) *if f is monotone, $f \in \mathscr{R}[a, b]$;*
(ii) *if f is continuous on $[a, b]$, $f \in \mathscr{R}[a, b]$.*

Proof For $n \in \mathbf{N}$, let $P_n = \{x_0, x_1, \ldots, x_n\} \in \mathscr{P}[a, b]$ be such that $\Delta x_j = (b - a)/n$ $(j = 1, 2, \ldots, n)$; plainly the sequence (P_n) has the property that $\lim_{n \to \infty} w(P_n) = 0$.
(i) Suppose f is increasing on $[a, b]$ (otherwise consider $-f$). Then $M_j = f(x_j), m_j = f(x_{j-1})$ $(j = 1, 2, \ldots, n)$ and

$$U(P_n, f) - L(P_n, f) = (b - a)n^{-1} \sum_{j=1}^{n} \left\{ f(x_j) - f(x_{j-1}) \right\}$$

$$= (b - a)n^{-1} \{f(b) - f(a)\} \to 0$$

as $n \to \infty$. Thus by Theorem 1.1.7, $f \in \mathscr{R}[a, b]$.
(ii) By Theorem 1.2.4, $f \in \mathscr{B}[a, b]$. Let $\varepsilon > 0$. By Theorem 1.2.3, f is uniformly continuous on $[a, b]$; hence there exists $\delta > 0$ such that if $s, t \in [a, b]$ and $|s - t| < \delta$, then we have $|f(s) - f(t)| < \varepsilon$. Suppose $n \in \mathbf{N}$ is such that $w(P_n) < \delta$. Then for each $j \in \{1, 2, \ldots, n\}$,

$$M_j - m_j = \sup \left\{ |f(s) - f(t)| : s, t \in [x_{j-1}, x_j] \right\}$$

(see Exercise 1.1.10/2), and hence

$$U(P_n, f) - L(P_n, f) = \sum_{j=1}^{n} (M_j - m_j)\Delta x_j \leq \varepsilon(b - a).$$

Since $w(P_n) < \delta$ for all save finitely many n, Theorem 1.1.7 and passage to the limit as $n \to \infty$ show that

$$0 \le \overline{\int_a^b} f - \underline{\int_a^b} f \le \varepsilon(b-a).$$

The final inequality being valid for all positive ε, it follows that $\overline{\int_a^b} f = \underline{\int_a^b} f$, that is, $f \in \mathscr{R}[a, b]$.   □

Note that there are Riemann-integrable functions which are neither continuous nor monotone: see Exercise 1.2.14/4 for one such example. This fact notwithstanding, although a Riemann-integrable function need not be continuous, it must have a point of continuity, indeed, infinitely many such. The next two lemmas are a preparation to prove this assertion.

Lemma 1.2.6 *Let $f \in \mathscr{R}[a, b]$ and let $a < c < d < b$. Then $f \in \mathscr{R}[c, d]$; more precisely, the restriction of f to $[c, d]$ belongs to $\mathscr{R}[c, d]$.*

Proof Let (P_n) be a sequence of partitions of $[a, b]$ such that $\{c, d\} \subset P_n$ $(n \in \mathbf{N})$ and $w(P_n) \to 0$. Let $Q_n = P_n \cap [c, d]$ $(n \in \mathbf{N})$. Then each $Q_n \in \mathscr{P}[c, d]$ and $w(Q_n) \to 0$. Since

$$U(Q_n, f) - L(Q_n, f) \le U(P_n, f) - L(P_n, f) \ (n \in \mathbf{N})$$

and $f \in \mathscr{R}[a, b]$, it follows that

$$0 \le \overline{\int_c^d} f - \underline{\int_c^d} f \le \overline{\int_a^b} f - \underline{\int_a^b} f = 0.$$

Thus $f \in \mathscr{R}[c, d]$.   □

Lemma 1.2.7 *Let $f \in \mathscr{R}[a, b]$ and $v > 0$. Then there exists a closed interval $[c, d] \subset [a, b]$ such that*

(i) $a < c < d < b$,
(ii) $d - c < v$,
(iii) $\mathrm{osc}(f; [c, d]) < v$.

Proof Since $f \in \mathscr{R}[a, b]$, by Exercise 1.1.10/6, there is a partition $P = \{x_0, x_1, \ldots, x_n\}$ of $[a, b]$ such that

$$U(P, f) - L(P, f) = \sum_{j=1}^n (M_j - m_j)\Delta x_j < v(b - a).$$

Appealing to Exercise 1.1.10/2, it follows that, for some i $(1 \le i \le n)$,

$$\operatorname{osc}(f; [x_{i-1}, x_i]) = M_i - m_i = \min_{1 \le j \le n} (M_j - m_j) < v.$$

Finally, any choice of closed interval $[c, d]$ such that $x_{i-1} < c < d < x_i$ and $d - c < v$ has the properties required.   □

Theorem 1.2.8 *Let $f \in \mathscr{R}[a, b]$. Then there is a real number c such that $a < c < b$ and f is continuous at c.*

Proof Since $f \in \mathscr{R}[a, b]$, by Lemma 1.2.7, there is a closed interval $[a_1, b_1] \subset [a, b]$ such that

$$a < a_1 < b_1 < b, \ b_1 - a_1 < 1 \text{ and } \operatorname{osc}(f; [a_1, b_1]) < 1.$$

In view of Lemma 1.2.6, $f \in \mathscr{R}[a_1, b_1]$.. Hence a further appeal to Lemma 1.2.7 shows that there is a closed interval $[a_2, b_2] \subset [a_1, b_1]$ such that

$$a_1 < a_2 < b_2 < b_1, \ b_2 - a_2 < 2^{-1} \text{ and } \operatorname{osc}(f; [a_2, b_2]) < 2^{-1}.$$

Continuing in this way, and allowing $a_0 := a$, $b_0 := b$, we see that there is a nested sequence $([a_n, b_n])$ of bounded, closed intervals such that, for each $n \in \mathbf{N}$,

(i) $a_{n-1} < a_n < b_n < b_{n-1}$,

(ii) $b_n - a_n < n^{-1}$, and

(iii) $\operatorname{osc}(f; [a_n, b_n]) < n^{-1}$.

Applying the Nested Intervals Principle (see the Appendix, Theorem A.4.15), we see that there exists $c \in \mathbf{R}$ such that $\{c\} = \cap_{n=1}^{\infty} [a_n, b_n]$. It remains to show that f is continuous at c. Let $\varepsilon > 0$. There exists $m \in \mathbf{N}$ such that $m\varepsilon > 1$ and, evidently, for all $x \in [a_m, b_m]$,

$$|f(x) - f(c)| \le \operatorname{osc}(f; [a_m, b_m]) < m^{-1} < \varepsilon.$$

With $\delta = \min\{c - a_m, b_m - c\}$, it follows that $|f(x) - f(c)| < \varepsilon$ whenever $|x - c| < \delta$. Thus f is continuous at c.   □

Definition 1.2.9 Let $\mathscr{I}$ be a non-empty interval in $\mathbf{R}$. A function $\phi : \mathscr{I} \to \mathbf{R}$ is said to satisfy a **Lipschitz condition (on $\mathscr{I}$)** if there exists $c > 0$ such that for all $s, t \in \mathscr{I}$,

$$|\phi(s) - \phi(t)| \le c \, |s - t| .$$

With this property, ϕ is also described as a **Lipschitz-continuous function (on $\mathscr{I}$)**; the number c is said to be a **Lipschitz constant for ϕ**.

Theorem 1.2.10 *Suppose that $f \in \mathscr{R}[a, b]$ and that $f([a, b]) \subset [\alpha, \beta]$; let $\phi : [\alpha, \beta] \to \mathbf{R}$ be a Lipschitz-continuous function on $[\alpha, \beta]$, and let $h = \phi \circ f$. Then $h \in \mathscr{R}[a, b]$.*

Proof Let c be a Lipschitz constant for ϕ. By Exercise 1.2.14/6, ϕ is continuous (uniformly continuous) on $[\alpha, \beta]$. Hence h is bounded. Let $P = \{x_0, x_1, \ldots, x_n\} \in \mathscr{P}[a, b]$ and write $I_j = [x_{j-1}, x_j]$ $(j = 1, 2, \ldots, n)$. By Exercise 1.1.10/2,

$$U(P, h) - L(P, h) = \sum_{j=1}^{n} \operatorname{osc}(h; I_j)\Delta x_j \leq c \sum_{j=1}^{n} \operatorname{osc}(f; I_j)\Delta x_j$$
$$= c\left(U(P, f) - L(P, f)\right).$$

Application of this inequality to the members of a sequence (P_n) of partitions of $[a, b]$ with $w(P_n) \to 0$ now shows, with the help of Theorem 1.1.7, that $h \in \mathscr{R}[a, b]$. $\square$

Corollary 1.2.11 *If f is in $\mathscr{R}[a, b]$, so are $|f|$ and f^2.*

Proof Put $K = \sup\{|f(x)| : a \leq x \leq b\}$ and let $\mathscr{I} = [-K, K]$. Then, for all $s, t \in \mathscr{I}$,

$$\left||s| - |t|\right| \leq |s - t|, \quad \left|s^2 - t^2\right| \leq 2K |s - t|.$$

The maps $t \longmapsto |t|$ and $t \longmapsto t^2$ each satisfy a Lipschitz condition on $\mathscr{I}$ and so appeal to Theorem 1.2.10 gives the result. $\square$

Theorem 1.2.10 enables us to generate new Riemann-integrable functions from functions known already to be Riemann-integrable. The next theorem goes further along this particular line and includes Theorem 1.2.10 as a special case. The condition that ϕ is a Lipschitz-continuous function is relaxed, simply requiring it to be continuous. The following lemma which, loosely speaking, asserts that every continuous real-valued function on a bounded, closed interval is 'almost' Lipschitz-continuous, paves the way for the relaxation.

Lemma 1.2.12 *Let $\phi : [\alpha, \beta] \to \mathbf{R}$ be continuous and let $\varepsilon > 0$. Then there exists $c > 0$ such that, for all $s, t \in [\alpha, \beta]$,*

$$|\phi(s) - \phi(t)| < \varepsilon + c |s - t|.$$

Proof To obtain a contradiction, we suppose the conclusion false. Then there exist $\varepsilon > 0$ and sequences (s_n), (t_n) in $[\alpha, \beta]$ such that for all $n \in \mathbf{N}$,

$$|\phi(s_n) - \phi(t_n)| \geq \varepsilon + n |s_n - t_n|.$$

By the Bolzano-Weierstrass theorem, there are points $s, t \in [\alpha, \beta]$ and subsequences $(s_{k(n)})$, $(t_{k(n)})$ of (s_n), (t_n) such that $\lim_{n \to \infty} s_{k(n)} = s$, $\lim_{n \to \infty} t_{k(n)} = t$. Evidently $\left|\phi(s_{k(n)}) - \phi(t_{k(n)})\right| \geq \varepsilon$; and since ϕ is continuous, we may let $n \to \infty$ and obtain $|\phi(s) - \phi(t)| \geq \varepsilon$, which implies that $s \neq t$. However, we then have

$$|\phi(s) - \phi(t)| = \lim_{n \to \infty} \left|\phi(s_{k(n)}) - \phi(t_{k(n)})\right| \geq \lim_{n \to \infty} k(n) \left|s_{k(n)} - t_{k(n)}\right|$$
$$= \lim_{n \to \infty} k(n) \lim_{n \to \infty} \left|s_{k(n)} - t_{k(n)}\right| = \infty,$$

which is impossible. The result claimed thus holds.   □

Theorem 1.2.13 *Suppose that* $f \in \mathscr{R}[a, b]$ *and that* $f([a, b]) \subset [\alpha, \beta]$; *let* $\phi :$ $[\alpha, \beta] \to \mathbf{R}$ *be continuous and set* $h = \phi \circ f$. *Then* $h \in \mathscr{R}[a, b]$.

Proof Since ϕ is continuous, h is bounded. Let $\varepsilon > 0$. By Lemma 1.2.12, there exists $c > 0$ such that, for all $s, t \in [\alpha, \beta]$,

$$|\phi(s) - \phi(t)| < \varepsilon + c\,|s - t|.$$

Let (P_n) be a sequence of partitions of $[a, b]$ with $w(P_n) \to 0$. Then for all $n \in \mathbf{N}$,

$$U(P_n, h) - L(P_n, h) \leq \varepsilon(b - a) + c\{U(P_n, f) - L(P_n, f)\},$$

and by letting $n \to \infty$ we obtain, since $f \in \mathscr{R}[a, b]$,

$$\overline{\int_a^b} h - \underline{\int_a^b} h \leq \varepsilon(b - a).$$

As this holds for all $\varepsilon > 0$, it follows that $\overline{\int_a^b} h = \underline{\int_a^b} h$, and the proof is complete. □

Note that Corollary 1.2.11 can be obtained from Theorem 1.2.13 even more directly than before.

Exercise 1.2.14

1. Let $f : [a, b] \to \mathbf{R}$ be continuous on $[a, b]$. Use the Bolzano-Weierstrass theorem to show directly that
   (i) f is bounded.
   (ii) f attains its bounds; that is, there exist $c, d \in [a, b]$ such that

   $$f(c) = \inf_{x \in [a,b]} f(x), \quad f(d) = \sup_{x \in [a,b]} f(x).$$

2. Let $f : (0, 1] \to \mathbf{R}$ be defined by $f(x) = \cos(\pi/x)$ $(0 < x \leq 1)$. Prove that f is continuous but not uniformly continuous on $(0, 1]$.
3. Let $f : (0, 1] \to \mathbf{R}$ be uniformly continuous on $(0, 1]$. Through either a proof or exhibition of a counterexample, decide whether or not f is necessarily bounded.
4. Let $f : [0, 1] \to \mathbf{R}$ be defined by $f(x) = x$ if $x = 1/n$ for some $n \in \mathbf{N}$, $f(x) = 0$ otherwise. Show that $f \in \mathscr{R}[0, 1]$ and that $\int_0^1 f = 0$. [Hint: partition the interval $[0, 1]$ into n^2 subintervals of equal length.]
5. Let $\mathscr{I}$ be a non-empty interval in $\mathbf{R}$ and let $f : \mathscr{I} \to \mathbf{R}$ be defined by

   $$f(x) = x^2.$$

   Show that f satisfies a Lipschitz condition on $\mathscr{I}$ if, and only if, $\mathscr{I}$ is bounded.

6. Let $\mathscr{I}$ be a non-empty interval in $\mathbf{R}$ and let $f : \mathscr{I} \to \mathbf{R}$ be a Lipschitz-continuous function on $\mathscr{I}$. Show that f is uniformly continuous on $\mathscr{I}$.

7. Let $f : [a, b] \to \mathbf{R}$ be differentiable. Show that f is Lipschitz-continuous on $[a, b]$ if, and only if, its derivative, f', is bounded on $[a, b]$.

8. (a) Give an example of a Lipschitz-continuous function on $[0, 1]$ which is not differentiable on $[0, 1]$.
   (b) Let $f : [0, 1] \to \mathbf{R}$ be defined by $f(x) = x^2 \sin(x^{-2})$ if $0 < x \leq 1$; $f(0) = 0$. Show that f does not satisfy a Lipschitz condition on $[0, 1]$.

9. Give an example of a function $f \in \mathscr{B}[0, 1] \backslash \mathscr{R}[0, 1]$ which has a point of continuity in the open interval $(0, 1)$.

10. Let $f \in \mathscr{R}[a, b]$ and let $x \in [a, b]$. Prove that there is a sequence (x_n) of distinct points in $[a, b]$ such that
    (i) $\lim_{n \to \infty} x_n = x$, and
    (ii) each x_n is a point of continuity of f.

1.3 Properties of the Integral

In this section we establish numerous useful properties of the Riemann integral. We begin with upper and lower integrals.

Theorem 1.3.1 *Let $f, g \in \mathscr{B}[a, b]$ and let $\lambda \in \mathbf{R}$. Then:*

(i) $\overline{\int_a^b} f + \overline{\int_a^b} g \geq \overline{\int_a^b} (f + g) \geq \underline{\int_a^b} (f + g) \geq \underline{\int_a^b} f + \underline{\int_a^b} g$;

(ii) *if $\lambda \geq 0$, then* $\overline{\int_a^b} \lambda f = \lambda \overline{\int_a^b} f$ *and* $\underline{\int_a^b} \lambda f = \lambda \underline{\int_a^b} f$;

(iii) *if $\lambda < 0$, then* $\overline{\int_a^b} \lambda f = \lambda \underline{\int_a^b} f$ *and* $\underline{\int_a^b} \lambda f = \lambda \overline{\int_a^b} f$;

(iv) *if $f(t) \geq g(t)$ for all $t \in [a, b]$, then*

$$\overline{\int_a^b} f \geq \overline{\int_a^b} g \text{ and } \underline{\int_a^b} f \geq \underline{\int_a^b} g;$$

(v) *if $f(t) = g(t)$ at all but a finite number of points of $[a, b]$, then*

$$\overline{\int_a^b} f = \overline{\int_a^b} g \text{ and } \underline{\int_a^b} f = \underline{\int_a^b} g;$$

(vi) *if $f(t) \geq 0$ for all $t \in [a, b]$ and $f(c) > 0$ at some point $c \in [a, b]$ at which f is continuous, then $\underline{\int_a^b} f > 0$.*

Proof Let (P_n) be a sequence in $\mathscr{P}[a, b]$ with $w(P_n) \to 0$.

(i) By Theorem 1.1.7,

$$\overline{\int_a^b} f + \overline{\int_a^b} g = \lim_{n\to\infty} U(P_n, f) + \lim_{n\to\infty} U(P_n, g) \geq \lim_{n\to\infty} U(P_n, f + g)$$

$$= \overline{\int_a^b}(f + g) \geq \underline{\int_a^b}(f + g) = \lim_{n\to\infty} L(P_n, f + g)$$

$$\geq \lim_{n\to\infty} L(P_n, f) + \lim_{n\to\infty} L(P_n, g) = \underline{\int_a^b} f + \underline{\int_a^b} g.$$

(ii) $\overline{\int_a^b} \lambda f = \lim_{n\to\infty} U(P_n, \lambda f) = \lambda \lim_{n\to\infty} U(P_n, f) = \lambda \overline{\int_a^b} f$; $\underline{\int_a^b} f$ is handled similarly.

(iii) $\overline{\int_a^b} \lambda f = \lim_{n\to\infty} U(P_n, \lambda f) = \lambda \lim_{n\to\infty} L(P_n, f) = \lambda \underline{\int_a^b} f$; $\underline{\int_a^b} \lambda f$ responds to similar treatment.

(iv) $\overline{\int_a^b} f = \lim_{n\to\infty} U(P_n, f) \geq \lim_{n\to\infty} U(P_n, g) = \overline{\int_a^b} g$; we proceed similarly with $\underline{\int_a^b} f$.

(v) Let $K \in \mathbf{R}$ be such that $\sup\{|f(t)| : a \leq t \leq b\}$, $\sup\{|g(t)| : a \leq t \leq b\} \leq K$, and suppose there are exactly k points of $[a, b]$ at which $f(t) \neq g(t)$. Each such point can lie in at most two of the intervals $[x_{j-1}, x_j]$ associated with the partition $P_n = \{x_0, x_1, \ldots, x_n\}$. Let $\widetilde{\sum}$ denote summation over those j such that $f(t) \neq g(t)$ for some $t \in [x_{j-1}, x_j]$; there are at most $2k$ such j. Then

$$|U(P_n, f) - U(P_n, g)| = \left| \widetilde{\sum} \{M_j(f) - M_j(g)\} \Delta x_j \right| \leq 2K.2kw(P_n)$$

$$\to 0 \text{ as } n \to \infty.$$

Hence $\overline{\int_a^b} f = \overline{\int_a^b} g$. That $\underline{\int_a^b} f = \underline{\int_a^b} g$ follows from a similar argument.

(vi) Since f is continuous at c, there is a closed interval I, with $c \in I \subset [a, b]$ and the length $l(I)$ of I positive, such that $f(t) \geq \frac{1}{2} f(c)$ for all $t \in I$. Define $g : [a, b] \to \mathbf{R}$ by $g(t) = \frac{1}{2} f(c)$ if $t \in I$, $g(t) = 0$ otherwise. Then $f \geq g$ and so, by (iv), $\underline{\int_a^b} f \geq \underline{\int_a^b} g = \frac{1}{2} f(c) l(I) > 0$. $\square$

We can now establish certain fundamental properties of the integral.

Theorem 1.3.2 *Let $f, g \in \mathscr{R}[a, b]$ and let $\lambda \in \mathbf{R}$. Then:*

(a) $f + g, \lambda f, fg \in \mathscr{R}[a, b]$; $\int_a^b (f + g) = \int_a^b f + \int_a^b g$, $\int_a^b \lambda f = \lambda \int_a^b f$;

(b) *if $f(t) \geq g(t)$ for all $t \in [a, b]$, then $\int_a^b f \geq \int_a^b g$;*

(c) $|f| \in \mathscr{R}[a, b]$ and

$$\left| \int_a^b f \right| \leq \int_a^b |f| \, ;$$

(d) if $f(t) \geq 0$ for all $t \in [a, b]$, f is continuous on $[a, b]$ and $\int_a^b f = 0$, then $f = 0$.

Proof (a) In view of Theorem 1.3.1, all that we have to do is to show that $fg \in \mathscr{R}[a, b]$; note that by fg is meant the product function $t \longmapsto f(t)g(t)$. By the first parts of (a), $f \pm g \in \mathscr{R}[a, b]$; by Corollary 1.2.11, $(f \pm g)^2 \in \mathscr{R}[a, b]$. Since $fg = \frac{1}{4}\{(f + g)^2 - (f - g)^2\}$, it follows from the first parts of (a) that $fg \in \mathscr{R}[a, b]$.

(b) This follows immediately from Theorem 1.3.1 (iv).

(c) By Corollary 1.2.11, $|f| \in \mathscr{R}[a, b]$. Let $\lambda = sgn \int_a^b f$. Then, since $\lambda f \leq |f|$,

$$\left| \int_a^b f \right| = \lambda \int_a^b f = \int_a^b \lambda f \leq \int_a^b |f| \, .$$

[Note that $sgn \, x := x/|x|$ if $x \in \mathbf{R}\{0\}$, $sgn \, 0 = 0$.]

(d) If f were not the zero function, then there would be a point $c \in [a, b]$ such that $f(c) > 0$; and then, by Theorem 1.3.1(vi), $\int_a^b f$ would be positive, giving a contradiction. □

Theorem 1.3.2 is particularly important: (a) shows that the family of all Riemann-integrable functions on a given interval $[a, b]$ is a real vector space when addition and multiplication by real numbers are defined in the obvious way; (b) is useful in the estimation of integrals of functions by integrals of simpler functions; and (c) will be so often used that recourse to it should become virtually automatic when faced with the problem of estimation of an integral.

Another most useful inequality is that of H. A. Schwarz:

Theorem 1.3.3 (Schwarz's inequality) *Let* $f, g \in \mathscr{R}[a, b]$. *Then*

$$\left(\int_a^b fg \right)^2 \leq \left(\int_a^b f^2 \right) \left(\int_a^b g^2 \right).$$

Proof For all $\lambda \in \mathbf{R}$, $(f + \lambda g)^2 \geq 0$; since $f^2, g^2, fg, (f + \lambda g)^2 \in \mathscr{R}[a, b]$ (by Theorem 1.3.2(a)) it follows from Theorem 1.3.2(b) that $\int_a^b (f + \lambda g)^2 \geq 0$ and hence

$$\int_a^b f^2 + \lambda^2 \int_a^b g^2 + 2\lambda \int_a^b fg \geq 0. \tag{1.3.1}$$

If $\int_a^b g^2 \neq 0$, the choice of $\lambda = -\int_a^b fg / \int_a^b g^2$ in (1.3.1) gives the result; if $\int_a^b g^2 = 0$, the choice of $|\lambda|$ sufficiently large shows that $\int_a^b fg = 0$ and the result again follows. □

It is next desirable to establish various results concerning the integrability of a function $f : [a, b] \to \mathbf{R}$ over subintervals $[c, d]$ of $[a, b]$. When dealing with objects such as the integral of f over $[c, d]$, for simplicity we shall use the notation $\int_c^d f$ rather than the more precise $\int_c^d g$, where g is the restriction of f to $[c, d]$.

Theorem 1.3.4 *Let $a < c < b$ and suppose that $f \in \mathscr{B}[a, b]$. Then*

$$\underline{\int_a^b} f = \underline{\int_a^c} f + \underline{\int_c^b} f, \quad \overline{\int_a^b} f = \overline{\int_a^c} f + \overline{\int_c^b} f.$$

Proof Let (P_n'), (P_n'') be sequences in $\mathscr{P}[a, c]$, $\mathscr{P}[c, b]$ respectively such that $w(P_n'), w(P_n'') \to 0$; let $P_n = P_n' \cup P_n''$ for all $n \in \mathbf{N}$. Then $P_n \in \mathscr{P}[a, b], w(P_n) \to 0$ and

$$\overline{\int_a^b} f = \lim_{n \to \infty} U(P_n, f) = \lim_{n \to \infty} U(P_n', f) + \lim_{n \to \infty} U(P_n'', f) = \overline{\int_a^c} f + \overline{\int_c^b} f.$$

The lower integrals are handled in a similar manner.   □

Corollary 1.3.5 *Suppose that $a = c_0 < c_1 < \ldots < c_m = b$ and let $f \in \mathscr{B}[a, b]$. Then*

$$\underline{\int_a^b} f = \sum_{j=1}^m \underline{\int_{c_{j-1}}^{c_j}} f \quad and \quad \overline{\int_a^b} f = \sum_{j=1}^m \overline{\int_{c_{j-1}}^{c_j}} f.$$

Proof Induction reduces the proof to that of the case $m = 2$, which is just Theorem 1.3.4.   □

Theorem 1.3.6 *Let $a < c < b$ and suppose that $f \in \mathscr{B}[a, b]$. Then $f \in \mathscr{R}[a, b]$ if, and only if, $f \in \mathscr{R}[a, c]$ and $f \in \mathscr{R}[c, b]$. Moreover,*

$$\int_a^b f = \int_a^c f + \int_c^b f$$

whenever one side of the equality exists.

Proof By Theorem 1.3.4,

$$\overline{\int_a^b} f - \underline{\int_a^b} f = \left(\overline{\int_a^c} f - \underline{\int_a^c} f \right) + \left(\overline{\int_c^b} f - \underline{\int_c^b} f \right);$$

thus by Corollary 1.1.8, $\overline{\int_a^b} f = \underline{\int_a^b} f$ if, and only if, $\overline{\int_a^c} f = \underline{\int_a^c} f$ and $\overline{\int_c^b} f = \underline{\int_c^b} f$. The rest follows from Theorem 1.3.4.   □

Corollary 1.3.7 *Let* $a \leq c < d \leq b$ *and suppose that* $f \in \mathscr{R}[a, b]$. *Then* $f \in \mathscr{R}[c, d]$.

Proof Although this result has already been established in Lemma 1.2.6, we give the following alternative proof. By Theorem 1.3.6, $f \in \mathscr{R}[a, b]$ implies that $f \in \mathscr{R}[a, d]$, which in turn implies that $f \in \mathscr{R}[c, d]$. □

Theorem 1.3.8 *Let* $f \in \mathscr{B}[a, b]$ *and suppose that* $f \in \mathscr{R}[c, d]$ *whenever* $a < c < d < b$. *Then* $f \in \mathscr{R}[a, b]$.

Proof Let K be such that $|f(x)| \leq K$ for all $x \in [a, b]$, and let $c, d \in (a, b), c < d$. Then since $f \in \mathscr{R}[c, d]$,

$$\left| \overline{\int_a^b} f - \underline{\int_a^b} f \right| = \left| \overline{\int_a^c} f - \underline{\int_a^c} f \right| + \left| \overline{\int_c^d} f - \underline{\int_c^d} f \right| + \left| \overline{\int_d^b} f - \underline{\int_d^b} f \right|$$

$$\leq 2K(c - a) + 0 + 2K(b - d).$$

As this is true for all $c, d \in (a, b)$ with $c < d$, we may let $c \to a, b \to d$ to obtain

$$\overline{\int_a^b} f = \underline{\int_a^b} f,$$

which means that $f \in \mathscr{R}[a, b]$. □

This last Theorem is a most useful test for integrability: the following examples give some idea of how it may be used.

Example 1.3.9

(i) Let $f : [0, 1] \to \mathbf{R}$ be defined by $f(x) = \sin(1/x)$ $(0 < x \leq 1)$, $f(0) = 10^{67}$. Despite the discontinuity at 0, $f \in \mathscr{R}[0, 1]$: for f is continuous on every subinterval $[c, d]$ of $[0, 1]$ with $c, d \in (0, 1)$ and $c < d$, and hence, by Theorem 1.2.5, $f \in \mathscr{R}[c, d]$ for all such c and d; f is also bounded on $[0, 1]$. By Theorem 1.3.8, $f \in \mathscr{R}[0, 1]$.

(ii) Let $f \in \mathscr{B}[a, b]$ be continuous at all points of $[a, b]$ save for a finite number. Then $f \in \mathscr{R}[a, b]$. To prove this, let the points of discontinuity of f lie among the points $a = c_0, c_1, \ldots, c_m = b$, where $c_0 < c_1 < \ldots < c_m$. By Theorem 1.3.8, $f \in \mathscr{R}[c_{j-1}, c_j]$ for $j = 1, \ldots, m$; by Corollary 1.3.5, $f \in \mathscr{R}[a, b]$.

Exercise 1.3.10

1. Let $\alpha_1, \ldots, \alpha_k \in \mathbf{R}$, suppose that $a = c_0 < c_1 < \ldots < c_k = b$ and let $f : [a, b] \to \mathbf{R}$ be such that $f(t) = \alpha_j$ if $t \in (c_{j-1}, c_j)$ $(j = 1, \ldots, k)$. Prove that $f \in \mathscr{R}[a, b]$ and that

$$\int_a^b f = \sum_{j=1}^k \alpha_j (c_j - c_{j-1})$$

irrespective of the values of f at the points $c_0, c_1, \ldots, c_k$.

2. Let $f : [0, 1] \to \mathbf{R}$ be defined by $f(x) = x^{-1/2} \sin x$ if $0 < x \leq 1$, $f(0) = 1$. Prove that $f \in \mathscr{R}[0, 1]$.

3. Let $\varepsilon > 0$ and suppose that $f \in \mathscr{R}[a, b]$ and $|f(t)| \geq \varepsilon$ for all $t \in [a, b]$. Prove that $1/f \in \mathscr{R}[a, b]$.

4. Let $f, g : [a, b] \to \mathbf{R}$ be increasing and decreasing respectively, and define

$$\phi(x) = f(x) - (b - a)^{-1} \int_a^b f(t)dt \quad (x \in [a, b]).$$

Show that there is a point c in $[a, b]$ such that $\phi(x) \leq 0$ if $a \leq x < c$, $\phi(x) \geq 0$ if $c < x \leq b$. By consideration of the identity

$$\int_a^b g(x)\phi(x)dx = \int_a^c g(x)\phi(x)dx + \int_c^b g(x)\phi(x)dx,$$

prove that $\int_a^b g(x)\phi(x)dx \leq 0$ and deduce that

$$\int_a^b f(x)g(x)dx \leq (b - a)^{-1} \left(\int_a^b f(x)dx \right) \int_a^b g(x)dx.$$

5. Let $c, \theta \in \mathbf{R}$ and suppose that $\theta > 0$; let $f \in \mathscr{B}[a, b]$. Show that

$$\overline{\int_a^b} f(t)dt = \theta \overline{\int_{(a-c)/\theta}^{(b-c)/\theta}} f(\theta t + c)dt, \quad \underline{\int_a^b} f(t)dt = \theta \underline{\int_{(a-c)/\theta}^{(b-c)/\theta}} f(\theta t + c)dt.$$

State and prove the corresponding result for $\theta < 0$.

6. Let $f : [0, 1] \to \mathbf{R}$ be defined as follows: $f(t) = 0$ if t is irrational, $f(t) = 1/q$ if $t = p/q$, where p and q are integers with no common factor greater than 1, and $q > 0$. Prove that f is continuous at irrational points but discontinuous at rational points. Show that $f \in \mathscr{R}[0, 1]$ and that

$$\int_0^1 f = 0.$$

1.4 Evaluation of Integrals: Integration and Differentiation

So far we have established that various types of functions have Riemann integrals and that the Riemann integral has properties which seem both natural and desirable. We now turn to the task of evaluation. Darboux's theorem suggests a direct approach, one that translates the problem into the evaluation of limits of sequences. Unfortunately, however, while this approach leads to systematic procedures for obtaining approximations to integrals, in practice it is seldom that these procedures facilitate the exact determination of a given integral. We present one such procedure below, together with an application.

Theorem 1.4.1 (The trapezium rule) *Let* $f \in \mathscr{R}[a, b]$ *and, for each* $n \in \mathbf{N}$, *let* $\{x_0, x_1, \ldots, x_n\}$ *be that partition which divides* $[a, b]$ *into* n *sub-intervals of equal length and let*

$$\tau_n = \frac{1}{2n} \sum_{r=1}^{n} \{f(x_{r-1}) + f(x_r)\}.$$

Then

$$\lim_{n \to \infty} \tau_n = \frac{1}{b-a} \int_a^b f.$$

Proof For each $r \in \{1, \ldots, n\}$ put $m_r = \inf\{f(x) : x_{r-1} \le x \le x_r\}$ and $M_r = \sup\{f(x) : x_{r-1} \le x \le x_r\}$. Then since

$$m_r \le \frac{1}{2}\{f(x_{r-1}) + f(x_r)\} \le M_r$$

it follows that

$$\frac{b-a}{n} \sum_{r=1}^{n} m_r \le \frac{b-a}{2n} \sum_{r=1}^{n} \{f(x_{r-1}) + f(x_r)\} \le \frac{b-a}{n} \sum_{r=1}^{n} M_r,$$

and so

$$\frac{b-a}{n} \sum_{r=1}^{n} m_r \le (b-a)\tau_n \le \frac{b-a}{n} \sum_{r=1}^{n} M_r.$$

The result now follows immediately from Theorem 1.1.7. $\qquad\qquad\square$

This is called the trapezium rule because $\frac{1}{2}\{f(x_{r-1}) + f(x_r)\}\frac{(b-a)}{n}$ is the area of the trapezium with vertices $(x_{r-1}, 0)$, $(x_r, 0)$, $(x_r, f(x_r))$, $(x_{r-1}, f(x_{r-1}))$ and $(b-a)\tau_n$ is the total area of these trapezia.

Example 1.4.2 Let $a \in \mathbf{R}$ and $|a| \neq 1$. We show that

$$\int_0^\pi \log(1 - 2a\cos x + a^2)dx = \begin{cases} 0 & \text{if } |a| < 1, \\ 2\pi \log|a| & \text{if } |a| > 1. \end{cases}$$

Proof The mapping $x \longmapsto \log(1 - 2a\cos x + a^2)$ is continuous and hence integrable over $[0, \pi]$. The trapezium rule and the identity

$$1 - a^{2n} = (1 - a^2) \prod_{r=1}^{n-1} \left(1 - 2a \cos \frac{r\pi}{n} + a^2\right)$$

show that the desired integral equals

$$\pi \lim_{n \to \infty} \frac{1}{2n} \left\{ \log(1-a)^2 + 2 \sum_{r=1}^{n-1} \log\left(1 - 2a \cos \frac{r\pi}{n} + a^2\right) + \log(1+a)^2 \right\}$$

$$= \pi \lim_{n \to \infty} \frac{1}{2n} \log \left\{ (1-a^2) \prod_{r=1}^{n-1} \left(1 - 2a \cos \frac{r\pi}{n} + a^2\right) \right\}^2$$

$$= \pi \lim_{n \to \infty} \frac{1}{n} \log \left| 1 - a^{2n} \right|.$$

The stated result follows. $\square$

The ease with which the trapezium rule copes with the above example should not beguile the reader. For the evaluation of

$$\int_0^1 \sqrt{x(1-x)} \, dx = \lim_{n \to \infty} \frac{1}{n} \sum_{r=1}^{n-1} \sqrt{\frac{r}{n}\left(1 - \frac{r}{n}\right)}$$

the rule is unhelpful, and commonly one finds exact evaluation through such rules impossible. A partial amelioration of this state of affairs comes about via the observation that given an integrable function, then rapid evaluation of its integral is possible if the function is recognisable as the derivative of another.

Definition 1.4.3 Let I be a non-degenerate interval in $\mathbf{R}$, let $f : I \to \mathbf{R}$ and suppose that $F : I \to \mathbf{R}$ is differentiable, with $F'(t) = f(t)$ for all $t \in I$. Then F is said to be a **primitive of** f (on I).

Theorem 1.4.4 (The first fundamental theorem of calculus) *Let $f \in \mathscr{R}[a, b]$ and suppose that F is a primitive of f. Then*

$$\int_a^b f = F(b) - F(a).$$

Proof Let $P = \{x_0, x_1, \ldots, x_n\} \in \mathscr{P}[a, b]$. By the mean-value theorem, there exists $t_j \in (x_{j-1}, x_j)$ such that $F(x_j) - F(x_{j-1}) = f(t_j)\Delta x_j$ $(j = 1, \ldots, n)$; hence

$$L(P, f) \le \sum_{j=1}^n f(t_j)\Delta x_j = \sum_{j=1}^n \left\{ F(x_j) - F(x_{j-1}) \right\}$$

$$= F(b) - F(a) \le U(P, f).$$

Thus

$$\underline{\int_a^b} f \leq F(b) - F(a) \leq \overline{\int_a^b} f,$$

and as $f \in \mathscr{R}[a, b]$, the result follows. □

Example 1.4.5

(i) To evaluate $\int_a^b x^2 dx$, we note that a primitive of the continuous (hence integrable) function f, where $f(x) = x^2$, is F, where $F(x) = x^3/3$. Thus by Theorem 1.4.4,

$$\int_a^b x^2 dx = (b^3 - a^3)/3.$$

(ii) The function $f : [-1, 1] \to \mathbf{R}$ defined by

$$f(x) = \begin{cases} 1 & \text{if } 0 < x \leq 1, \\ 0 & \text{if } x = 0, \\ -1 & \text{if } -1 \leq x < 0 \end{cases}$$

is in $\mathscr{R}[-1, 1]$ as it is continuous on $[-1, 1]$ except at 0; but it has no primitive on $[-1, 1]$ as it does not have the intermediate-value property enjoyed by continuous functions and functions which are derivatives (see Exercise 1.4.15/11).

(iii) The function $F : [0, 1] \to \mathbf{R}$ defined by

$$F(x) = \begin{cases} x^2 \sin(1/x^2) & \text{if } 0 < x \leq 1, \\ 0 & \text{if } x = 0 \end{cases}$$

has derivative $f : [0, 1] \to \mathbf{R}$ given by $f(x) = 2x \sin(1/x^2) - 2x^{-1} \cos(1/x^2)$ $(0 < x \leq 1)$, $f(0) = 0$. As f is unbounded on $[0, 1]$, it is not in $\mathscr{R}[0, 1]$. This gives an example of a function which has a primitive but is not Riemann-integrable. It is possible to construct a function which is bounded on $[0, 1]$, has a primitive but is not in $\mathscr{R}[0, 1]$; this task is much harder. The first published example is believed to be that of Volterra [18]; see also [7, p. 107], [8, p. 210], [12, pp. 37–39] and the discussion in [9], 9.3.

Examples (ii) and (iii) exhibit the force of the hypotheses in Theorem 1.4.4.

We can now give the familiar integration-by-parts method for the evaluation of integrals.

Theorem 1.4.6 *Let $f, g \in \mathscr{R}[a, b]$ and suppose there are differentiable functions $F, G : [a, b] \to \mathbf{R}$ such that $F' = f$ and $G' = g$. Then*

$$\int_a^b Fg = F(b)G(b) - F(a)G(a) - \int_a^b Gf.$$

Proof Let $H = FG$ (the product function). Then $H' = Fg + Gf$; by Theorem 1.3.2(a), $H' \in \mathscr{R}[a, b]$; and by Theorem 1.4.4,

$$\int_a^b (Fg + Gf) = F(b)G(b) - F(a)G(a),$$

which gives the result.   □

Example 1.4.7

(i) To evaluate $\int_0^1 xe^x dx$, take $F(x) = x$, $G(x) = e^x$; then

$$\int_0^1 xe^x dx = e - \int_0^1 e^x dx = 1.$$

(ii) The evaluation of $\int_1^2 \log x dx$ proceeds by taking $F(x) = \log x$, $G(x) = x$; then

$$\int_1^2 \log x dx = 2 \log 2 - \int_1^2 dx = 2 \log 2 - 1.$$

Remark 1.4.8 Theorems 1.4.4 and 1.4.6 may be slightly varied to give useful results. For example, Theorem 1.4.4 may be revised as follows: suppose that $f \in \mathscr{R}[a, b]$ and that there is a continuous function $F : [a, b] \to \mathbf{R}$ such that F is differentiable on (a, b) and $F'(t) = f(t)$ for all $t \in (a, b)$. Then $\int_a^b f = F(b) - F(a)$.

The proof is the same as that of Theorem 1.4.4. Theorem 1.4.6 may be similarly revised.

Associated with Theorem 1.4.4 is a natural question of existence: which functions in $\mathscr{R}[a, b]$ have a primitive? The second fundamental theorem of calculus, given below, provides a partial result in this connection. Bearing upon the question, note that Theorems 1.3.6 and 1.4.4 show that $f \in \mathscr{R}[a, b]$ has a primitive if, and only if, the mapping $x \longmapsto \int_a^x f$ is differentiable and has derivative f.

Theorem 1.4.9 (The second fundamental theorem of calculus) *Let J be a non-degenerate interval in $\mathbf{R}$, let $a \in J$ and suppose that $f : J \to \mathbf{R}$ belongs to $\mathscr{R}(I)$ for every closed, bounded, non-degenerate interval $I \subset J$. Define $F : J \to \mathbf{R}$ by*

$$F(x) = \begin{cases} \int_a^x f & \text{if } x > a, \ x \in J, \\ 0 & \text{if } x = a, \\ -\int_x^a f & \text{if } x < a, \ x \in J. \end{cases}$$

Then F is continuous. If f is right- (left-) continuous at $x_0 \in J$, then F is right- (left-) differentiable at x_0 and

$$F'_+(x_0) = f(x_0) \quad (F'_-(x_0) = f(x_0)).$$

In particular, if f is continuous at x_0, then F is differentiable at x_0 and $F'(x_0) = f(x_0)$.

Proof Suppose that $b \in J, b > a$. Then $f \in \mathcal{R}[a, b]$ and there exists a real number M, depending on b, such that $|f(t)| < M$ if $a \le t \le b$. Let $\varepsilon > 0$. By Theorems 1.3.2 and 1.3.6, if $a \le x < y \le b$, then

$$|F(y) - F(x)| = \left| \int_x^y f \right| \le \int_x^y |f| \le M(y - x) < \varepsilon$$

if $|y - x| < \varepsilon/M$. Hence F is continuous at each point of (a, b), is right-continuous at a and left-continuous at b. A similar discussion shows that F is continuous at points to the left of a.

Now suppose that f is right-continuous at $x_0 \in J$ and let $\varepsilon > 0$. Then there exists $\delta > 0$ such that $|f(x) - f(x_0)| < \varepsilon$ if $0 \le x - x_0 < \delta$. Thus if $x_0 < x < x_0 + \delta$,

$$\left| \frac{F(x) - F(x_0)}{x - x_0} - f(x_0) \right| = \left| (x - x_0)^{-1} \int_{x_0}^x (f(t) - f(x_0))dt \right|$$

$$\le (x - x_0)^{-1} \int_{x_0}^x \varepsilon dt = \varepsilon.$$

Hence $F'_+(x_0) = f(x_0)$. The rest is now clear.   □

Remark 1.4.10

(i) It is convenient to extend the use of the symbol $\int_a^x f$. So far defined for $x > a$ and $f \in \mathcal{R}[a, x]$, we define it to be zero if $x = a$, and to be $-\int_x^a f$ if $x < a$ and $f \in \mathcal{R}[x, a]$. Given this extension, it is immediate from Theorem 1.4.9 that, if $f : J \to \mathbf{R}$ is continuous, then

$$\frac{d}{dx} \left(\int_a^x f \right) = f(x) \ (x \in J). \tag{1.4.1}$$

(ii) Suppose the hypothesis concerning f in Theorem 1.4.9 is relaxed, simply to require that $f \in \mathcal{B}(I)$ for every closed, bounded, non-degenerate interval $I \subset J$. Then, adopting an extension of symbols analogous to that of (i), the maps

$$x \longmapsto \underline{\int_a^x} f \text{ and } x \longmapsto \overline{\int_a^x} f$$

are easily proved to have properties identical to those derived for the function F of the theorem.

Next we give an application of Theorem 1.4.9 which yields the valuable technique of integration known as 'integration by change of variable' or 'integration by substitution'.

Theorem 1.4.11 (The change of variable theorem) *Let $\phi : [a, b] \to \mathbf{R}$ be continuously differentiable, let J be a non-degenerate interval in $\mathbf{R}$ such that $\phi([a, b]) \subset J$, and suppose that $f : J \to \mathbf{R}$ is continuous. Then*

$$\int_a^b f(\phi(t))\phi'(t)dt = \int_{\phi(a)}^{\phi(b)} f(x)dx.$$

(Note that $\phi(a)$ need not be less than $\phi(b)$; the convention introduced in Remark 1.4.10 is to be used in such cases.)

Proof Let $F(x) = \int_{\phi(a)}^x f$ ($x \in J$). By Theorem 1.4.9, F is continuously differentiable on J and $F' = f$. Hence by the chain rule $F \circ \phi$ is continuously differentiable on $[a, b]$ and

$$(F \circ \phi)'(t) = f(\phi(t))\phi'(t) \text{ for } a \le t \le b.$$

Since $(F \circ \phi)'$ is continuous on $[a, b]$ it is in $\mathcal{R}[a, b]$, and so by Theorem 1.4.4,

$$\int_a^b f(\phi(t))\phi'(t)dt = F(\phi(b)) - F(\phi(a))$$

$$= \int_{\phi(a)}^{\phi(b)} f(x)dx,$$

as required.   □

Note that if ϕ is differentiable but ϕ' is not continuous, then the same proof shows that the result still holds under the additional hypothesis that $\int_a^b f(\phi(t))\phi'(t)dt$ exists.

Example 1.4.12

(i) To evaluate $\int_0^1 \sqrt{x(1-x)}dx$, note that $f : x \longmapsto \sqrt{x(1-x)}$ is continuous on $[0, 1]$; also $\phi : t \longmapsto \frac{1}{2}(1+\sin t)$ is continuously differentiable on $[-\pi/2, \pi/2]$ and $\phi([-\pi/2, \pi/2]) = [0, 1]$. Hence by Theorem 1.4.11,

$$\int_0^1 \sqrt{x(1-x)}dx = \int_{-\pi/2}^{\pi/2} \frac{1}{4}\cos^2 t\, dt = \frac{\pi}{8}.$$

(ii) Suppose $0 < u < 1$. The map $f : x \longmapsto (1+x^2)^{-1}$ is continuous on $[u, 1]$ and $\phi : t \longmapsto t^{-1}$ is continuously differentiable on $[1, u^{-1}]$. By Theorem 1.4.11,

$$\int_1^{u^{-1}} (1+t^2)^{-1}dt = \int_u^1 (1+x^2)^{-1}dx;$$

by Theorem 1.4.9,

$$\lim_{u \to 0^+} \int_u^1 (1+x^2)^{-1}dx = \int_0^1 (1+x^2)^{-1}dx = \frac{\pi}{4}.$$

Hence, as $v \to \infty$,

$$\tan^{-1} v = \int_0^v (1+x^2)^{-1}dx \to 2\int_0^1 (1+x^2)^{-1}dx = \frac{\pi}{2}.$$

We conclude this section by giving two 'mean-value' theorems for integrals of products of functions; these are useful in the estimation of integrals which are difficult to evaluate directly.

Theorem 1.4.13 (The first mean-value theorem for integrals) *Let $f : [a, b] \to \mathbf{R}$ be continuous, and suppose that $g \in \mathscr{R}[a, b]$ and $g(t) \geq 0$ for $a \leq t \leq b$. Then there exists $\xi \in [a, b]$ such that*

$$\int_a^b fg = f(\xi)\int_a^b g.$$

Proof Let m and M be respectively the minimum and maximum value of f on $[a, b]$. Then for all $t \in [a, b]$, $mg(t) \leq f(t)g(t) \leq Mg(t)$, and so $m\int_a^b g \leq \int_a^b fg \leq M\int_a^b g$. Hence by the intermediate-value theorem, the result follows. □

Note that the special case in which $g = 1$ shows that

$$\int_a^b f = (b-a)f(\xi).$$

Theorem 1.4.14 (The second mean-value theorem for integrals) *Suppose that $f : [a, b] \to \mathbf{R}$ is monotone, differentiable and such that $f' \in \mathscr{R}[a, b]$; let $g : [a, b] \to \mathbf{R}$ be continuous. Then there exists $\xi \in [a, b]$ such that*

$$\int_a^b fg = f(a)\int_a^\xi g + f(b)\int_\xi^b g.$$

If in addition f is decreasing and $f(t) \geq 0$ for all $t \in [a, b]$, then there exists $\zeta \in [a, b]$ such that

$$\int_a^b fg = f(a)\int_a^\zeta g.$$

Proof Let $G(x) = \int_a^x g$ for $a \le x \le b$; by Theorem 1.4.9, $G'(x) = g(x)$ on $[a, b]$. Integration by parts (Theorem 1.4.6) now shows that

$$\int_a^b fg = f(b)G(b) - \int_a^b Gf'.$$

As G is continuous and either $f'(t) \ge 0$ for all $t \in [a, b]$ or $f'(t) \le 0$ for all $t \in [a, b]$, it follows from Theorem 1.4.13 that there exists $\xi \in [a, b]$ such that

$$\int_a^b fg = f(b)G(b) - G(\xi) \int_a^b f' = f(b)G(b) + \{f(a) - f(b)\} G(\xi)$$

$$= f(a) \int_a^\xi g + f(b) \int_\xi^b g.$$

For the second part, note that either $G(b) \ge G(\xi)$, in which case

$$f(a)G(b) \ge f(b)G(b) + \{f(a) - f(b)\} G(\xi) \ge f(a)G(\xi);$$

or $G(b) \le G(\xi)$, so that

$$f(a)G(\xi) \ge f(b)G(b) + \{f(a) - f(b)\} G(\xi) \ge f(a)G(b).$$

Whichever is the case, the intermediate-value theorem shows that there exists $\zeta \in [a, b]$ such that

$$f(a) \int_a^\zeta g = f(a)G(\zeta) = f(b)G(b) + \{f(a) - f(b)\} G(\xi) = \int_a^b fg.$$

□

Exercise 1.4.15

1. Let $f \in \mathcal{R}[a, b]$ and suppose that for each $n \in \mathbf{N}$ and each $r \in \{1, \ldots, n\}$, real numbers $\xi_r^{(n)}$ are chosen so that

$$a + (r - 1)(b - a)/n \le \xi_r^{(n)} \le a + r(b - a)/n.$$

Prove that

$$\lim_{n \to \infty} \frac{1}{n} \sum_{r=1}^n f(\xi_r^{(n)}) = (b - a)^{-1} \int_a^b f.$$

2. Using the result of the preceding exercise, show that

$$(i) \int_0^\pi \sin x dx = 2, \quad (ii) \int_0^1 e^{-x} dx = 1 - e^{-1}.$$

3. Prove that

$$\frac{\pi}{4} = \int_0^1 \frac{1}{1+t^2} dt = \lim_{n \to \infty} n \sum_{r=1}^n \frac{1}{n^2 + r^2}.$$

4. (*Simpson's rule*) Let $f \in \mathcal{R}[a, b]$ and, for each $n \in \mathbf{N}$, let $\{x_0, x_1, \ldots, x_{2n}\}$ be that partition which divides $[a, b]$ into $2n$ sub-intervals of equal length and let

$$\theta_n = \frac{1}{6n} \sum_{r=1}^n \{f(x_{2r-2}) + 4f(x_{2r-1}) + f(x_{2r})\}.$$

Show that

$$\lim_{n \to \infty} \theta_n = \frac{1}{(b-a)} \int_a^b f.$$

5. Let $f : [a, b] \to \mathbf{R}$ be differentiable and suppose that $f' \in \mathcal{R}[a, b]$. Using integration by parts, show that

$$\lim_{\lambda \to \infty} \int_a^b f(t) \cos(\lambda t) dt = 0, \quad \lim_{\lambda \to \infty} \int_a^b f(t) \sin(\lambda t) dt = 0.$$

(This is a special case of the *Riemann-Lebesgue lemma*: see Exercise 1.7.17/16.)

6. Let $f : \mathbf{R} \to \mathbf{R}$ be continuous and define $G : \mathbf{R} \to \mathbf{R}$ by

$$G(x) = \int_0^{\sin x} f(t) dt \quad (x \in \mathbf{R}).$$

Show that G is differentiable on $\mathbf{R}$ and compute G'.

7. For each $\lambda > 0$ let $I_\lambda(x)$ be defined by

$$I_\lambda(x) = \int_x^\pi \frac{1}{2} \sin(\lambda t) \operatorname{cosec}(t/2) dt \quad (0 < x < 2\pi).$$

Prove that

$$\lim_{\lambda \to \infty} I_\lambda(x) = 0 \quad (0 < x < 2\pi).$$

By considering $I_{n+\frac{1}{2}}(x) - I_{n-\frac{1}{2}}(x)$, show that the series $\sum_{n=1}^\infty \frac{\sin nx}{n}$ converges for $0 < x < 2\pi$ and find its sum. Deduce that

$$\frac{\pi}{4} = 1 - \frac{1}{3} + \frac{1}{5} - \frac{1}{7} + \ldots .$$

8. Let $a > 0$ and suppose that $f \in \mathscr{R}[0, a]$. Prove that

$$\int_0^a f(x)dx = \int_0^a f(a - x)dx = \frac{1}{2}\int_0^a (f(x) + f(a - x))dx.$$

Show that

$$\int_0^a \log(1 + \tan a \tan x)dx = a \log(\sec a) \quad (0 < a < \pi/2).$$

9. For all real $\alpha > 0$ and all integers $n \geq 0$, let

$$I_n = I_n(\alpha) = \int_{-1}^1 (1 - x^2)^n \cos(\alpha x)dx.$$

Show that if $n \geq 2$,

$$\alpha^2 I_n = 2n(2n - 1)I_{n-1} - 4n(n - 1)I_{n-2},$$

and deduce that for all $n \geq 1$,

$$\alpha^{2n+1} I_n(\alpha) = n!\{P_n(\alpha) \cos \alpha + Q_n(\alpha) \sin \alpha\},$$

where P_n and Q_n are polynomials of degree less than $2n + 1$ with integer coefficients.
Show that if $\alpha = \pi/2$ and if $\pi/2$ were equal to b/a for some positive integers a and b, then

$$J_n = b^{2n+1}I_n(\pi/2)/n!$$

would be an integer. By considering what happens to J_n as $n \to \infty$, prove that π is irrational.

10. Let $J(n) = \int_0^{\pi/2} \sin^n \theta d\theta$ $(n = 0, 1, 2, ...)$. Prove that

$$J(2n) = \frac{(2n)!\pi}{(n!)^2 2^{2n+1}}, \quad J(2n + 1) = \frac{(n!)^2 2^{2n}}{(2n + 1)!}$$

and that, for $n \geq 1$,

$$J(2n)J(2n + 1) < J^2(2n) < J(2n)J(2n - 1).$$

Deduce *Wallis's inequality*:

$$\frac{1}{\left(n + \frac{1}{2}\right)\pi} < \frac{((2n)!)^2}{(n!)^4 \, 2^{4n}} < \frac{1}{n\pi} \quad (n \in \mathbf{N})$$

and hence obtain *Wallis's product:*

$$\pi = \lim_{n \to \infty} \frac{2^{4n}(n!)^4}{n((2n)!)^2}.$$

11. Let $f : [a, b] \to \mathbf{R}$ be differentiable and let $f'(a) < \gamma < f'(b)$. Prove that there exists $c \in (a, b)$ such that $f'(c) = \gamma$. (A similar result holds, of course, if $f'(a) > f'(b)$.) [Hint: consider $g(x) = f(x) - \gamma x$.]

12. Give an example of a function $f : [0, 1] \to \mathbf{R}$ which has a primitive, is in $\mathscr{R}[0, 1]$ and is not continuous.

13. Using Theorem 1.4.14, or otherwise, show that if $0 < a < b$, then

$$\left| \int_a^b \frac{\sin x}{x} dx \right| \leq \frac{2}{a}.$$

14. Let $I = [0, 1]$ and let $g : I \times I \to \mathbf{R}$ be defined by

$$g(x, y) = \begin{cases} 1 - y & \text{if } x \text{ is rational,} \\ y & \text{if } x \text{ is irrational.} \end{cases}$$

Show that

$$\int_0^1 \left(\int_0^1 g(x, y) dx \right) dy \neq \int_0^1 \left(\int_0^1 g(x, y) dy \right) dx.$$

15. Let $a \in \mathbf{R}$, $|a| \neq 1$, and let

$$I(a) = \int_0^\pi \log(1 - 2a \cos x + a^2) dx$$

(see Example 1.4.2). By splitting the domain of integration into $[0, \pi/2]$ and $[\pi/2, \pi]$, and then making the substitution $x = \pi - y$ in the integral over $[\pi/2, \pi]$, show that $I(a) = \frac{1}{2}I(a^2)$. Hence calculate $I(a)$.

16. (Van der Corput's theorem) Let $k \in \mathbf{N}$, $\lambda > 0$ and suppose that $f : [a, b] \to \mathbf{R}$ has derivatives of all orders and $\left| f^{(k)}(x) \right| \geq 1$ for all $x \in [a, b]$. Show that there is a constant c_k, independent of f and λ, such that

$$\left| \int_a^b \cos(\lambda f(x)) dx \right| \leq c_k \lambda^{-1/k}$$

provided that either (i) $k \geq 2$, or (ii) $k = 1$ and f' is monotone (increasing or decreasing).

[Hint: For (ii), write

$$\int_a^b \cos(\lambda f(x))dx = \int_a^b \frac{1}{\lambda f'(x)} \frac{d}{dx} (\sin(\lambda f(x)))\, dx$$

and use integration by parts. For (i), use induction: Let $P(k)$ be the proposition that the desired inequality holds for some k, assume that $P(k)$ holds, suppose that $f^{(k+1)}(x) \geq 1$ for all $x \in [a, b]$, and let c be the unique point in $[a, b]$ at which $\left| f^{(k)} \right|$ assumes its minimum. If $f^{(k)}(c) = 0$, consider the intervals $[a, c - \delta]$, $[c - \delta, c + \delta]$, $[c + \delta, b]$ for suitable $\delta > 0$; if $f^{(k)}(c) \neq 0$, note that c is either a or b.]

1.5 Applications

Here we give a variety of results to illustrate the use of integrals in different parts of elementary analysis.

1.5.1 The Integral Formula for the Logarithmic Function

To begin, recall that the exponential function $\exp : \mathbf{R} \to \mathbf{R}^+$ defined by $\exp x = \sum_{n=0}^{\infty} \frac{x^n}{n!}$ ($x \in \mathbf{R}$) has various pleasant properties: it is differentiable and $(\exp x)' = \exp x$ for all $x \in \mathbf{R}$; $\exp(x).\exp(-x) = 1$ for all $x \in \mathbf{R}$; $\exp x > 0$ for all $x \in \mathbf{R}$ and $\exp$ is strictly increasing; $\exp(x + y) = \exp(x).\exp(y)$ for all $x, y \in \mathbf{R}$; for all $n \in \mathbf{Z}$, $x^n \exp x \to \infty$ as $x \to \infty$.

Since the exponential function is strictly increasing and has everywhere a non-zero derivative, its inverse, $\log : \mathbf{R}^+ \to \mathbf{R}$ is also strictly increasing and differentiable. This inverse, the logarithmic function, is defined by

$$\log(\exp x) = x \text{ for all } x \in \mathbf{R};$$

or, equivalently, by

$$\exp(\log y) = y \text{ for all } y \in \mathbf{R}^+.$$

From the properties of $\exp$ mentioned above it follows directly that for all $u, v > 0$,

$$\log(uv) = \log u + \log v, \quad (\log u)' = 1/u;$$

and that

$$\lim_{u \to \infty} \log u = \infty, \quad \lim_{u \to 0+} \log u = -\infty.$$

Since

$$1 = (\log u)' \, |_{u=1} = \lim_{h \to 0} \frac{\log(1+h) - \log 1}{h} = \lim_{h \to 0} \frac{1}{h} \log(1+h)$$

$$= \lim_{n \to \infty} n \log \left(1 + \frac{1}{n} \right) = \lim_{n \to \infty} \log \left(1 + \frac{1}{n} \right)^n,$$

use of the continuity of exp shows that

$$\exp(1) = \exp \left(\lim_{n \to \infty} \log \left(1 + \frac{1}{n} \right)^n \right) = \lim_{n \to \infty} \exp \left(\log \left(1 + \frac{1}{n} \right)^n \right)$$

$$= \lim_{n \to \infty} \left(1 + \frac{1}{n} \right)^n = e.$$

Hence

$$e = \exp(1) = \sum_{n=0}^{\infty} \frac{1}{n!} \text{ and } \log e = 1.$$

As for powers, induction shows that for all $n \in \mathbf{Z}$ and all $t > 0$,

$$t^n = (\exp(\log t))^n = \exp(n \log t).$$

Since

$$\left(\exp \left(\frac{1}{m} \log t \right) \right)^m = \exp(\log t) = t,$$

it follows that

$$t^{1/m} = \exp \left(\frac{1}{m} \log t \right)$$

for all $m \in \mathbf{N}$ and all $t > 0$. Consideration of these results shows that for all $r \in \mathbf{Q}$ and all $t > 0$,

$$t^r = \exp(r \log t).$$

This makes it natural to *define* t^α, for any real number α and any $t > 0$, by

$$t^\alpha = \exp(\alpha \log t).$$

It is now easy to verify that the usual laws of indices hold, that the function $t \longmapsto t^{\alpha}$ is differentiable on $(0, \infty)$ and that $(t^{\alpha})' = \alpha t^{\alpha-1}$. Thus a primitive of the function $t \longmapsto t^{\alpha}$ on $(0, \infty)$ is the function with values $(1 + \alpha)^{-1} t^{1+\alpha}$ $(\alpha \neq -1)$ and $\log t$ $(\alpha = -1)$. The function $t \longmapsto t^{\alpha}$ is continuous and is therefore in $\mathscr{R}[x, y]$ if $0 < x < y < \infty$; by Theorem 1.4.4,

$$\int_x^y t^{\alpha} dt = \begin{cases} (1+\alpha)^{-1} \left(y^{1+\alpha} - x^{1+\alpha} \right) & \text{if } \alpha \neq -1, \\ \log y - \log x & \text{if } \alpha = -1. \end{cases}$$

Evidently $\int_1^x t^{-1} dt = \log x$ $(1 < x < \infty)$, $\int_x^1 t^{-1} dt = -\log x$ $(0 < x < 1)$; hence

$$\log x = \int_1^x t^{-1} dt \quad \text{if } 0 < x < \infty.$$

This integral representation of the logarithmic function is often taken as the *definition* of $\log x$; various properties of the function follow in a simple way from it. Thus if $x, y > 0$, then

$$\int_1^{xy} t^{-1} dt = \int_1^x t^{-1} dt + \int_x^{xy} t^{-1} dt,$$

and use of the substitution $s = t/x$ in the second integral on the right-hand side gives

$$\int_1^{xy} t^{-1} dt = \int_1^x t^{-1} dt + \int_1^y s^{-1} ds,$$

so that

$$\log(xy) = \log x + \log y.$$

Information about the growth of $\log x$ with x can be obtained very easily from the integral formula. For if $0 < \varepsilon < \alpha$ and $x > 1$, then

$$x^{-\alpha} \log x = x^{-\alpha} \int_1^x t^{-1} dt < x^{-\alpha} \int_1^x t^{\varepsilon-1} dt = x^{-\alpha} \left(\frac{x^{\varepsilon} - 1}{\varepsilon} \right)$$
$$< x^{\varepsilon-\alpha}/\varepsilon;$$

hence $x^{-\alpha} \log x \to 0$ as $x \to \infty$, provided that $\alpha > 0$.

Series expansions can also be obtained by use of the integral representation. The identity

$$(1+t)^{-1} - \sum_{r=1}^{n} (-t)^{r-1} = (1+t)^{-1} (-t)^n \quad (t \neq -1)$$

(obtained by summation of the geometric series $\sum_{r=1}^{n}(-t)^{r-1}$) shows that if $-1 < x \leq 1$, then

$$\left| \int_0^x (1+t)^{-1} dt - \sum_{r=1}^{n} \int_0^x (-t)^{r-1} dt \right| = \left| \int_0^x (1+t)^{-1}(-t)^n dt \right|$$

$$\leq \begin{cases} \dfrac{x^{n+1}}{(n+1)}, & \text{if } 0 \leq x \leq 1, \\[2ex] \dfrac{|x^{n+1}|}{(n+1)(1+x)}, & \text{if } -1 < x < 0 \end{cases}$$

$$\to 0 \text{ as } n \to \infty.$$

We thus have

$$\log(1+x) = \sum_{r=1}^{\infty} \frac{(-1)^{r-1}}{r} x^r \text{ if } -1 < x \leq 1.$$

In particular, this gives

$$\log 2 = \sum_{r=1}^{\infty} \frac{(-1)^{r-1}}{r}.$$

Lastly, yet another use of the integral representation of the logarithm leads to a very simple proof of the famous *arithmetic-geometric mean inequality*: this asserts that given any natural number n, any positive numbers $a_1, \ldots, a_n$ and any positive numbers $p_1, \ldots, p_n$ such that $p_1 + \ldots + p_n = 1$, then

$$G_n := \prod_{r=1}^{n} a_r^{p_r} \leq \sum_{r=1}^{n} p_r a_r =: A_n,$$

with equality if, and only if, the a_r are all equal. To prove this, note that for all $x > 0$,

$$x - 1 - \log x = \int_1^x \left(1 - t^{-1}\right) dt \geq 0;$$

moreover, equality holds if, and only if, $x = 1$. If the a_k are all equal, then plainly $G_n = A_n$; and if they are not all equal, then $G_n < A_n$ since

$$A_n G_n^{-1} - 1 = \sum_{k=1}^{n} p_k \left(a_k G_n^{-1} - 1\right) > \sum_{k=1}^{n} p_k \log\left(a_k G_n^{-1}\right) = 0.$$

1.5.2 The Integral Test for Convergence of Series

Theorem 1.5.1 *Let $r \in \mathbf{N}$, let $\phi : [r, \infty) \to [0, \infty)$ be decreasing and put*

$$a_n = \sum_{k=r}^{n} \phi(k) - \int_r^{n+1} \phi \quad (n \in \mathbf{N}, n \geq r).$$

Then there exists $a \in \mathbf{R}$ such that $a_n \uparrow a$ as $n \uparrow \infty$, and $0 \leq a \leq \phi(r)$. Moreover, the series $\sum_{n=r}^{\infty} \phi(n)$ converges if, and only if, the sequence $\left(\int_r^n \phi\right)_{n \geq r}$ converges.

Proof Since ϕ is monotone, it is Riemann-integrable over every closed, bounded interval contained in $[r, \infty)$. For $k \geq r$, $k \in \mathbf{N}$,

$$\phi(k+1) \leq \phi(t) \leq \phi(k) \text{ if } k \leq t \leq k+1,$$

and hence

$$\phi(k+1) \leq \int_k^{k+1} \phi \leq \phi(k).$$

Thus

$$\sum_{k=r}^{n} \phi(k+1) \leq \int_r^{n+1} \phi \leq \sum_{k=r}^{n} \phi(k),$$

which implies that

$$0 \leq \sum_{k=r}^{n} \phi(k) - \int_r^{n+1} \phi \leq \phi(r) - \phi(n+1) \leq \phi(r).$$

It follows that (a_n) is an increasing sequence bounded above by $\phi(r)$, and so there is a real number a with $a_n \uparrow a$ as $n \uparrow \infty$. The rest follows directly from the definition of a_n.   □

Example 1.5.2 Suppose that $\alpha \geq 0$ and let $\phi(t) = t^{-\alpha} \; (1 \leq t < \infty)$. Then

$$\int_1^m t^{-\alpha} dt = \begin{cases} (1-\alpha)^{-1}(m^{1-\alpha} - 1) & \text{if } \alpha \neq 1, \\ \log m & \text{if } \alpha = 1. \end{cases}$$

By Theorem 1.5.1, there is a real number A_α such that as $m \to \infty$,

$$\sum_{n=1}^{m-1} n^{-\alpha} - \int_1^m t^{-\alpha} dt \to A_\alpha.$$

Moreover, since

$$\lim_{m \to \infty} \int_1^m t^{-\alpha} dt = \begin{cases} \infty & \text{if } \alpha \le 1, \\ (\alpha - 1)^{-1} & \text{if } \alpha > 1, \end{cases}$$

we see that $\sum_{n=1}^{\infty} n^{-\alpha}$ converges if $\alpha > 1$ and diverges if $\alpha \le 1$. (When $\alpha \le 0$ the series diverges as the nth term fails to converge to 0 as $n \to \infty$.) The choice of $\alpha = 1$ shows that

$$A_1 = \lim_{m \to \infty} \left(1 + \frac{1}{2} + ... + \frac{1}{m-1} - \log m \right) < \infty;$$

this constant A_1 is known as Euler's constant, is usually denoted by γ, and is approximately equal to 0.5772.

1.5.3 Taylor's Theorem and the Binomial Series

Decidedly the most useful version of Taylor's theorem, from the point of view of estimation of the remainder, is that which expresses the remainder as an integral. We present this version here, and then illustrate its use in obtaining the binomial theorem for an arbitrary real exponent.

Theorem 1.5.3 (Taylor's theorem) *Let J be an interval in $\mathbf{R}$, let α and β be distinct points of J, let f and its first n derivatives (for some $n \in \mathbf{N}$) be real-valued functions defined on J, and suppose that $f^{(n)} \in \mathcal{R}(I)$ for each non-degenerate, closed, bounded interval $I \subset J$. Then*

$$f(\beta) = \sum_{k=0}^{n-1} \frac{f^{(k)}(\alpha)}{k!} (\beta - \alpha)^k + \frac{1}{(n-1)!} \int_\alpha^\beta (\beta - t)^{n-1} f^{(n)}(t) dt.$$

Proof When $n = 1$, the result follows directly from the fundamental theorem of calculus (Theorem 1.4.4). When $n > 1$, put

$$E_k(\beta) = \frac{1}{(k-1)!} \int_\alpha^\beta (\beta - t)^{k-1} f^{(k)}(t) dt \quad (1 \le k \le n).$$

By Theorem 1.4.4, $E_1(\beta) = f(\beta) - f(\alpha)$; and by Theorem 1.4.6,

$$E_k(\beta) - E_{k+1}(\beta) = \frac{(\beta - \alpha)^k}{k!} f^{(k)}(\alpha) \quad (1 \le k \le n - 1).$$

Summation of this over k from 1 to $n - 1$ then gives

$$f(\beta) - f(\alpha) - \sum_{k=1}^{n-1} \frac{(\beta - \alpha)^k}{k!} f^{(k)}(\alpha) = E_n(\beta),$$

as required.   □

Theorem 1.5.4 (The binomial series theorem) *Let* $a \in \mathbf{R}$. *Then*

$$(1 + x)^a = 1 + \sum_{n=1}^{\infty} \binom{a}{n} x^n \quad (-1 < x < 1),$$

where

$$\binom{a}{n} = \frac{a(a - 1)...(a - n + 1)}{n!} \quad (n \in \mathbf{N}).$$

Proof Let $x \in \mathbf{R}$, $|x| < 1$. Application of Theorem 1.5.3 with $\alpha = 0$, $\beta = x$ and $f(t) = (1 + t)^a$ gives

$$(1 + x)^a = 1 + \sum_{k=1}^{n} \binom{a}{k} x^k + E_{n+1}(x),$$

where

$$E_{n+1}(x) = \int_0^x \frac{(x - t)^n}{n!} a(a - 1)...(a - n)(1 + t)^{a-n-1} dt.$$

It remains to prove that $\lim_{n \to \infty} E_{n+1}(x) = 0$. Put

$$C_a(x) = \begin{cases} (1 + x)^{a-1} & \text{if } a \geq 1, x \geq 0, \\ 1 & \text{if } a \geq 1, x \leq 0, \\ 1 & \text{if } a \leq 1, x \geq 0, \\ (1 + x)^{a-1} & \text{if } a \leq 1, x \leq 0. \end{cases}$$

Note that $(1 + t)^{a-1} \leq C_a(x)$ for all t between 0 and x, and hence

$$|E_{n+1}(x)| \leq C_a(x) \frac{|a(a - 1)...(a - n)|}{n!} \left| \int_0^x \left| \frac{x - t}{1 + t} \right|^n dt \right|.$$

Since $\left| \frac{x-t}{1+t} \right| \leq |x|$ for all t between 0 and x,

$$|E_{n+1}(x)| \leq u_n(x),$$

where

$$u_n(x) = C_a(x) \frac{|a(a-1)...(a-n)|}{n!} |x|^{n+1}.$$

The ratio test now shows that $\sum u_n(x)$ converges if $|x| < 1$: hence $\lim_{n\to\infty} u_n(x) = 0$, and so $\lim_{n\to\infty} E_{n+1}(x) = 0$. □

1.5.4 Approximations to Integrals

To conclude this section we return to the topic of evaluation. At the beginning of Sect. 1.4 we noted the relative failure of procedures, such as the trapezium rule, to effect exact evaluation of the integral of a given function. Nevertheless, these procedures do have merit from the point of view of approximation. Provided that a given function is sufficiently regular, in principle its integral may be computed to any pre-assigned degree of accuracy. Theorem 1.5.6 below substantiates this remark for the trapezium rule; the following lemma prepares the ground.

Lemma 1.5.5 *Let $h > 0$, let $f : [-h, h] \to \mathbf{R}$ be twice differentiable and suppose that $f^{(2)} \in \mathscr{R}[-h, h]$. Then*

$$\left| \int_{-h}^{h} f(x)dx - h\{f(-h) + f(h)\} \right| \le \frac{2}{3}h^3 \sup_{-h \le t \le h} \left| f^{(2)}(t) \right|.$$

Proof Let $L : \mathscr{R}[-h, h] \to \mathbf{R}$ be defined by

$$L(u) = \int_{-h}^{h} u(x)dx - h\{u(-h) + u(h)\}.$$

Whenever u is continuously differentiable,

$$L(u) = -\int_{-h}^{h} x u^{(1)}(x)dx = -\int_{-h}^{h} x \left\{ u^{(1)}(x) - u^{(1)}(0) \right\} dx.$$

Plainly the mapping L is linear; moreover, $L(p) = 0$ whenever p is a polynomial of degree at most 1.

Let $M := \sup_{-h \le t \le h} \left| f^{(2)}(t) \right|$. Evidently

$$L(f) = -\int_{-h}^{h} x \left(\int_{0}^{x} f^{(2)}(t)dt \right) dx,$$

and hence

$$|L(f)| \le \int_{-h}^{h} \left| x \int_{0}^{x} f^{(2)}(t)dt \right| dx \le M \int_{-h}^{h} x^2 dx = \frac{2}{3} M h^3.$$

$\square$

Theorem 1.5.6 (Error estimates for the trapezium rule) *Let* $f : [a, b] \to \mathbf{R}$ *be twice differentiable and suppose that* $f^{(2)} \in \mathscr{R}[a, b]$. *Let* $\{x_0, x_1, \ldots, x_n\}$ *be that partition of* $[a, b]$ *which divides it into n intervals of equal length and let*

$$\tau_n = \frac{1}{2n} \sum_{r=1}^{n} \{f(x_{r-1}) + f(x_r)\}.$$

Then

$$\left| (b-a)^{-1} \int_{a}^{b} f(x)dx - \tau_n \right| \le \frac{1}{12} \left(\frac{b-a}{n} \right)^2 \sup_{a \le t \le b} \left| f^{(2)}(t) \right|.$$

Proof Put $M = \sup_{a \le t \le b} \left| f^{(2)}(t) \right|$ and $h = (b-a)/2n$. For $1 \le r \le n$, let $g_r : [-h, h] \to \mathbf{R}$ be given by

$$g_r(t) = f(t + (x_{r-1} + x_r)/2).$$

Then g_r is twice differentiable, $g_r^{(2)} \in \mathscr{R}[-h, h]$ and

$$\left| \int_{-h}^{h} g_r(t)dt - h\{g_r(-h) + g_r(h)\} \right| \le \frac{2}{3} M h^3.$$

It follows that, for $1 \le r \le n$,

$$\left| \int_{x_{r-1}}^{x_r} f(t)dt - \frac{(b-a)}{2n} \{f(x_{r-1}) + f(x_r)\} \right| \le \frac{1}{12} M \left(\frac{b-a}{n} \right)^3.$$

$\square$

Exercise 1.5.7

1. Let $\phi : [a, b] \to \mathbf{R}$ be continuous and such that $\phi(t) > 0$ for all $t \in [a, b]$.
   Prove that
   $$\frac{1}{b-a} \int_{a}^{b} \log \phi(t)dt \le \log \left\{ \frac{1}{b-a} \int_{a}^{b} \phi(t)dt \right\}.$$

2. Prove that $\sum_{n=2}^{\infty} \frac{1}{n(\log n)^{\alpha}}$ converges if $\alpha > 1$ and diverges if $\alpha \le 1$; and that $\sum_{n=3}^{\infty} \frac{1}{n \log n \log \log n}$ diverges.

3. Prove that as $n \to \infty$, $\sum_{r=1}^{n} \left(\frac{s}{r} - \frac{1}{r^s} \right) - s \log n$ ($s > 1$) tends to a limit $\psi(s)$, where $0 \le \psi(s) + \frac{1}{s-1} \le s - 1$.

4. Let $n \in \mathbf{N}$ and let $f : [a, b] \to \mathbf{R}$ be such that $f^{(n)}$ is continuous on $[a, b]$. Show that if α and β are distinct points of $[a, b]$, then

$$f(\beta) - \sum_{k=0}^{n-1} \frac{f^{(k)}(\alpha)}{k!} (\beta - \alpha)^k = R_n(\beta),$$

where

$$R_n(\beta) = \frac{f^{(n)}(\gamma)}{n!} (\beta - \alpha)^n = \frac{f^{(n)}(\lambda)}{(n-1)!} (\beta - \lambda)^{n-1}(\beta - \alpha)$$

for suitable γ and λ between α and β.

5. Let J be a non-degenerate interval in $\mathbf{R}$, and let $a \in J$. Let $f : J \to \mathbf{R}$ be a function possessing a primitive, and suppose that $f \in \mathscr{R}(I)$ for each non-degenerate, closed, bounded interval $I \subset J$. For $n \in \mathbf{N}$, let $f_n : J \to \mathbf{R}$ be defined by

$$f_1(x) = \int_a^x f(t)dt; \quad f_n(x) = \int_a^x f_{n-1}(t)dt \ (n \ge 2).$$

Prove that

$$f_n(x) = \int_a^x \frac{(x-t)^{n-1}}{(n-1)!} f(t)dt \ (x \in J, n \in \mathbf{N}).$$

6. (i) Let $d_n = \log n! - \left(n + \frac{1}{2} \right) \log n + n$ ($n \in \mathbf{N}$). Prove that

$$d_n - d_{n+1} = \left(n + \frac{1}{2} \right) \log \left(1 + \frac{1}{n} \right) - 1$$

$$= \frac{1}{3(2n+1)^2} + \frac{1}{5(2n+1)^4} + \cdots$$

$$< \frac{1}{3\{(2n+1)^2 - 1\}} = \frac{1}{12n} - \frac{1}{12(n+1)}.$$

Let $c_n = d_n - \frac{1}{12n}$ ($n \in \mathbf{N}$). Show that the sequences (c_n) and (d_n) are increasing and decreasing, respectively, and that each is convergent. Let $\lim_{n \to \infty} d_n = \lambda$.

(ii) Let $a_n = n! e^n n^{-(n+\frac{1}{2})}$ ($n \in \mathbf{N}$). Show that $\lim_{n \to \infty} a_n = e^\lambda$ and that $e^\lambda = \lim_{n \to \infty} (a_n^2 / a_{2n}) = \sqrt{2\pi}$. Deduce *Stirling's formula:*

$$n! \sim (2\pi n)^{1/2}(n/e)^n \text{ as } n \to \infty;$$

that is,

$$\lim_{n \to \infty} n!(2\pi n)^{-1/2}(n/e)^{-n} = 1.$$

7. (Error-estimates for Simpson's rule-see Exercise 1.4.15/4.)

(i) Let $h > 0$. Let $L : \mathscr{R}[-h, h] \to \mathbf{R}$ be defined by

$$L(u) = \int_{-h}^{h} u - \frac{h}{3} \{u(-h) + 4u(0) + u(h)\}.$$

Prove that L is a linear mapping and that its kernel contains the class of polynomials of degree at most 3.
Let $f : [-h, h] \to \mathbf{R}$ be four times differentiable and suppose that $f^{(4)} \in \mathscr{R}[-h, h]$. Prove that

$$|L(f)| \le \frac{h^5}{90} \sup_{-h \le t \le h} \left| f^{(4)}(t) \right|.$$

(ii) Let $f : [a, b] \to \mathbf{R}$ be four times differentiable and suppose that $f^{(4)} \in \mathscr{R}[a, b]$. Let $\{x_0, x_1, \ldots, x_{2n}\}$ be that partition of $[a, b]$ which divides it into $2n$ intervals of equal length, and let

$$\theta_n = \frac{1}{6n} \sum_{r=1}^{n} \{f(x_{2r-2}) + 4f(x_{2r-1}) + f(x_{2r})\}.$$

Prove that

$$\left| (b-a)^{-1} \int_a^b f - \theta_n \right| \le \frac{(b-a)^4}{2880n^4} \sup_{a \le t \le b} \left| f^{(4)}(t) \right|.$$

[Hint for part (i): Taylor's theorem shows that

$$L(f) = L\left(x \mapsto \int_0^x \frac{(x-t)^3}{3!} f^{(4)}(t)dt\right);$$

exercise 5 above shows that

$$\frac{d}{dx}\left(\int_0^x \frac{(x-t)^4}{4!} f^{(4)}(t)dt\right) = \int_0^x \frac{(x-t)^3}{3!} f^{(4)}(t)dt.]$$

1.6 The Improper Riemann Integral

The theory so far developed requires that the functions to be integrated should be *bounded* and that the integration should take place over a *closed, bounded* interval. These conditions mean that we are unable to attach a meaning to symbols such as $\int_0^1 x^{-1/2}dx$, $\int_0^\infty e^{-x}dx$ or $\int_1^\infty (x-1)^{-1/2}e^{-x^2}dx$; and our object here is to relax these constraints so that the notion of an integral is more widely available, and in particular may be able to deal with problems to be encountered in Complex Analysis: see Chap. 3.

Definition 1.6.1 Let J be an interval in $\mathbf{R}$. By $\mathcal{R}_{loc}(J)$ we mean the family of all real-valued functions f, with domain containing J (and depending upon the particular function f), such that for every closed, bounded, non-degenerate interval $I \subset J$, $f|_I \in \mathcal{R}(I)$. For simplicity we shall write $\mathcal{R}_{loc}(a, b)$ for $\mathcal{R}_{loc}((a, b))$, $\mathcal{R}_{loc}[a, b)$ for $\mathcal{R}_{loc}([a, b))$, etc.

Note that $\mathcal{R}_{loc}(J_1) \subset \mathcal{R}_{loc}(J_2)$ if $J_1 \supset J_2$. Corollary 1.3.7 shows that if J is non-degenerate, closed and bounded, then $\mathcal{R}(J) = \mathcal{R}_{loc}(J)$; otherwise this equality fails to hold since $\mathcal{R}(J)$ is not defined. As an example of the kind of functions we have in mind consider the function $f : (0, 1] \to \mathbf{R}$ defined by $f(x) = x^{-1}$ $(0 < x \le 1)$: this is continuous on (0,1] and so is Riemann-integrable over every closed interval $[a, b] \subset (0, 1]$; hence it is in $\mathcal{R}_{loc}(0, 1]$.

Definition 1.6.2 Let $a \in \mathbf{R} \cup \{-\infty\}$, $b \in \mathbf{R} \cup \{\infty\}$, $a < b$, and suppose that $f \in \mathcal{R}_{loc}(a, b)$; assume also that there exists $c \in (a, b)$ such that

$$\lim_{u \to a+} \int_u^c f \text{ and } \lim_{v \to b-} \int_c^v f \text{ exist in } \mathbf{R}. \tag{1.6.1}$$

Then f is said to be **improperly Riemann-integrable** over (a, b), and the **improper Riemann integral of** f **over** (a, b), denoted by $(IR) \int_a^b f$ or $\int_a^b f$, is defined to be

$$\lim_{u \to a+} \int_u^c f + \lim_{v \to b-} \int_c^v f. \tag{1.6.2}$$

Remark 1.6.3

(i) If $a = -\infty$, by $\lim_{u \to a+}$ we mean $\lim_{u \to -\infty}$; similarly if $b = \infty$, $\lim_{v \to b-}$ means $\lim_{v \to \infty}$.
(ii) The definition of the improper Riemann integral is independent of the choice of c in (a, b), for given that the limits in (1.6.1) hold for some particular $c \in (a, b)$, then for all $d \in (a, b)$,

$$\lim_{u \to a+} \int_u^c f + \lim_{v \to b-} \int_c^v f = \lim_{u \to a+} \int_u^d f + \int_d^c f + \lim_{v \to b-} \int_c^v f$$

$$= \lim_{u \to a+} \int_u^d f + \lim_{v \to b-} \int_d^v f.$$

Example 1.6.4 Let $a = -\infty$, $b = \infty$, $f : \mathbf{R} \to \mathbf{R}$ where $f(x) = (1 + x^2)^{-1}$ ($x \in \mathbf{R}$). As f is continuous, $f \in \mathscr{R}_{loc}(\mathbf{R})$. Moreover, taking $c = 0$ we see that

$$\int_0^v (1 + x^2)^{-1} dx = \tan^{-1} v \to \frac{\pi}{2} \text{ as } v \to \infty,$$

and

$$\int_u^0 (1 + x^2)^{-1} dx = -\tan^{-1} u \to \frac{\pi}{2} \text{ as } u \to -\infty,$$

so that

$$(IR) \int_{-\infty}^{\infty} (1 + x^2)^{-1} dx = \pi.$$

Lemma 1.6.5 *Let $a, b \in \mathbf{R}$, $a < b$, and suppose that $f \in \mathscr{R}[a, b]$. Then f is improperly Riemann-integrable over (a, b) and*

$$(IR) \int_a^b f = \int_a^b f.$$

Proof By Corollary 1.3.7, $f \in \mathscr{R}_{loc}[a, b]$. By Theorem 1.4.9, given any $c \in (a, b)$, the function $u \longmapsto \int_c^u f$ is continuous on $[a, b]$. Hence

$$\lim_{v \to b-} \int_c^v f = \int_c^b f, \quad \lim_{u \to a+} \int_u^c f = \int_a^c f,$$

and so

$$(IR) \int_a^b f = \int_a^c f + \int_c^b f = \int_a^b f.$$

□

This result shows that the improper Riemann integral is an extension of the Riemann integral. It also means that there will be no confusion if we denote the improper Riemann integral by $\int_a^b f$, and we shall do this from time to time.

We now give a number of results which make it easier to determine whether or not a given function is improperly Riemann-integrable over a given interval.

Lemma 1.6.6 *Suppose that* $-\infty < a < b \leq \infty$ *and that* $f \in \mathcal{R}_{loc}[a, b)$. *Then* f *is improperly Riemann-integrable over* (a, b) *if, and only if,*

$$\lim_{v \to b-} \int_a^v f \quad \text{exists in } \mathbf{R}, \tag{1.6.3}$$

where the integral in (1.6.3) *is a Riemann integral. If this limit exists, then*

$$(IR) \int_a^b f = \lim_{v \to b-} \int_a^v f.$$

Proof Suppose that f is improperly Riemann-integrable over (a, b). Then there exist $c \in (a, b)$ and $\alpha \in \mathbf{R}$ such that $\lim_{v \to b-} \int_c^v f = \alpha$. Since

$$\int_a^v f = \int_a^c f + \int_c^v f$$

for all $v \in [a, b)$, it follows that

$$\lim_{v \to b-} \int_a^v f = \alpha + \int_a^c f.$$

Conversely, let $\beta = \lim_{v \to b-} \int_a^v f$ (note that β is finite) and let $c \in (a, b)$. Then $\lim_{v \to b-} \int_c^v f = \beta - \int_a^c f$ and as in the proof of Lemma 1.6.5 we see that $\lim_{u \to a+} \int_u^c f = \int_a^c f$. Hence f is improperly Riemann-integrable over (a, b) and

$$(IR) \int_a^b f = \int_a^c f + \beta - \int_a^c f = \beta.$$

$\square$

Remark 1.6.7 The same style of proof shows that if $-\infty \leq a < b < \infty$ and $f \in \mathcal{R}_{loc}(a, b]$, then f is improperly Riemann-integrable over (a, b) if, and only if, $\lim_{u \to a+} \int_u^b f$ exists in $\mathbf{R}$; if this limit exists, then

$$(IR) \int_a^b f = \lim_{u \to a+} \int_u^b f.$$

Example 1.6.8 Let $\alpha \in \mathbf{R}$, $a > 0$, $f(x) = x^\alpha$. Then:

(i) f is improperly Riemann-integrable over (a, ∞) if, and only if, $\alpha < -1$; and if $\alpha < -1$, then $\int_a^\infty f = -a^{1+\alpha}/(1+\alpha)$;

(ii) f is improperly Riemann-integrable over $(0, a)$ if, and only if, $\alpha > -1$; and if $\alpha > -1$, then $\int_0^a f = a^{1+\alpha}/(1+\alpha)$;

(iii) f is not improperly Riemann-integrable over $(0, \infty)$.

To establish these results, note first that as f is continuous on $(0, \infty)$, the function f is in $\mathscr{R}_{loc}(0, \infty)$ no matter what α is. Moreover, if $v, c \in (0, \infty)$, then

$$\int_c^v f = \begin{cases} (1+\alpha)^{-1}(v^{1+\alpha} - c^{1+\alpha}) & \text{if } \alpha \neq -1, \\ \log v - \log c & \text{if } \alpha = -1. \end{cases} \tag{1.6.4}$$

Since $\lim_{v\to\infty} \int_c^v f$ exists in $\mathbf{R}$ if, and only if, $\alpha < -1$; and $\lim_{u\to 0+} \int_u^c f$ exists in $\mathbf{R}$ if, and only if, $\alpha > -1$, (iii) follows immediately. From (1.6.4), with $c = a$, and Lemma 1.6.6 we obtain (i); (ii) follows similarly with the aid of Remark 1.6.7.

Lemma 1.6.9 *Let $a, b \in \mathbf{R}$, $a < b$; let $f : (a, b) \to \mathbf{R}$ be bounded and in $\mathscr{R}_{loc}(a, b)$; suppose that $g : [a, b] \to \mathbf{R}$ is such that $g \mid_{(a,b)} = f$. Then $g \in \mathscr{R}[a, b]$, f is improperly Riemann-integrable over (a, b) and*

$$(IR) \int_a^b f = \int_a^b g.$$

Proof By Theorem 1.3.8, $g \in \mathscr{R}[a, b]$. The rest follows from Lemma 1.6.5. $\quad\square$

Example 1.6.10 Let $f : (0, 1) \to \mathbf{R}$ be given by $f(x) = (\log x) \log(1 - x)$ for $0 < x < 1$. We claim that f is improperly Riemann-integrable over $(0, 1)$. For since

$$\lim_{x\to 0+} f(x) = \lim_{x\to 0+} \left\{ x \log x \cdot \frac{\log(1 - x)}{x} \right\} = \lim_{x\to 0+} (x \log x) \lim_{x\to 0+} \left\{ \frac{\log(1 - x)}{x} \right\}$$
$$= 0$$

and

$$\lim_{x\to 1-} f(x) = \lim_{x\to 1-} \left\{ \frac{\log x}{1 - x} \cdot (1 - x) \log(1 - x) \right\} = 0,$$

it follows that f is bounded. Thus by Lemma 1.6.9, f is improperly Riemann-integrable over $(0, 1)$.

Theorem 1.6.11 *Let $-\infty < a < b \leq \infty$ and let $f : [a, b) \to \mathbf{R}$ be non-negative and in $\mathscr{R}_{loc}[a, b)$. Then $\int_a^b f$ exists if, and only if, there is a constant $K \in \mathbf{R}$ such that $\int_a^v f \leq K$ for all $v \in (a, b)$.*

Proof If $\int_a^b f$ exists, the result is immediate. For the converse, let $\varepsilon > 0$ and observe that $F(v) := \int_a^v f \leq K$ if $v \in (a, b)$; put $\overline{F} = \sup\{F(v) : v \in (a, b)\}$. There exists $v_1 \in (a, b)$ such that $\overline{F} - \varepsilon < F(v_1) \leq \overline{F}$, and since F is increasing, $\left| F(v) - \overline{F} \right| < \varepsilon$ whenever $v_1 < v < b$. Hence $\overline{F} = \lim_{v\to\infty} F(v) = \int_a^b f$. $\quad\square$

This theorem is a particularly useful one, for it enables a comparison test for integrals to be obtained, similar to that for series with non-negative terms. For example, to show that the improper Riemann integral $\int_0^\infty \frac{|\cos x|}{1+x^2} dx$ exists, it is enough to observe that for all $v \in (0, \infty)$,

$$\int_0^v \frac{|\cos x|}{1+x^2} dx \leq \int_0^v \frac{1}{1+x^2} dx \leq \frac{\pi}{2},$$

for then Theorem 1.6.11 does the trick. However, it applies only to *non-negative* functions. Nevertheless, Theorem 1.6.11 coupled with the following result enables us to cope with a wide variety of situations.

Theorem 1.6.12 *Let* $-\infty < a < b \leq \infty$ *and suppose that* $f \in \mathcal{R}_{loc}[a, b)$. *Then if* $\int_a^b |f|$ *exists, so does* $\int_a^b f$.

Proof Let $v \in (a, b)$. Then as $f \in \mathcal{R}[a, v]$, so do f^+, f^- and $|f|$ (recall that $f^+ = \frac{1}{2}(f + |f|)$, $f^- = \frac{1}{2}(|f| - f)$). Also

$$0 \leq \int_a^v f^-, \int_a^v f^+ \leq \int_a^v |f| \leq \int_a^b |f|.$$

By Lemma 1.6.6, f^+ and f^- are improperly Riemann-integrable over (a, b); hence

$$\lim_{v \to b-} \int_a^v f = \lim_{v \to b-} \int_a^v (f^+ - f^-) = \lim_{v \to b-} \int_a^v f^+ - \lim_{v \to b-} \int_a^v f^-,$$

and the result follows. $\qquad\qquad\qquad\qquad\qquad\qquad\qquad\qquad\qquad\qquad\qquad\square$

As an illustration of the usefulness of this result, note that it shows directly that the function $f : (0, \infty) \to \mathbf{R}$ given by $f(x) = \frac{\cos x}{1+x^2}$ $(0 < x < \infty)$ is improperly Riemann-integrable over $(0, \infty)$, since we know that $\int_0^\infty \frac{|\cos x|}{1+x^2} dx$ exists.

Remark 1.6.13 The converse of Theorem 1.6.12 is *false*: if f is improperly Riemann-integrable over (a, b) it does *not* follow that $(IR) \int_a^b |f| dx$ exists. To illustrate this let $f : [0, \infty) \to \mathbf{R}$ be defined by $f(x) = \frac{\sin x}{x}$ $(x > 0)$, $f(0) = 1$. We claim that f is improperly Riemann-integrable over $(0, \infty)$ but that $|f|$ is not. To justify this, note that for each $m \in \mathbf{N}, m > 2$,

$$\int_\pi^{m\pi} \frac{|\sin x|}{x} dx = \sum_{n=1}^{m-1} \int_{n\pi}^{(n+1)\pi} \frac{|\sin x|}{x} dx \geq \sum_{n=1}^{m-1} \frac{1}{(n+1)\pi} \int_{n\pi}^{(n+1)\pi} |\sin x| dx$$

$$= \sum_{n=1}^{m-1} \frac{1}{(n+1)\pi} \int_0^\pi \sin x dx = \frac{2}{\pi} \sum_{n=1}^{m-1} \frac{1}{(n+1)}$$

$$= \frac{2}{\pi} \sum_{k=2}^m \frac{1}{k}.$$

Since $\sum_2^\infty \frac{1}{k}$ diverges, $\lim_{v \to \infty} \int_\pi^v \frac{|\sin x|}{x} dx = \infty$, and hence $|f|$ is not improperly Riemann-integrable over $(0, \infty)$. However, for each $v \in (\pi, \infty)$,

$$\int_0^v f = \int_0^\pi f + \int_\pi^v f.$$

Integration by parts gives

$$\int_\pi^v f = -\frac{1}{\pi} - v^{-1}\cos v - \int_\pi^v \frac{\cos x}{x^2}dx;$$

and since $x \longmapsto x^{-2}|\cos x|$ is in $\mathscr{R}_{loc}[\pi, \infty)$ and

$$\int_\pi^v \frac{|\cos x|}{x^2}dx \le \int_\pi^v \frac{1}{x^2}dx \le 1/\pi,$$

it follows from Theorem 1.6.11 that $x \longmapsto x^{-2}|\cos x|$ is improperly Riemann-integrable over (π, ∞); by Theorem 1.6.12, so is $x \longmapsto x^{-2}\cos x$. Accordingly, as $\lim_{v\to\infty}\int_0^v f$ exists in $\mathbf{R}$, it follows that f is improperly Riemann-integrable over $(0, \infty)$.

Various results for improper Riemann integrals, companion to those developed in Sects. 1.3 and 1.4 for the Riemann integral, may now be established without difficulty. However, a technical matter worth noting arises in the course of this procedure: this relates to the additivity of the integral. Generalisation of Theorem 1.3.6 requires a modification of Definition 1.6.2 to enlarge the class of improperly Riemann-integrable functions.

Definition 1.6.14 Let f be a real-valued function defined on (a, b) save perhaps at a finite number of points of this interval, and suppose there are finitely many points $a_1, \ldots, a_p \in (a, b)$, with $a_1 < a_2 < \ldots < a_p$, such that f is improperly Riemann-integrable over each subinterval (a_{i-1}, a_i) $(i = 1, \ldots, p+1)$, where $a_0 = a, a_{p+1} = b$. Then f is said to be **improperly Riemann-integrable over** (a, b), and we define the **improper Riemann integral of** f **over** (a, b) to be $\sum_{i=1}^{p+1}\int_{a_{i-1}}^{a_i} f$, denoted by $\int_a^b f$.

Note that while the choice of points a_i in this definition is not unique, nevertheless it may easily be shown that if $\int_a^b f$ exists, then it does not depend upon the particular a_i chosen.

Exercise 1.6.15

1. Let $a \in \mathbf{R}$ and suppose that $f \in \mathscr{R}_{loc}[a, \infty)$. Show that f is improperly Riemann-integrable over (a, ∞) if, and only if, given any $\varepsilon > 0$ there exists $x_0 \in \mathbf{R}$ such that

$$\left|\int_x^y f\right| < \varepsilon \text{ whenever } y > x \ge x_0.$$

2. Let $a \in \mathbf{R}$, $g \in \mathscr{R}_{loc}[a, \infty)$, $g(x) \ge 0$ for all $x \in [a, \infty)$ and suppose that g is improperly Riemann-integrable over (a, ∞). Let $K \in \mathbf{R}$, $f \in \mathscr{R}_{loc}[a, \infty)$ and

suppose that

$$|f(x)| \leq K g(x) \text{ when } a \leq x < \infty.$$

Prove that f is improperly Riemann-integrable over (a, ∞).

3. Show that the following improper integral exists:

$$\Gamma(p) = \int_0^\infty x^{p-1} e^{-x} dx \quad (p > 0).$$

Prove that $\Gamma(p + 1) = p \Gamma(p) \ (p > 0)$. Deduce that $\Gamma(n + 1) = n! \ (n \in \mathbf{N})$.

4. Determine those values of $\alpha \in \mathbf{R}$ for which the following integrals exist as improper Riemann integrals:

(a) $\displaystyle\int_0^\infty \frac{x^{\alpha-1}}{1+x} dx,$ (b) $\displaystyle\int_0^1 \frac{1}{x^\alpha (1 - \log x)} dx,$

(c) $\displaystyle\int_0^\pi x^{-\alpha} \sin x dx,$ (d) $\displaystyle\int_0^\pi \frac{\sin^\alpha x}{\log(1 + x)} dx.$

5. Let $f \in \mathscr{R}_{loc}(\mathbf{R})$ and let f be periodic with period $\tau > 0 : f(t + \tau) = f(t)$ $(t \in \mathbf{R})$. Prove that

$$\int_x^{x+\tau} f = \int_0^\tau f \ (x \in \mathbf{R}).$$

6. Prove that if $u_n = \int_0^{\pi/2} \sin 2nx \cot x dx$ and $v_n = \int_0^{\pi/2} \frac{\sin 2nx}{x} dx$, then $u_n = \pi/2$ and $\lim_{n \to \infty} v_n = \int_0^\infty \frac{\sin x}{x} dx$. Further, show that $\lim_{n \to \infty} (u_n - v_n) = 0$, and deduce that

$$\int_0^\infty \frac{\sin x}{x} dx = \frac{\pi}{2}.$$

7. Show that if $\theta > 1$, then the improper Riemann integral

$$\int_0^{1/e} x^{-1} (\log(1/x))^{-\theta} dx$$

exists, and evaluate it.

8. Let $\lambda \in \mathbf{R}$. Show that the improper Riemann integral $\int_0^\infty e^{-x} \sin(\lambda x) dx$ exists and equals $\lambda/(1 + \lambda^2)$.

9. Show that both $\int_1^\infty \frac{\cos x}{x} dx$ and $\int_2^\infty \frac{\cos x}{\log x} dx$ exist as improper Riemann integrals.

10. Show that $\int_{-\infty}^\infty \sin(e^x) dx$ exists as an improper Riemann integral, and evaluate it.

11. Noting that

$$1 + x^4 = (1 - \sqrt{2}x + x^2)(1 + \sqrt{2}x + x^2),$$

express $(1 + x^4)^{-1}$ in terms of partial fractions and show that the improper Riemann integral $\int_0^\infty (1 + x^4)^{-1} dx$ exists and equals $\pi/2\sqrt{2}$.

1.7 Uniform Convergence

Let (f_n) be a sequence of Riemann-integrable functions on an interval $[a, b]$, and suppose that there is a function $f : [a, b] \to \mathbf{R}$ such that for each $x \in [a, b]$, $\lim_{n\to\infty} f_n(x) = f(x)$. A natural question to ask is whether $f \in \mathcal{R}[a, b]$, and if so, whether

$$\lim_{n\to\infty} \int_a^b f_n(x)dx = \int_a^b f(x)dx.$$

In general the answer is 'no': if $a = 0$, $b = 1$, $f_n(x) = n^2 x e^{-nx}$ then $f(x) = 0$, $\lim_{n\to\infty} \int_0^1 f_n(x)dx = \lim_{n\to\infty}(-ne^{-n} - e^{-n} + 1) = 1$, while $\int_0^1 f(x)dx = 0$. In this example, the limit function f is Riemann-integrable, but even this need not be so, as is evident from the case $a = 0$, $b = 1$ and

$$f_n(x) = \begin{cases} 1 & \text{if } x = k2^{-n} \text{ for some } k \in \mathbf{Z} \text{ with } 0 \le k \le 2^n, \\ 0 & \text{otherwise.} \end{cases}$$

Here

$$f(x) = \begin{cases} 1 & \text{if } x = k2^{-n} \text{ for some } n \in \mathbf{N}, k \in \mathbf{Z} \text{ with } 0 \le k \le 2^n, \\ 0 & \text{otherwise,} \end{cases}$$

and $f \notin \mathcal{R}[0, 1]$. However, when $a = 0$, $b = 1$, $f_n(x) = x^n/n$, we have $f(x) = 0$ and

$$\lim_{n\to\infty} \int_0^1 f_n(x)dx = \lim_{n\to\infty} \frac{1}{n(n + 1)} = 0 = \int_0^1 f(x)dx,$$

and all is well.

The concept of uniform convergence which we now introduce enables us to distinguish between these cases.

Definition 1.7.1 Let S be a non-empty set, and for each $n \in \mathbf{N}$ let $f_n : S \to \mathbf{R}$. If there is a function $f : S \to \mathbf{R}$ such that $\lim_{n\to\infty} f_n(s) = f(s)$ for all $s \in S$, the sequence (f_n) is said to **converge pointwise on S to f**; if there is a function $f : S \to \mathbf{R}$ such that given any $\varepsilon > 0$, there exists $N \in \mathbf{N}$ such that $\sup_{s\in S} |f_n(s) - f(s)| < \varepsilon$ if $n \ge N$, it is said to **converge uniformly on S to f**.

Equivalently, (f_n) converges uniformly on S if there is a function $f : S \to \mathbf{R}$ such that given any $\varepsilon > 0$, there exists $N \in \mathbf{N}$ such that for all $s \in S$, $|f_n(s) - f(s)| < \varepsilon$ if $n \geq N$; while for pointwise convergence on S we require that given any $s \in S$ and any $\varepsilon > 0$, there exists $N \in \mathbf{N}$ such that $|f_n(s) - f(s)| < \varepsilon$ if $n \geq N$. This trivial reformulation of the definition helps to underline the basic difference between these two forms of convergence: for uniform convergence on S, the integer N depends on ε and does not depend upon the particular $s \in S$; while for pointwise convergence N may well depend on the particular s chosen as well as on ε. Evidently uniform convergence on S implies pointwise convergence on S. The converse is false, as we see from consideration of the sequence (f_n), where $f_n : \mathbf{R} \to \mathbf{R}$ and

$$f_n(s) = \begin{cases} ns, & \text{if } 0 \leq s \leq n^{-1}, \\ 2 - ns, & \text{if } n^{-1} < s \leq 2n^{-1}, \\ 0, & \text{otherwise.} \end{cases}$$

Then $f_n(s) \to 0$ for each $s \in \mathbf{R}$, but since $f_n(n^{-1}) = 1$ for all $n \in \mathbf{N}$, the convergence cannot be uniform on $\mathbf{R}$, for there can be no $N \in \mathbf{N}$ such that for all $n > N$, $\sup_{x \in \mathbf{R}} |f_n(x)| < 1/2$.

Example 1.7.2

(i) Let $f_n(x) = (1 + nx)^{-1}$ for $0 < x \leq 1$ and $n \in \mathbf{N}$. For all $x \in (0, 1]$, $\lim_{n \to \infty} f_n(x) = 0$. However, the convergence is not uniform on $(0, 1]$, as for all $n \in \mathbf{N}$, $f_n(n^{-1}) = 1/2$.

(ii) Let $g_n(x) = x(1 + nx)^{-1}$ for $0 < x \leq 1$ and $n \in \mathbf{N}$. Given any $x \in (0, 1]$ and any $\varepsilon > 0$, $0 < g_n(x) < 1/n$ if $n > 1/\varepsilon$: hence (g_n) converges uniformly on $(0, 1]$ to 0.

Cauchy's general principle of convergence (Appendix, Theorem A.4.14) has a natural analogue for uniform convergence which we give next.

Theorem 1.7.3 (Cauchy's general principle of uniform convergence) *Let (f_n) be a sequence of real-valued functions defined on a non-empty set S. Then (f_n) converges uniformly on S if, and only if, given any $\varepsilon > 0$, there exists $N \in \mathbf{N}$ such that $\sup_{s \in S} |f_n(s) - f_m(s)| < \varepsilon$ if $m, n \geq N$.*

Proof Suppose that (f_n) converges uniformly on S to a function f, and let $\varepsilon > 0$. Then there exists $N \in \mathbf{N}$ such that for all $n \geq N$ and all $s \in S$, $|f_n(s) - f(s)| < \varepsilon/2$. Hence for all $s \in S$ and all $m, n \geq N$,

$$|f_n(s) - f_m(s)| \leq |f_n(s) - f(s)| + |f(s) - f_m(s)| < \varepsilon,$$

which gives the desired result.

Conversely, suppose that Cauchy's criterion holds. Then for each $s \in S$, $(f_n(s))$ is a Cauchy sequence of real numbers, and hence converges, to $f(s)$, say; this defines a function $f : S \to \mathbf{R}$. Let $\varepsilon > 0$. Then there exists $N \in \mathbf{N}$ such that for all $m, n \geq N$

and for all $s \in S$ we have $|f_n(s) - f_m(s)| < \varepsilon/2$. Let $m \to \infty$: then for all $n \geq N$, $\sup_{s \in S} |f_n(s) - f(s)| \leq \varepsilon/2 < \varepsilon$. The proof is complete. $\square$

Now that uniform convergence of sequences has been treated we may pass without difficulty to the uniform convergence of series by the natural consideration of partial sums.

Definition 1.7.4 Let (f_n) be a sequence of real-valued functions on a non-empty set S and put $u_n(s) = \sum_{k=1}^{n} f_k(s)$ for all $s \in S$ and all $n \in \mathbf{N}$. The symbol $\sum_{k=1}^{\infty} f_k$, referred to as the **series generated by** (f_n), is used to denote the sequence (u_n). If the sequence (u_n) converges pointwise on S to f, we say that the series $\sum_{k=1}^{\infty} f_k$ **converges pointwise on** S **to** f and write $f = \sum_{k=1}^{\infty} f_k$. By context one understands whether $\sum_{k=1}^{\infty} f_k$ represents the series itself or the limit function f. If (u_n) is uniformly convergent on S we say that $\sum_{k=1}^{\infty} f_k$ **converges uniformly on** S. The function u_n is called the nth **partial sum** of the series $\sum_{k=1}^{\infty} f_k$.

There are various tests for detecting uniform convergence of series, but the simplest, and perhaps the most useful, is the following, due to Weierstrass.

Theorem 1.7.5 (The Weierstrass M-test) *Let* (f_n) *be a sequence of real-valued functions defined on a non-empty set* S, *and suppose that* (M_n) *is a sequence of non-negative real numbers, with* $\sum_{n=1}^{\infty} M_n$ *convergent, such that for all* $n \in \mathbf{N}$, $\sup_{s \in S} |f_n(s)| \leq M_n$. *Then* $\sum_{n=1}^{\infty} f_n$ *is uniformly convergent on* S.

Proof Let $\varepsilon > 0$. Since $\sum_1^{\infty} M_n$ is convergent, there exists $N \in \mathbf{N}$ such that for all $m, n \in \mathbf{N}$ with $m > n \geq N$, we have $\sum_{r=n+1}^{m} M_r < \varepsilon$. Thus for all $s \in S$ and all $m, n > N$,

$$\left| \sum_{r=n+1}^{m} f_r(s) \right| \leq \sum_{r=n+1}^{m} |f_r(s)| \leq \sum_{r=n+1}^{m} M_r < \varepsilon.$$

The result now follows from Theorem 1.7.3. $\square$

Example 1.7.6 The series $\sum_1^{\infty} \frac{1}{(n+x^2)(n+1+x^2)}$ converges uniformly on $\mathbf{R}$, since for all $x \in \mathbf{R}$ and all $n \in \mathbf{N}$,

$$\frac{1}{(n + x^2)(n + 1 + x^2)} \leq \frac{1}{n^2},$$

and $\sum_1^{\infty} \frac{1}{n^2}$ is convergent; Theorem 1.7.5 can then be applied with $M_n = \frac{1}{n^2}$.

We now turn to continuous functions. Let S be a non-empty subset of the real line and, replacing I by S in Definition 1.2.1, let $C(S)$ be the family of all real-valued continuous functions on S. If (f_n) is a sequence in $C(S)$ which converges pointwise on S to $f : S \to \mathbf{R}$, then f need not be continuous. To illustrate this, let $S = \mathbf{R}$ and define f_n by

$$f_n(t) = \begin{cases} 1, & t \geq n^{-1}, \\ nt, & |t| < n^{-1}, \\ -1, & t \leq -n^{-1}. \end{cases}$$

Clearly (f_n) converges pointwise on $\mathbf{R}$ to f, where

$$f(t) = \begin{cases} 1, & t > 0, \\ 0, & t = 0, \\ 1, & t < 0. \end{cases}$$

Evidently each f_n is continuous on $\mathbf{R}$, but $f \notin C(\mathbf{R})$. If the convergence is *uniform* on S this behaviour is impossible, as the following theorem shows: continuity is preserved by uniform convergence.

Theorem 1.7.7 *Let $S \subset \mathbf{R}$, $S \neq \emptyset$ and suppose (f_n) is a sequence of real-valued functions on S which converges uniformly on S to $f : S \to \mathbf{R}$. If $s \in S$ is a point of continuity of each f_n, then f is continuous at s. In particular, if each f_n belongs to $C(S)$, then $f \in C(S)$.*

Proof Let $\varepsilon > 0$. Since (f_n) converges uniformly on S to f, there exists $N \in \mathbf{N}$ such that for all $n \geq N$ and all $t \in S$, $|f_n(t) - f(t)| < \varepsilon/3$. Since f_N is continuous at s, there exists $\delta > 0$ such that $|f_N(t) - f_N(s)| < \varepsilon/3$ if $t \in S$ and $|s - t| < \delta$. Thus if $t \in S$ and $|s - t| < \delta$, then

$$|f(t) - f(s)| \leq |f(t) - f_N(t)| + |f_N(t) - f_N(s)| + |f_N(s) - f(s)| < \varepsilon.$$

Hence f is continuous at s.   □

As an immediate Corollary we have

Corollary 1.7.8 *Let (f_n) be a sequence in $C(S)$ and suppose $\sum_{n=1}^{\infty} f_n$ converges uniformly on S to f. Then $f \in C(S)$.*

Proof Simply consider the partial sums $u_n = \sum_{k=1}^{n} f_k$ and use Theorem 1.7.7.   □

Example 1.7.9 To determine whether or not the series $\sum_{n=0}^{\infty} \frac{x^2}{(1+x^2)^n}$ converges uniformly on $\mathbf{R}$, note that $f(x) := \sum_{n=0}^{\infty} \frac{x^2}{(1+x^2)^n}$ is given by $f(x) = 1 + x^2$ if $x \neq 0$, $f(0) = 0$: thus f is not continuous on $\mathbf{R}$. Since $x \longmapsto \frac{x^2}{(1+x^2)^n}$ is continuous on $\mathbf{R}$, for all n, it follows from Corollary 1.7.8 that the series cannot be uniformly convergent on $\mathbf{R}$.

One of the most useful and interesting results concerning uniform convergence is the Weierstrass approximation theorem, which asserts that any real-valued function which is continuous on a closed, bounded interval can be uniformly approximated on that interval, as closely as we please, by a polynomial. We give a short direct proof of this important result, a proof which has the virtue that it may be refined, for suitably differentiable functions, to yield simultaneous approximation of these functions and their derivatives: see [4, pp. 112–114]. Another proof, using integration theory, will be found in Exercise 1.7.17/14.

Theorem 1.7.10 (The Weierstrass approximation theorem) *Let $a, b \in \mathbf{R}$, with $a < b$, and let $f \in C(I)$, where $I = [a, b]$. Then there is a sequence (p_n) of polynomials which converges uniformly on I to f.*

Proof We may, without loss of generality, suppose that $a = 0$, $b = 1$. For if the theorem has been proved in this special case, then if we let $g : [0, 1] \to [a, b]$ be given by $g(x) = a + x(b - a)$, it will follow that there is a sequence (p_n) of polynomials which converges uniformly on $[0, 1]$ to $f \circ g$: thus $\left(p_n \circ g^{-1}\right)$ is a sequence of polynomials which converges uniformly on $[a, b]$ to f.

To prove the theorem when $a = 0$ and $b = 1$, consider the identity

$$(x + y)^n = \sum_{k=0}^{n} \binom{n}{k} x^k y^{n-k}, \tag{1.7.1}$$

differentiate both sides with respect to x and then multiply by x:

$$nx(x + y)^{n-1} = \sum_{k=0}^{n} k \binom{n}{k} x^k y^{n-k}. \tag{1.7.2}$$

Now differentiate (1.7.1) twice with respect to x and multiply the result by x^2 :

$$n(n - 1)x^2(x + y)^{n-2} = \sum_{k=0}^{n} k(k - 1) \binom{n}{k} x^k y^{n-k}. \tag{1.7.3}$$

Put $r_k(x) = \binom{n}{k} x^k (1 - x)^{n-k}$ and set $y = 1 - x$: (1.7.1)–(1.7.3) then become

$$\sum_{k=0}^{n} r_k(x) = 1, \quad \sum_{k=0}^{n} kr_k(x) = nx, \quad \sum_{k=0}^{n} k(k - 1)r_k(x) = n(n - 1)x^2. \tag{1.7.4}$$

Thus

$$\sum_{k=0}^{n} (k - nx)^2 r_k(x) = n^2 x^2 - 2n^2 x^2 + n(n - 1)x^2 + nx \tag{1.7.5}$$

$$= nx(1 - x).$$

Since f is continuous on $[0, 1]$, it is bounded: there exists $M \in \mathbf{R}$ such that for all $x \in [0, 1]$, $|f(x)| \le M$. Moreover, by Theorem 1.2.3, f is uniformly continuous on $[0, 1]$. Let $\varepsilon > 0$: then there exists $\delta > 0$ such that $|f(x) - f(y)| < \varepsilon$ whenever $x, y \in [0, 1]$ and $|x - y| < \delta$. Hence for all $x \in [0, 1]$, use of (1.7.4) shows that

$$\left| f(x) - \sum_{k=0}^{n} f\left(\frac{k}{n}\right) r_k(x) \right| = \left| \sum_{k=0}^{n} \left(f(x) - f\left(\frac{k}{n}\right) \right) r_k(x) \right|$$
$$\leq |S_1| + |S_2|,$$

where S_1 denotes the sum of $\left(f(x) - f\left(\frac{k}{n}\right)\right) r_k(x)$ over those k in $[0, n] \cap \mathbf{N}_0$ such that $|k - nx| < n\delta$, and S_2 is the same except that the summation is over those k such that $|k - nx| \geq n\delta$. The uniform continuity of f now shows that, noting that $r_k(x) \geq 0$,

$$|S_1| \leq \varepsilon \sum_{k=0}^{n} r_k(x) = \varepsilon;$$

while (1.7.5) gives

$$|S_2| \leq 2M \sum_{|k-nx| \geq n\delta} r_k(x) \leq \frac{2M}{(n\delta)^2} \sum_{k=0}^{n} (k - nx)^2 r_k(x)$$
$$= \frac{2Mx(1-x)}{n\delta^2} \leq \frac{M}{2n\delta^2}.$$

Thus there exists $N \in \mathbf{N}$ such that for all $x \in [0, 1]$ and all $n \geq N$, $|S_2| < \varepsilon$. It follows that

$$\left| f(x) - \sum_{k=0}^{n} f\left(\frac{k}{n}\right) r_k(x) \right| \leq 2\varepsilon$$

for all $x \in [0, 1]$ and all $n \geq N$; and since $p_n := \sum_{k=0}^{n} f\left(\frac{k}{n}\right) r_k$ is a polynomial the theorem follows. $\qquad \square$

As an immediate application of Weierstrass's theorem we give the following.

Theorem 1.7.11 *Suppose* $f : [0, 1] \to \mathbf{R}$ *is continuous and such that for all* $n \in \mathbf{N}_0$,

$$\int_0^1 t^n f(t) dt = 0.$$

Then $f = 0$.

Proof Let $\varepsilon > 0$. By Theorem 1.7.10, there is a polynomial p, with $p(t) = a_0 + a_1 t + \ldots + a_n t^n$ say, such that for all $t \in [0, 1]$, $|f(t) - p(t)| < \varepsilon$. Hence

$$\int_0^1 |f(t)|^2\, dt = \int_0^1 f(t)\,(f(t) - p(t))\, dt + \int_0^1 f(t) p(t)\, dt$$

$$\leq \varepsilon \int_0^1 |f(t)|\, dt$$

since $\int_0^1 f(t) p(t)\, dt = \sum_{k=0}^n a_k \int_0^1 t^k f(t)\, dt = 0$. As this is true for all $\varepsilon > 0$, $\int_0^1 |f(t)|^2\, dt = 0$. Since $|f|^2$ is continuous, Theorem 1.3.2(d) now shows that f must be the zero function.   □

To conclude this section, we return to the topic with which we began it, namely uniform convergence and integration. From now on, we suppose that $a, b \in \mathbf{R}$, $a < b$ and $I = [a, b]$.

Theorem 1.7.12   *Let (u_n) be a sequence of functions in $\mathcal{R}(I)$ which converges uniformly on I to a function u. Then $u \in \mathcal{R}(I)$ and*

$$\int_a^b u = \lim_{n \to \infty} \int_a^b u_n.$$

Proof Since (u_n) converges uniformly on I to u,

$$\varepsilon_n := \sup\{|u(t) - u_n(t)| : t \in I\} \to 0 \text{ as } n \to \infty;$$

and as

$$\sup_I |u(t)| \leq \varepsilon_n + \sup_I |u_n(t)| < \infty,$$

u is bounded on I. Moreover,

$$-\varepsilon_n + u_n(t) \leq u(t) \leq u_n(t) + \varepsilon_n$$

for all $t \in I$ and all $n \in \mathbf{N}$; hence

$$-\varepsilon_n(b - a) + \int_a^b u_n \leq \underline{\int_a^b} u \leq \overline{\int_a^b} u \leq \varepsilon_n(b - a) + \int_a^b u_n,$$

and thus

$$\left| \overline{\int_a^b} u - \underline{\int_a^b} u \right| \leq 2\varepsilon_n(b - a) \to 0 \text{ as } n \to \infty.$$

It follows that

$$\overline{\int_a^b} u = \underline{\int_a^b} u;$$

that is, $u \in \mathscr{R}(I)$.

Finally,

$$\left| \int_a^b u - \int_a^b u_n \right| \leq \int_a^b |u - u_n| \leq \int_a^b \varepsilon_n = \varepsilon_n(b-a) \to 0 \text{ as } n \to \infty.$$

$\square$

Corollary 1.7.13 *Let* (f_n) *be a sequence in* $\mathscr{R}(I)$ *and suppose the series*

$$f(t) = \sum_{n=1}^{\infty} f_n(t)$$

converges uniformly on I. *Then* $f \in \mathscr{R}(I)$ *and*

$$\int_a^b f = \sum_{n=1}^{\infty} \int_a^b f_n.$$

Proof Put $u_n = \sum_{k=1}^n f_k$ $(n \in \mathbf{N})$; (u_n) converges uniformly on I to f. Thus by Theorem 1.7.12,

$$\int_a^b f = \lim_{n \to \infty} \int_a^b u_n = \lim_{n \to \infty} \sum_{k=1}^n \int_a^b f_k = \sum_{k=1}^{\infty} \int_a^b f_k.$$

$\square$

Theorem 1.7.14 *Suppose that* $\sum_{n=0}^{\infty} a_n x^n$ *has non-zero radius of convergence* R, *and define* $f : (-R, R) \to \mathbf{R}$ *by* $f(x) = \sum_{n=0}^{\infty} a_n x^n$. *Then* $\sum_{n=0}^{\infty} \frac{a_n}{n+1} x^{n+1}$ *has radius of convergence* R, f *is continuous on* $(-R, R)$ *and given any* a, b *with* $-R < a < b < R$,

$$\int_a^b f = \sum_{n=0}^{\infty} \frac{a_n}{n+1} \left(b^{n+1} - a^{n+1} \right).$$

Proof Let $\sum_{n=0}^{\infty} \frac{a_n}{n+1} x^{n+1}$ have radius of convergence R_1. If $|x| < R$, then there exists $x_0 \in \mathbf{R}$, with $|x| < |x_0| < R$, such that $\sum_{n=0}^{\infty} a_n x_0^n$ is convergent; hence there exists $M \in \mathbf{R}$ such that for all $n \in \mathbf{N}_0 := \mathbf{N} \cup \{0\}$, $\left| a_n x_0^n \right| \leq M$. Thus

$$\left| \frac{a_n}{n+1} x^{n+1} \right| \leq M |x| |x/x_0|^n$$

and so comparison with $\sum |x/x_0|^n$ shows that $\sum_0^{\infty} \frac{a_n}{n+1} x^{n+1}$ is convergent. Hence $R_1 \geq R$.

Now suppose that $|y| < R_1$. Then there exists $y_0 \in \mathbf{R}$ such that $|y| < |y_0| < R_1$ and $\sum_0^\infty \frac{a_n}{n+1} y_0^{n+1}$ is convergent. Hence there exists $K \in \mathbf{R}$ with $\left| \frac{a_n}{n+1} y_0^{n+1} \right| \le K$ for all $n \in \mathbf{N}_0$, which implies that

$$|a_n y^n| \le K \left(\frac{n+1}{|y_0|} \right) \left| \frac{y}{y_0} \right|^n :$$

thus comparison with $\sum (n + 1) |y/y_0|^n$ shows that $\sum a_n y^n$ is convergent, so that $R \ge R_1$. Thus $R_1 = R$.

Let $-R < a < b < R$ and put $c = \max(|a|, |b|)$. Since $c < R$, $\sum a_n c^n$ is absolutely convergent. As $|a_n x^n| \le |a_n c^n|$ for all $x \in [a, b]$ and all $n \in \mathbf{N}_0$, Weierstrass's M-test shows that $\sum a_n x^n$ converges uniformly on $[a, b]$. By Corollary 1.7.8, f is continuous on $[a, b]$, and as a and b may be chosen arbitrarily close to $-R$ and R respectively, f is continuous on $(-R, R)$. The proof is now completed by appeal to Corollary 1.7.13.   □

To conclude this section we observe that in Theorem 1.7.12, the strong hypothesis of uniform convergence of the sequence (u_n) on the interval I is used to show that the limit function $u \in \mathscr{R}(I)$ and that

$$\int_a^b u = \lim_{n \to \infty} \int_a^b u_n.$$

The same conclusion as regards the convergence of the integrals can be reached under conditions weaker than that of uniform convergence, provided that the limit function is known to belong to $\mathscr{R}(I)$. We present such a result below, beginning with a theorem that implies the desired conclusion for monotone sequences.

Theorem 1.7.15 *Let (f_n) be a decreasing sequence of functions in $\mathscr{B}(I)$ that converges pointwise on I to 0. Then*

$$\lim_{n \to \infty} \int_a^b f_n = 0.$$

Proof Let $\varepsilon > 0$. By Exercise 1.7.17/17, given any $n \in \mathbf{N}$, there is a continuous function g_n on I such that $0 \le g_n \le f_n$ and

$$\int_a^b f_n < \int_a^b g_n + 2^{-n} \varepsilon.$$

Put $h_n = \min(g_1, \ldots, g_n)$: thus $0 \le h_n \le g_n \le f_n$ and (h_n) is a sequence of continuous functions on I that decreases to zero everywhere on I. By Dini's theorem (Exercise 1.7.17/18), (h_n) converges uniformly to zero on I, and hence, by Theorem 1.7.12,

$$\lim_{n \to \infty} \int_a^b h_n = 0.$$

We claim that for all $n \in \mathbf{N}$,

$$0 \le \int_a^b f_n \le \int_a^b h_n + \varepsilon(1 - 2^{-n});$$

granted this, it is clear that the theorem holds. To justify the claim, note that if $i \in \{1, \ldots, n\}$,

$$0 \le g_n = g_i + (g_n - g_i) \le g_i + \max(g_i, \ldots, g_n) - g_i$$

$$\le g_i + \sum_{j=1}^{n-1} (\max(g_j, \ldots, g_n) - g_j),$$

and hence

$$0 \le g_n \le h_n + \sum_{j=1}^{n-1} (\max(g_j, \ldots, g_n) - g_j).$$

Moreover, since $\max(g_j, \ldots, g_n) \le \max(f_j, \ldots, f_n) = f_j$, we have

$$\int_a^b f_j \ge \int_a^b (\max(g_j, \ldots, g_n) - g_j) + \int_a^b g_j,$$

which gives

$$\int_a^b (\max(g_j, \ldots, g_n) - g_j) \le \int_a^b f_j - \int_a^b g_j \le 2^{-j} \varepsilon.$$

Thus

$$\int_a^b g_n \le \int_a^b h_n + \sum_{j=1}^{n-1} 2^{-j} \varepsilon = \int_a^b h_n + \varepsilon(1 - 2^{-(n-1)}),$$

and so

$$\int_a^b f_n < \int_a^b g_n + 2^{-n} \varepsilon \le \int_a^b h_n + \varepsilon(1 - 2^{-(n-1)} + 2^{-n}).$$

The claim follows immediately. $\square$

Theorem 1.7.16 (Arzelà's theorem) *Let* (f_n) *be a sequence in* $\mathscr{R}(I)$ *that converges pointwise on* I *to* $f \in \mathscr{R}(I)$; *assume also that there exists* $M > 0$ *such that for all* $x \in I$ *and all* $n \in \mathbf{N}$, $|f_n(x)| \le M$. *Then*

$$\int_a^b f = \lim_{n \to \infty} \int_a^b f_n.$$

Proof First suppose that $f = 0$ and $0 \le f_n(x) \le M$ for all $x \in I$ and all $n \in \mathbf{N}$. Put $g_n(x) = \sup\{f_{n+k}(x) : k \in \mathbf{N}_0\}$ $(x \in I, n \in \mathbf{N})$. Then $0 \le f_n \le g_n$, (g_n) is decreasing and

$$0 = \lim_{n \to \infty} f_n(x) = \limsup_{n \to \infty} f_n(x) = \lim_{n \to \infty} g_n(x) \ (x \in I).$$

By Theorem 1.7.15,

$$\lim_{n \to \infty} \int_a^b g_n = 0,$$

and hence

$$0 \le \lim_{n \to \infty} \int_a^b f_n \le \lim_{n \to \infty} \int_a^b g_n = 0,$$

as required. In the general case, application of what has been proved to the functions $|f_n - f|$, which have pointwise limit 0 and are bounded above by $2M$, shows that

$$\lim_{n \to \infty} \int_a^b |f_n - f| = 0.$$

Since

$$\left| \int_a^b f_n - \int_a^b f \right| \le \int_a^b |f_n - f|,$$

the result is immediate.   □

The proof follows that given in [11]. Note that Arzelà [1] proved this result in 1885, well before the creation of the theory of the Lebesgue integral. That the hypothesis $f \in \mathscr{R}(I)$ cannot be removed is shown by the following example. Let $\{r_n : n \in \mathbf{N}\}$ be the set of all rationals in $I := [0, 1]$ and for each $n \in \mathbf{N}$ define $f_n : I \to \mathbf{R}$ by

$$f_n(x) = \begin{cases} 1, & x \in \{r_i : i = 1, \ldots, n\}, \\ 0, & \text{otherwise.} \end{cases}$$

As each f_n is bounded and has only finitely many discontinuities in I, it belongs to $\mathscr{R}(I)$. The sequence (f_n) is monotonic increasing and has limit f, where $f(x) = 1$ $(x \in \mathbf{Q} \cap [0, 1])$, $f(x) = 0$ otherwise. Thus $f \notin \mathscr{R}(I)$.

From this and Example 1.4.5 (iii) we see two deficiencies of the Riemann integral:
(i) it does not behave particularly well with respect to limit processes;
(ii) a function f may have a bounded derivative on a bounded interval I yet that derivative may not itself be in $\mathscr{R}(I)$; Riemann integration and differentiation are not completely reversible.

The Lebesgue integral, developed in 1902, is not only more general than that of Riemann but is also superior to it in these respects. For example, the Lebesgue analogue of Theorem 1.7.16 holds without the assumption that the limit function f be integrable: this property follows from the other hypotheses. Moreover, if $f : [a, b] \to \mathbf{R}$ has a bounded derivative f', then f' is Lebesgue integrable and $\int_a^b f' = f(b) - f(a)$: integration and differentiation are reversible if the derivative is bounded. Nevertheless, there are functions that are improperly Riemann-integrable but not Lebesgue integrable: one such is the function $f : (0, \infty) \to \mathbf{R}$ given by $f(x) = (\sin x)/x$.

Exercise 1.7.17

1. Determine whether or not the sequence (f_n) converges uniformly on $[0, 1]$, where
   (a) $f_n(x) = n^3 x/(1 + n^4 x^2)$, (b) $f_n(x) = x^n(1 - x^n)$,
   (c) $f_n(x) = x^n/(1 + x^n)$,   (d) $f_n(x) = x^n/n$.

2. Let $p \in \mathbf{R}$ and for each $n \in \mathbf{N}$ let $f_n : [0, 1] \to \mathbf{R}$ be defined by

$$f_n(x) = n^p x(1 - x)^n \ (0 \le x \le 1).$$

   Prove that $\lim_{n\to\infty} f_n(x) = 0 \ (0 \le x \le 1)$. Prove also that the sequence (f_n) converges uniformly on $[0, 1]$ to the zero function if, and only if, $p < 1$; and that $\lim_{n\to\infty} \int_0^1 f_n = 0$ if, and only if, $p < 2$.

3. Show that $\sum_{n=1}^{\infty} \frac{x^n}{1+x^n}$ converges for all $x \in [0, 1)$ and that it converges uniformly on $[0, a]$ for each $a \in (0, 1)$. Does it converge uniformly on $[0, 1]$?

4. Prove that

$$\int_0^{\pi/2} \frac{\log\left(1 - \frac{1}{4}\sin^2\theta\right)}{\sin\theta}\,d\theta = -\frac{1}{2}\sum_{n=0}^{\infty} \frac{(n!)^2}{(2n+2)!}.$$

5. By expanding $(1 + \cos\theta \cos x)^{-1}$ in ascending powers of $\cos\theta \cos x$ and then integrating with respect to x, prove that if $0 < \theta < \pi$, then

$$\operatorname{cosec}\theta = 1 + \sum_{k=1}^{\infty} \frac{(2k)!}{2^{2k}(k!)^2}(\cos\theta)^{2k}.$$

6. (i) Prove that $\sum_{n=1}^{\infty} \frac{\cos nx}{n^2}$ is uniformly convergent on $\mathbf{R}$.

   (ii) Let $0 < \varepsilon < \pi$. Prove that $\sum_{n=1}^{\infty} \frac{\sin nx}{n}$ is uniformly convergent on $[\varepsilon, 2\pi - \varepsilon]$.

   (iii) From Exercise 1.4.15/7 it is known that

   $$\frac{1}{2}(\pi - x) = \sum_{n=1}^{\infty} \frac{\sin nx}{n} \qquad (0 < x < 2\pi).$$

   Prove that

   $$\frac{1}{4}(\pi - x)^2 = \sum_{n=1}^{\infty} \frac{\cos nx - \cos n\pi}{n^2} \qquad (0 \le x \le 2\pi).$$

   By integrating this last equality over $[0, 2\pi]$, deduce that

   $$\frac{1}{4}(\pi - x)^2 - \frac{\pi^2}{12} = \sum_{n=1}^{\infty} \frac{\cos nx}{n^2} \qquad (0 \le x \le 2\pi).$$

   In particular, show that

   $$\sum_{n=1}^{\infty} \frac{1}{n^2} = \frac{\pi^2}{6}.$$

   [Hints for part (ii): Let $A_n(x) = \sum_{k=1}^{n} \sin kx$, $P_n(x) = \sum_{k=1}^{n} \frac{\sin kx}{k}$.

   (a) There exists $K = K(\varepsilon) \in \mathbf{R}$ such that

   $$|A_n(x)| \le K \text{ if } \varepsilon \le x \le 2\pi - \varepsilon \text{ and } n \in \mathbf{N}.$$

   (b) $P_n(x) = \sum_{k=1}^{n} \left(k^{-1} - (k+1)^{-1}\right) A_k(x) + (n+1)^{-1} A_n(x)$.
   (c) $|P_m(x) - P_n(x)| \le 2K(n+1)^{-1}$ if $\varepsilon \le x \le 2\pi - \varepsilon$ and $m > n$.]

7. For each $x \in \mathbf{R}$ let $[x]$ denote the integer such that $x - 1 < [x] \le x$ and let $\{x\} = x - [x]$ denote the fractional part of x. What discontinuities does the mapping $x \longmapsto \{x\}$ have?

   Let $f : \mathbf{R} \to \mathbf{R}$ be defined by

   $$f(x) = \sum_{n=1}^{\infty} \frac{\{nx\}}{n^3} \qquad (x \in \mathbf{R}).$$

Prove that

(i)   the series defining f converges uniformly on $\mathbf{R}$;
(ii) $\{x \in \mathbf{R} : f \text{ is discontinuous at } x\} = \mathbf{Q}$, the set of all rationals;
(iii) f is Riemann-integrable over every non-degenerate, closed, bounded interval in $\mathbf{R}$.

8. Let $g : \mathbf{R} \to \mathbf{R}$ be defined by $g(x) = |x|$ if $|x| \leq 2$, $g(x+4m) = g(x)$ if $x \in \mathbf{R}$ and $m \in \mathbf{Z}$; define $f_n : \mathbf{R} \to \mathbf{R}$ by $f_n(x) = 4^{-n}g(4^n x)$ $(x \in \mathbf{R}, n \in \mathbf{N})$ and put $f(x) = \sum_{n=1}^{\infty} f_n(x)$ $(x \in \mathbf{R})$. Show that the series defining f is uniformly convergent on $\mathbf{R}$, and deduce that f is continuous on $\mathbf{R}$. Prove that f is not differentiable at any point of $\mathbf{R}$.

9. Let $f : [0, 1] \to \mathbf{R}$ be defined by $f(x) = \sum_{n=1}^{\infty} \frac{x^n(1-x)}{\sqrt{n}}$ $(0 \leq x \leq 1)$. Prove that f is continuous on $[0, 1]$.

10. Show that the function $f : \mathbf{R} \to \mathbf{R}$ defined by $f(x) = \sum_{n=1}^{\infty} (2n)! \sin^3(x/n!)$ $(x \in \mathbf{R})$ is continuous on $\mathbf{R}$.

11. Prove that if $-1 \leq t \leq 1$, then

$$\int_0^1 \frac{(1-x)}{(1-tx^3)} dx = \sum_{n=0}^{\infty} \frac{t^n}{(3n+1)(3n+2)}.$$

Deduce that

$$\frac{\pi}{3\sqrt{3}} = \frac{1}{1.2} + \frac{1}{4.5} + \frac{1}{7.8} + \frac{1}{10.11} + \dots .$$

12. For each $n \in \mathbf{N}$, define $g_n : \mathbf{R} \to \mathbf{R}$ by $g_n(x) = (1 + x^{2n})^{1/2n}$. Prove that as $n \to \infty$, g_n converges uniformly on $\mathbf{R}$ to g, where $g(x) = 1$ if $|x| < 1$, $g(x) = |x|$ if $|x| \geq 1$.

13. For each $n \in \mathbf{N}$ the function $f_n : [0, 1] \to \mathbf{R}$ is defined by $f_n(x) = -n^{-1}x^n \log x$ if $0 < x \leq 1$, $f_n(0) = 0$. Prove that $\sum_{n=1}^{\infty} f_n$ converges uniformly on $[0, 1]$ and deduce that

$$\int_0^1 (\log x) \log(1 - x) dx = \sum_{n=1}^{\infty} \frac{1}{n(n+1)^2}.$$

14. Let $f : \mathbf{R} \to \mathbf{R}$ be continuous and such that $f(t) = 0$ if $t \leq 0$ or $t \geq 1$. For each $n \in \mathbf{N}$ define $q_n : \mathbf{R} \to \mathbf{R}$ by $q_n(t) = c_n(1 - t^2)^n$ $(t \in \mathbf{R})$, where c_n is so chosen that

$$\int_{-1}^1 q_n(t) dt = 1.$$

Show that for all $n \in \mathbf{N}$, $c_n < n$, and deduce that $q_n \to 0$ uniformly on $[-1, -\delta] \cup [\delta, 1]$ for any $\delta \in (0, 1]$. Put

$$p_n(s) = \int_0^1 f(t)q_n(t - s)dt \quad (n \in \mathbf{N}, s \in \mathbf{R}).$$

Noting the conditions on f, prove that

$$p_n(s) = \int_{-1}^1 f(u + s)q_n(u)du \quad (n \in \mathbf{N}, s \in [0, 1])$$

and deduce that the sequence (p_n) of polynomials converges uniformly on $[0, 1]$ to f. Hence obtain Weierstrass's polynomial approximation theorem.

15. Let $f \in \mathscr{R}[a, b]$ and $\varepsilon > 0$. Prove that there is a function ψ which is continuous on $[a, b]$ and is such that

$$\int_a^b |f - \psi| < \varepsilon.$$

Deduce that there is a polynomial p such that

$$\int_a^b |f - p| < \varepsilon.$$

[Hint: Let $\{x_0, x_1, \ldots, x_n\}$ be a partition of $[a, b]$. Consider the continuous function ψ defined by

$$\psi(t) = (x_r - x_{r-1})^{-1} \{f(x_r)(t - x_{r-1}) + f(x_{r-1})(x_r - t)\}$$

whenever $x_{r-1} \leq t \leq x_r$ and $r \in \{1, 2, \ldots, n\}$.]

16. Use the above in conjunction with Exercise 1.4.15/5 to establish the *Riemann-Lebesgue lemma* for Riemann-integrable functions: if $f \in \mathscr{R}[a, b]$, then for all $\theta \in \mathbf{R}$,

$$\lim_{\lambda \to \infty} \int_a^b f(t) \cos(\lambda t + \theta)dt = 0.$$

17. Let $f \in \mathscr{B}[a, b]$, $f \geq 0$ and $\varepsilon > 0$. Show that there are a partition $P = \{x_0, x_1, \ldots, x_n\} \in \mathscr{P}[a, b]$ and non-negative numbers $m_1, \ldots m_n$ such that the function

$$s := \sum_{r=1}^{n-1} m_r \chi_{[x_{r-1}, x_r)} + m_n \chi_{[x_{n-1}, x_n]}$$

in terms of characteristic functions of intervals satisfies

$$0 \le s \le f \text{ and } \underline{\int_a^b f} \le \int_a^b s + \varepsilon/2.$$

Hence show that there is a function g that is continuous on $[a, b]$ and satisfies

$$0 \le g \le f, \ \underline{\int_a^b f} \le \int_a^b g + \varepsilon.$$

[Hint: remove any discontinuities of s by means of line segments. More formally, with $0 < \delta < \frac{1}{3}\min\{x_r - x_{r-1} : 1 \le r \le n\}$, define $g : [a, b] \to \mathbf{R}$ in each of the intervals $[x_{r-1}, x_r]$ $(r = 1, 2, \ldots, n)$ as follows:

$$g(x) = \begin{cases} \delta^{-1}m_r(x - x_{r-1}), & x_{r-1} \le x \le x_{r-1} + \delta, \\ m_r, & x_{r-1} + \delta \le x \le x_r - \delta, \\ \delta^{-1}m_r(x_r - x), & x_r - \delta \le x \le x_r. \end{cases}$$

Then g is continuous, $0 \le g \le s$ and $\int_a^b (s - g) < \varepsilon/2$, so that $0 \le g \le f$ and $\int_a^b f \le \int_a^b g + \varepsilon$.]

18. (Dini's theorem) Let (f_n) be a monotone sequence of continuous functions on $[a, b]$ that converges to f pointwise on $[a, b]$, where f is a continuous function on $[a, b]$. Prove that (f_n) converges uniformly on $[a, b]$ to f. [Hint: if not, and $(f_n(x))$ is decreasing, there exist $\varepsilon > 0$ and a sequence (x_n) in $[a, b]$ such that $f_n(x_n) - f(x_n) \ge \varepsilon$ for all $n \in \mathbf{N}$.]

Chapter 2
Metric Spaces

Here we give the elements of the theory of metric spaces: the ideas developed in this chapter will be extensively used in the rest of the book.

A metric space is simply a non-empty set X such that to each $x, y \in X$ there corresponds a non-negative number called the distance between x and y. To make the theory sufficiently rich this distance is supposed to have certain properties, such as symmetry and the triangle inequality, that are familiar from Euclidean geometry. As we shall see, the previous chapter offers many examples of such spaces. The idea of a metric space was introduced in 1906 by Fréchet and was significantly developed further in 1914 by Hausdorff, who introduced the term 'metric space'. Further impetus was provided from 1920 onwards by the fundamental work of the Polish school led by Banach: this was largely concerned with the case in which X was a linear space and was of great significance in the establishment of functional analysis as an important part of mathematics. Here we shall not assume that X has any linear structure as neither the results given nor the applications to complex analysis made in the next chapter need this property.

In this chapter we introduce some basic terminology and discuss in detail the fundamental properties of completeness, compactness and connectedness which such spaces may possess; further, special attention is paid to various forms of homotopy and to simple-connectedness. These properties not only have intrinsic interest but also are essential for later work surrounding such central results of complex analysis as, for example, the general version of the famous theorem due to Cauchy. Quite apart from the elegance of metric space theory, it is remarkably useful in that often a single theorem may be applied to handle seemingly different problems. Applications include a proof of the existence of a continuous, nowhere differentiable function, justification of differentiation under the integral sign, and establishment of a solution of an initial-value problem for a certain type of differential equation.

R. H. Dyer and D. E. Edmunds, *From Real to Complex Analysis*,
Springer Undergraduate Mathematics Series, DOI: 10.1007/978-3-319-06209-9_2,
© Springer International Publishing Switzerland 2014

2.1 Basic Definitions

First we recall certain fundamental properties of real numbers: for all $x, y, z \in \mathbf{R}$,

(i) $|x - y| \geq 0$; $|x - y| = 0$ if, and only if, $x = y$;
(ii) $|x - y| = |y - x|$;
(iii) $|x - y| \leq |x - z| + |z - y|$.

The quantity $|x - y|$ is naturally thought of as the distance between the real numbers x and y. We seek to generalise all this, replacing $\mathbf{R}$ by an arbitrary non-empty set and $|x - y|$ by a function of x and y which satisfies axioms based on (i), (ii) and (iii). This is done, not simply as an exercise in the axiomatic approach, but because the structure obtained will enable us to solve many apparently different problems with the same technique.

Definition 2.1.1 Let X be a non-empty set and let $d : X \times X \to \mathbf{R}$ be such that for all $x, y, z \in X$,

(i) $d(x, y) \geq 0$; $d(x, y) = 0$ if, and only if, $x = y$;
(ii) $d(x, y) = d(y, x)$ (the symmetry property);
(iii) $d(x, y) \leq d(x, z) + d(z, y)$ (the triangle inequality).

The function d is called a **metric** or **distance function** on X; the pair (X, d) is called a **metric space**; when no ambiguity is possible we shall, for simplicity, often refer to X, rather than (X, d), as a metric space.

To illustrate this definition we give a variety of examples.

Example 2.1.2

(i) $X = \mathbf{R}, d(x, y) = |x - y|$: this was our prototype; d is called the **usual metric** on $\mathbf{R}$.
(ii) $X = \mathbf{R}, d(x, y) = |x - y| / (1 + |x - y|)$. To check that the triangle inequality holds, we observe that for all $x, y, z \in \mathbf{R}$,

$$d(x, y) = 1 - \frac{1}{1 + |x - y|} \leq 1 - \frac{1}{1 + |x - z| + |z - y|}$$

$$= \frac{|x - z| + |z - y|}{1 + |x - z| + |z - y|} \leq d(x, z) + d(z, y).$$

As the other properties required of a metric obviously hold, d is a metric.
(iii) Let $n \in \mathbf{N}$ and take

$$X = \mathbf{R}^n = \{x = (x_1, ..., x_n) = (x_i) : x_i \in \mathbf{R} \text{ for } i = 1, ..., n\}.$$

Various metrics can be defined on this set in a natural way: some of the most common are $d_p (1 \leq p < \infty)$ and d_∞, where

$$d_p(x, y) = \left(\sum_{i=1}^{n} |x_i - y_i|^p \right)^{1/p}, \quad 1 \leq p < \infty,$$

$$d_\infty(x, y) = \max \{|x_i - y_i| : i = 1, ..., n\}.$$

The metric d_2 is usually referred to as the **Euclidean** metric on $\mathbf{R}^n$; when $n = 1$, all these metrics coincide. That each makes $\mathbf{R}^n$ into a metric space is clear, save perhaps for the proof of the triangle inequality for d_p, $1 \leq p < \infty$. This follows from the Minkowski inequality (see Exercise 2.1.45/1), in view of which we see that for all $x, y, z \in \mathbf{R}^n$,

$$d_p(x, y) = \left(\sum_{i=1}^{n} |(x_i - z_i) + (z_i - y_i)|^p \right)^{1/p}$$

$$\leq \left(\sum_{i=1}^{n} |x_i - z_i|^p \right)^{1/p} + \left(\sum_{i=1}^{n} |z_i - y_i|^p \right)^{1/p}$$

$$= d_p(x, z) + d_p(z, y).$$

We repeat that these examples illustrate the important fact that the same set may be endowed with different metrics. Note that for all $x, y \in \mathbf{R}^n$, $d_\infty(x, y) = \lim_{p \to \infty} d_p(x, y)$: this follows from the obvious inequalities

$$d_\infty(x, y) \leq d_p(x, y) \leq n^{1/p} d_\infty(x, y), \quad 1 \leq p < \infty.$$

(iv) Let X be any non-empty set and define $d : X \times X \to \mathbf{R}$ by the rule that

$$d(x, y) = \begin{cases} 1, & x \neq y, \\ 0, & x = y. \end{cases}$$

It is easy to check that d is a metric: (X, d) is called the **discrete metric space associated with** X, d being the **discrete metric** on X. This example is not only simple and a little surprising (going against our intuition about distances), but is also most useful as a source of counterexamples to rash conjectures about metric spaces.

In what follows, if $\mathbf{R}^n$ is referred to as a metric space without any metric being specified, then the Euclidean metric is to be assumed. When $n = 1$, identifying $\mathbf{R}^1$ and $\mathbf{R}$, this is the usual or standard metric.

(v) Let $a, b \in \mathbf{R}$, $a < b$, $I = [a, b]$, and let $X = C(I)$, the set of all continuous real-valued functions on I; for each $f, g \in C(I)$ define

$$\rho_1(f, g) = \int_a^b |f(t) - g(t)|\, dt,$$

$$\rho_\infty(f, g) = \max\{|f(t) - g(t)| : t \in [a, b]\}.$$

First note that since $|f - g|$ is a continuous, real-valued function on the closed, bounded interval I, both $\rho_1(f, g)$ and $\rho_\infty(f, g)$ are well-defined real numbers. It is now routine to check that both ρ_1 and ρ_∞ satisfy all the axioms (i), (ii) and (iii) of Definition 2.1.1: note in particular that in view of Theorem 1.3.2 (d), $\rho_1(f, g) = 0$ implies that $f = g$. Hence ρ_1 and ρ_∞ are metrics on $C(I)$. For details of a whole scale of metrics ρ_p $(1 \leq p < \infty)$ on $C(I)$ see Exercise 2.1.45/2.

(vi) Next we give an example similar to $(\mathbf{R}^n, d_p)$ but in which the elements of the space are certain infinite sequences. That is, we let

$$X = \left\{ x = (x_i)_{i \in \mathbf{N}} : x_i \in \mathbf{R} \text{ for all } i \in \mathbf{N}, \ \sum_1^\infty |x_i|^p < \infty \right\}, \ 1 \leq p < \infty,$$

and define d by

$$d(x, y) = \left(\sum_1^\infty |x_i - y_i|^p \right)^{1/p} \qquad \text{for all } x, y \in X.$$

To show that (X, d) is a metric space, it is first necessary to verify that d is well-defined; that is, that $d(x, y) < \infty$ for all $x, y \in X$: in previous examples this has been rather obvious. For each $n \in \mathbf{N}$ we have, by Minkowski's inequality,

$$\left(\sum_1^n |x_i - y_i|^p \right)^{1/p} \leq \left(\sum_1^n |x_i|^p \right)^{1/p} + \left(\sum_1^n |y_i|^p \right)^{1/p}$$

$$\leq \left(\sum_1^\infty |x_i|^p \right)^{1/p} + \left(\sum_1^\infty |y_i|^p \right)^{1/p}.$$

Hence $d(x, y) < \infty$ and

$$\left(\sum_1^\infty |x_i - y_i|^p\right)^{1/p} \le \left(\sum_1^\infty |x_i|^p\right)^{1/p} + \left(\sum_1^\infty |y_i|^p\right)^{1/p}.$$

The triangle inequality now follows from this generalised version of Min-kowski's inequality. To verify the remaining axioms is trivial.

This particular set X is usually referred to as ℓ_p.

(vii) Let $a, b \in \mathbf{R}$, $a < b$, $I = [a, b]$ and let $X = \mathscr{B}(I)$, the set of all bounded, real-valued functions on I, with d defined by

$$d(f, g) = \sup\{|f(t) - g(t)| : t \in I\} \text{ when } f, g \in \mathscr{B}(I).$$

It is easy to verify that d is a metric on $\mathscr{B}(I)$: axioms (i) and (ii) obviously hold, and if $f, g, h \in \mathscr{B}(I)$, then

$$\begin{aligned}
d(f, g) &= \sup\{|f(t) - h(t) + h(t) - g(t)| : t \in I\} \\
&\le \sup\{|f(t) - h(t)| + |h(t) - g(t)| : t \in I\} \\
&\le \sup\{|f(t) - h(t)| : t \in I\} + \sup\{|h(t) - g(t)| : t \in I\} \\
&= d(f, h) + d(h, g),
\end{aligned}$$

so that axiom (iii) also holds.

Since $C(I) \subset \mathscr{R}(I) \subset \mathscr{B}(I)$, we may regard $C(I)$ and $\mathscr{R}(I)$ as metric spaces, each with the metric inherited from $\mathscr{B}(I)$, that is, with the metrics $d \mid_{C(I) \times C(I)}$ and $d \mid_{\mathscr{R}(I) \times \mathscr{R}(I)}$ respectively.

(viii) Let (X, d) be a metric space and let Y be any non-empty subset of X; let d_Y be the restriction of d to $Y \times Y$. Then (Y, d_Y) is a metric space. Example 2.1.2 (vii) illustrates this most useful principle. In the case of any subset Y of $\mathbf{R}^n$, we shall for simplicity adopt the convention that if no metric is specified, Y is assumed to be endowed with the Euclidean metric inherited from $\mathbf{R}^n$.

(ix) Let $(X_1, d_1), \ldots, (X_n, d_n)$ be metric spaces. The product space

$$X_1 \times \ldots \times X_n = \prod_{i=1}^n X_i = \{(x_1, \ldots, x_n) : x_i \in X_i \text{ for } i = 1, \ldots, n\}$$

may be made into a metric space by endowing it with the metric d, where

$$d(x, y) = \left\{\sum_{i=1}^n d_i^2(x_i, y_i)\right\}^{1/2} \quad \text{if } x = (x_i), y = (y_i).$$

This may be established just as it was shown in Example 2.1.2 (iii) that $(\mathbf{R}^n, d_2)$ is a metric space.

(x) Let X be a vector space over $\mathbf{R}$. Let $\|\cdot\| : X \to \mathbf{R}$ be a map such that

(a) $\|x\| = 0$ if, and only if, $x = 0$;

(b) $\|\alpha x\| = |\alpha| \, \|x\|$ for all $x \in X$ and all $\alpha \in \mathbf{R}$;

(c) $\|x + y\| \le \|x\| + \|y\|$ for all $x, y \in X$.

Such a map is said to be a **norm** on X. Any such norm generates a metric d on X given by
$$d(x, y) = \|x - y\| \, .$$

In several of the examples of metric spaces given above, namely (i), (iii), (v), (vi) and (vii), the underlying set X may be viewed as a real linear space and the metric is generated by a norm given by
$$\|x\| = d(x, 0).$$

We now introduce some particularly important subsets of a metric space.

Definition 2.1.3 Let (X, d) be a metric space. Given any $x \in X$ and any $r > 0$, let $B(x, r) = \{y \in X : d(x, y) < r\} : B(x, r)$ is called the **open ball in X with centre x and radius** r. A subset G of X is called **open** if given any $x \in G$, there exists $r > 0$ (depending upon x) such that $B(x, r) \subset G$.

Example 2.1.4

(i) Take $X = \mathbf{R}$ and let d be the usual metric given by $d(x, y) = |x - y|$ $(x, y \in \mathbf{R})$, so that $B(x, r) = (x - r, x + r)$. Then $(0, 1)$ is open, for given any $x \in (0, 1), B(x, \min\{x, 1 - x\}) \subset (0, 1)$; similarly, (a, b) is open for all $a \in \{-\infty\} \cup \mathbf{R}$ and all $b \in \mathbf{R} \cup \{+\infty\}$ with $a < b$. However, if $a, b \in \mathbf{R}$ and $a < b$, then $[a, b]$ is not open, for no matter what $r > 0$ we choose, $B(a, r)$ is not contained in $[a, b]$; similarly, $[a, b)$ and $(a, b]$ are not open.

(ii) In any metric space (X, d), X is plainly open; so is $\emptyset$, for since $\emptyset$ has no points, the statement 'for all $x \in \emptyset, B(x, r) \subset \emptyset$ for all $r > 0$' is true!

(iii) Let (X, d) be any metric space, let $x \in X$ and $r > 0$. Then $B(x, r)$ is open: this justifies our description of $B(x, r)$ as the *open* ball with centre x and radius r. To prove this, let $y \in B(x, r)$ and put $\varepsilon = r - d(x, y) > 0$. Then $B(y, \varepsilon) \subset B(x, r)$, for if $z \in B(y, \varepsilon)$, then
$$d(z, x) \le d(z, y) + d(y, x) < \varepsilon + d(x, y) = r.$$

(iv) Let $X = \mathbf{R}^2$ and let d be the metric d_2 of Example 2.1.2 (iii); that is,
$$d((x_1, x_2), (y_1, y_2)) = \left\{ (x_1 - y_1)^2 + (x_2 - y_2)^2 \right\}^{1/2}.$$

The set

$$S = \{(x_1, x_2) : x_2 > x_1\}$$

is open in $(\mathbf{R}^2, d)$, for given any $(x_1, x_2) \in S$, $B((x_1, x_2), (x_2 - x_1)/\sqrt{2}) \subset S$.

(v) Let $X = \mathbf{R}^2$ and consider the metrics d_1, d_2, d_∞ of Example 2.1.2 (iii) on $\mathbf{R}^2$. In $(\mathbf{R}^2, d_\infty)$, the open ball with centre $(0, 0)$ and radius 1 is

$$\{(x_1, x_2) : \max\{|x_1|, |x_2|\} < 1\}.$$

In $(\mathbf{R}^2, d_1)$ the same open ball is $\{(x_1, x_2) : |x_1| + |x_2| < 1\}$, while in $(\mathbf{R}^2, d_2)$ this open ball has the more familiar specification $\{(x_1, x_2) : x_1^2 + x_2^2 < 1\}$. The reader is invited to sketch these three open balls.

(vi) In $(\mathbf{R}^2, d_2)$ the set $\mathbf{Q} \times \mathbf{Q} = \{(q_1, q_2) : q_1, q_2 \text{ rational}\}$ is not open, for given any $r > 0$, $\left(\sqrt{2}/n, 0\right) \in B\left((0, 0), r\right)$ for all sufficiently large $n \in \mathbf{N}$.

Some basic properties of open sets are given by the following Lemma.

Lemma 2.1.5 *Let (X, d) be a metric space.*

(i) *Every union of open subsets of X is open.*

(ii) *The intersection of every finite family of open subsets of X is open.*

(iii) *Let Y be a non-empty subset of X and let d_Y be the restriction of d to $Y \times Y$. Then U is an open subset of (Y, d_Y) if, and only if, there is an open subset V of (X, d) such that $U = V \cap Y$.*

Proof

(i) Let $\mathcal{U}$ be any family of open subsets of X and put $G = \bigcup \mathcal{U}$. If $G = \emptyset$ there is nothing to prove. Suppose $G \neq \emptyset$ and let $x \in G$. Then $x \in U$ for some $U \in \mathcal{U}$, and as U is open, there exists $r > 0$ such that $B(x, r) \subset U \subset G$. Hence G is open.

(ii) Let $\mathcal{U}$ be a finite family of open sets and put $F = \bigcap \mathcal{U}$. If $\mathcal{U} = \emptyset$, then $F = X$ and there is nothing to prove; again there is nothing to prove if $F = \emptyset$. Suppose $\mathcal{U} \neq \emptyset$, $F \neq \emptyset$ and let $x \in F$. Then $x \in U$ for all $U \in \mathcal{U}$; hence there exists $r_U > 0$ such that $B(x, r_U) \subset U$ for all $U \in \mathcal{U}$. Put $r = \min\{r_U : U \in \mathcal{U}\} : r > 0$ as $\mathcal{U}$ is a finite family, and so $B(x, r) \subset U$ for all $U \in \mathcal{U}$. Thus $B(x, r) \subset F$, and hence F is open.

(iii) If U is open in (Y, d_Y) then given any $u \in U$, there exists $r_u > 0$ such that $\{x \in Y : d(u, x) < r_u\} \subset U$; thus

$$U = \bigcup_{u \in U} \{x \in Y : d(u, x) < r_u\} = V \cap Y,$$

where

$$V = \bigcup_{u \in U} \{x \in X : d(u, x) < r_u\}$$

is open in (X, d).

Conversely, suppose $U = V \cap Y$, where V is open in (X, d). Then given any $u \in U$, there exists $r_u > 0$ such that

$$\{x \in Y : d(x, u) < r_u\} = Y \cap \{x \in X : d(x, u) < r_u\} \subset Y \cap V,$$

and so U is open in (Y, d_Y).   □

Note that the intersection of infinitely many open sets need not be open. For if $X = \mathbf{R}$ and d is the usual metric on $\mathbf{R}$ (so that $d(x, y) = |x - y|$ for all $x, y \in \mathbf{R}$), then $\bigcap_{n=1}^{\infty} (-1/n, 1/n) = \{0\}$, which is not open in $(\mathbf{R}, d)$.

In general, not all subsets of a metric space are open: see Example 2.1.4 (i). We can, however, associate with each set a largest open subset: $(0, 1)$ is the largest open subset of $[0, 1)$ in $\mathbf{R}$, endowed with the usual metric, for instance.

Definition 2.1.6 Let (X, d) be a metric space and let $A \subset X$. The **interior** of A is defined to be the set

$$\overset{o}{A} = \bigcup \{G : G \subset A \text{ and } G \text{ is open in } X\}.$$

A point is said to be an **interior point** of A if it belongs to $\overset{o}{A}$.

Note that $\overset{o}{A}$ is the union of all the open sets contained in A; in view of Lemma 2.1.5 (i), it is plainly the largest open subset of A.

Example 2.1.7 Let $X = \mathbf{R}$ and let d be the usual metric on $\mathbf{R}$. The interior of $[0, 1] \cup \{67\}$ is $(0, 1)$; that of $\mathbf{N}$ is $\emptyset$.

Lemma 2.1.8 *A subset A of a metric space (X, d) is open if, and only if, $A = \overset{o}{A}$.*

Proof If A is open, then $A \subset \overset{o}{A} \subset A$, and so $A = \overset{o}{A}$. Conversely, if $A = \overset{o}{A}$, then since $\overset{o}{A}$ is open so is A.   □

Dual to the notion of an open set is that of a closed set.

Definition 2.1.9 A subset A of a metric space X is **closed** if $X \backslash A$ is open.

Example 2.1.10

(i) In any metric space (X, d), both X and $\emptyset$ are closed (and open!). Moreover, given any $a \in X$, $\{a\}$ is closed, for given any $b \in X \backslash \{a\}$, $B\left(b, \frac{1}{2}d(a, b)\right) \subset X \backslash \{a\}$, so that $X \backslash \{a\}$ is open.

(ii) In $\mathbf{R}$, with the usual metric, $[a, b]$ is closed, for

$$\mathbf{R} \backslash [a, b] = (-\infty, a) \cup (b, \infty)$$

is open, as it is the union of two open sets.

(iii) The set $A = \{y \in X : d(x, y) \leq r\}$ (x being a given point of X and r being a given positive number) is called the **closed ball with centre** x **and radius** r. This set is closed, for if $z \in X\backslash A$, then $B\,(z, d\,(z, x) - r) \subset X\backslash A$, which shows that $X\backslash A$ is open.

Lemma 2.1.11 *In any metric space, arbitrary intersections and finite unions of closed sets are closed.*

Proof Let $\Im$ be a collection of closed sets. Then by De Morgan's rules and Lemma 2.1.5,

$$^c(\underset{F \in \Im}{\cap}\, F) = \underset{F \in \Im}{\cup}\, {}^cF$$

is open; hence $\underset{F \in \Im}{\cap} F$ is closed. Let $F_1, \ldots, F_n$ be closed sets. Then

$$^c\,(F_1 \cup \cdots \cup F_n) = {}^cF_1 \cap \cdots \cap {}^cF_n,$$

a finite intersection of open sets. Thus $^c\,(F_1 \cup \cdots \cup F_n)$ is open, by Lemma 2.1.5; hence $(F_1 \cup \cdots \cup F_n)$ is closed. $\qquad\square$

Dual to the notion of the interior is that of the closure of a set.

Definition 2.1.12 The **closure** $\overline{A}$ of a subset A of a metric space X is the intersection of all closed sets in X which contain A.

In view of Lemma 2.1.11, $\overline{A}$ is the smallest closed set which contains A. Two simple, but useful, lemmas now follow.

Lemma 2.1.13 *Let A be a subset of a metric space. Then A is closed if, and only if, $A = \overline{A}$. Moreover, $\overline{\overline{A}} = \overline{A}$.*

Proof If $A = \overline{A}$, then since $\overline{A}$ is closed, so is A. Conversely, if A is closed then it is the smallest closed set which contains A, and hence $A = \overline{A}$. Since $\overline{A}$ is closed, it now follows that $\overline{\overline{A}} = \overline{A}$. $\qquad\square$

Lemma 2.1.14 *Let A be a subset of a metric space X. Then*

$$^c(\overset{o}{A}) = \overline{{}^cA} \quad and \quad {}^c(\overline{A}) = \overset{o}{\overline{{}^cA}}\,.$$

Proof A point x belongs to $^c(\overset{o}{A})$ if, and only if, x fails to belong to any open set $G \subset A$; and this is so if, and only if, $x \in F$ for all closed $F \supset {}^c A$, which is equivalent to the statement that $x \in \overline{{}^cA}$.

The second identity follows from the first on replacing A by ^cA. Alternatively, note that

$$^c(\overline{A}) = {}^c\left(\cap_{F \supset A, F \text{ closed}}\, F\right) = \cup_{F \supset A, F \text{ closed}}{}^cF = \cup_{O \text{ open, } O \subset {}^cA}\, O = \overset{o}{\overline{{}^cA}}. \qquad\square$$

For economy of expression we need the following definition.

Definition 2.1.15 Let X be a metric space and let $a \in X$. Any open set containing a will be called a **neighbourhood** of a.

A simple example of a neighbourhood of a is given by the open ball $B(a, r)$ centred at a and with radius r; every neighbourhood of a contains such a ball. In terms of neighbourhoods we can give a useful characterisation of the points of the closure of a set.

Lemma 2.1.16 *Let A be a subset of a metric space X. Then $x \in \overline{A}$ if, and only if, every neighbourhood V of x has non-empty intersection with A.*

Proof The statement that for all neighbourhoods V of x we have $V \cap A \neq \emptyset$ is equivalent to saying that $x \notin \overset{o}{\widehat{^cA}} = {}^c(\overline{A})$, by Lemma 2.1.14. $\qquad\qquad\square$

Definition 2.1.17 Let A be a subset of a metric space X. The **boundary** ∂A of A is defined to be $\overline{A} \backslash \overset{o}{A}$.

Note that by Lemma 2.1.14,

$$\partial A = \overline{A} \cap \overline{{}^cA} = \overline{{}^cA} \backslash \overset{o}{\widehat{{}^cA}} .$$

Example 2.1.18

(i) Let $X = \mathbf{R}$, endowed with the usual metric. The boundary of $[0, 1]$ is $\{0, 1\}$, that of $\mathbf{Q}$ and $\mathbf{R} \backslash \mathbf{Q}$ is $\mathbf{R}$.

(ii) Let X be a discrete metric space and let $x \in X$. Then $\partial B(x, r) = \emptyset$ for all $r > 0$. This contrasts sharply with the situation in $\mathbf{R}^n$, equipped with the Euclidean metric: in this setting $\partial B(x, r) = \left\{ y \in \mathbf{R}^n : \sum_{i=1}^{n} (x_i - y_i)^2 = r^2 \right\}$.

Now that we have introduced the basic ideas concerning subsets of a metric space we turn to the convergence of sequences.

Definition 2.1.19 A sequence (x_n) in a metric space (X, d) is said to **converge to a point** $x \in X$ if, and only if, given any $\varepsilon > 0$, there exists $N \in \mathbf{N}$ such that $d(x, x_n) < \varepsilon$ if $n \geq N$; we write this as $x_n \to x$, $\lim_{n \to \infty} x_n = x$ or $d(x, x_n) \to 0$ as $n \to \infty$. A sequence (x_n) in X is said to be **convergent** if, and only if, there exists $x \in X$ such that $x_n \to x$; we also say in this case that (x_n) has **limit** x.

Note that $x_n \to x$ if, and only if, given any neighbourhood V of x, there exists $N \in \mathbf{N}$ such that $x_n \in V$ for all $n \geq N$.

Lemma 2.1.20 *Let (x_n) be a sequence in a metric space (X, d). Then (x_n) converges to at most one point.*

Proof Suppose that $x_n \to x$ and $x_n \to y$. Then

$$d(x, y) \leq d(x, x_n) + d(x_n, y) \to 0 \text{ as } n \to \infty.$$

Hence $d(x, y) = 0$, and so $x = y$. $\qquad\qquad\square$

Of course, a sequence may well not converge to any point.

Example 2.1.21

(i) Let $X = \mathbf{R}^n$ and let d be the Euclidean metric on $\mathbf{R}^n$ (see Example 2.1.2 (iii)); let $(x^{(m)})_{m \in \mathbf{N}}$ be a sequence in $\mathbf{R}^n$, with $x^{(m)} = (x_1^{(m)}, \ldots, x_n^{(m)})$. The sequence $(x^{(m)})$ converges to $x = (x_i)$ in $\mathbf{R}^n$ if, and only if,

$$\sum_{i=1}^{n} (x_i - x_i^{(m)})^2 \to 0 \text{ as } m \to \infty;$$

from this it is clear that $(x^{(m)})$ converges to x in $\mathbf{R}^n$ if, and only if, $(x_i^{(m)})$ converges to x_i as $m \to \infty$, for each $i \in \{1, \ldots, n\}$.

(ii) Let (X, d) be a discrete metric space. Then a sequence (x_n) in X is convergent if, and only if, it is eventually constant; that is if, and only if, there exists $N \in \mathbf{N}$ such that $x_n = x_N$ for all $n \geq N$. For if (x_n) is convergent in X, there exists $x \in X$ and $N \in \mathbf{N}$ such that $d(x, x_n) < 1$ for all $n \geq N$, so that $x_n = x$ for all $n \geq N$. The converse is obvious.

Convergent sequences of real numbers are bounded. Given an appropriate extension of the definition of boundedness, the same is true in a general metric space.

Definition 2.1.22 Let (X, d) be a metric space. A non-empty set $A \subset X$ is said to be **bounded** if there is a real number M such that

$$d(x, y) \leq M \ (x, y \in A);$$

otherwise, A is said to be **unbounded**. The extended real number

$$\text{diam}(A) := \sup\{d(x, y) : x, y \in A\}$$

is called the **diameter** of A.

Note that the set A is bounded if, and only if, $\text{diam}(A) < \infty$.

Lemma 2.1.23 *Let (x_n) be a convergent sequence in a metric space (X, d). Then $\{x_n : n \in \mathbf{N}\}$ is bounded.*

Proof Suppose that $\lim_{n \to \infty} x_n = x$. Then there exists $N \in \mathbf{N}$ such that for all $n \geq N$, $d(x, x_n) < 1$. Put $r = \max\{1, d(x, x_1), \ldots, d(x, x_{N-1})\}$. Then $d(x, x_n) \leq r$ for all $n \in \mathbf{N}$; further,

$$d(x_m, x_n) \leq d(x_m, x) + d(x, x_n) \leq 2r \ (m, n \in \mathbf{N}).$$

Thus $\{x_n : n \in \mathbf{N}\}$ is bounded. □

We can now give a most useful characterisation of the closure of a subset of a metric space.

Lemma 2.1.24 *Let A be any subset of a metric space X. Then $x \in \overline{A}$ if, and only if, there is a sequence (x_n) of points of A such that $\lim_{n \to \infty} x_n = x$.*

Proof Suppose there is a sequence (x_n) in A such that $x_n \to x$ as $n \to \infty$. Then for all $r > 0$, $A \cap B(x, r) \neq \emptyset$; and so by Lemma 2.1.16, $x \in \overline{A}$.

Conversely, suppose that $x \in \overline{A}$. Then by Lemma 2.1.16, for all $n \in \mathbf{N}$ we have $B(x, \frac{1}{n}) \cap A \neq \emptyset$. Appeal to the countable axiom of choice (Axiom A.5.2) gives the existence of a sequence (x_n) with $x_n \in B(x, \frac{1}{n}) \cap A$ for all $n \in \mathbf{N}$; plainly $x_n \to x$.   □

To conclude this rapid discussion of sequences we introduce the notion of a point of accumulation of a set.

Definition 2.1.25 Let A be a subset of a metric space X. A point $x \in X$ is called an **accumulation point of** A, or a **limit point of** A, if given any neighbourhood V of x, there exists $a \in A \cap V$ with $a \neq x$.

Note the difference between a point of accumulation of A and a point in $\overline{A}$: every point of accumulation of A is evidently in $\overline{A}$, but the converse is false. For example, with $X = \mathbf{R}$ endowed with the usual metric and $A = (0, 1) \cup \{2\}$, the point 2 belongs to $\overline{A}$ but is not a point of accumulation of A as $B(2, 1)$ contains no point of A distinct from 2.

Lemma 2.1.26 *Let $A \subset X$. Then x is a point of accumulation of A if, and only if, there is a sequence (x_n) of distinct points of A with $x_n \to x$ as $n \to \infty$.*

Proof Let x be a point of accumulation of A. Then given any $n \in \mathbf{N}$, there exists $x_n \in A \cap B(x, \frac{1}{n})$, $x_n \neq x$; this gives a sequence (x_n) of points of A which converges to x, with each $x_n \neq x$. The difficulty is that the points of this sequence may not be distinct, and to overcome this we proceed as follows, noting that there must be infinitely many distinct points in the sequence, for otherwise the sequence could not converge to x. Define $m : \mathbf{N} \to \mathbf{N}$ by $m(1) = 1$, $m(k + 1) =$ least integer p such that $x_p \notin \{x_{m(1)}, x_{m(2)}, \ldots, x_{m(k)}\}$ $(k \geq 1)$; thus $m(2) =$ least p such that $x_p \neq x_1$. Then $(x_{m(n)})_{n \in \mathbf{N}}$ is a subsequence of (x_n) consisting of distinct points of A, and $\lim_{n \to \infty} x_{m(n)} = x$. The converse is obvious.   □

2.1.1 Continuous Functions

Definition 2.1.27 Let (X_1, d_1), (X_2, d_2) be metric spaces. A map $f : X_1 \to X_2$ is said to be **continuous at** $x \in X_1$ if given any $\varepsilon > 0$, there exists $\delta > 0$ such that $d_2(f(y), f(x)) < \varepsilon$ if $d_1(x, y) < \delta$. (In general, δ depends upon x and ε.) If f is continuous at each point of X_1, it is said to be **continuous (on X_1)**. If given any $\varepsilon > 0$, there exists $\delta > 0$ (depending only on ε) such that $d_2(f(y), f(x)) < \varepsilon$ whenever $d_1(x, y) < \delta$, then f is called **uniformly continuous on X_1**.

This definition is the obvious extension of the ε, δ definition of continuity and uniform continuity for maps from subsets of $\mathbf{R}$ to $\mathbf{R}$ given in Chap. 1. However, in the wider context of metric spaces it is desirable to have other characterisations of continuity, and we now deal with this, beginning with the local property (that is, continuity at a point) and then turning to the global position (continuity on the whole space).

Lemma 2.1.28 *Let (X_1, d_1) and (X_2, d_2) be metric spaces, let $f : X_1 \to X_2$ and let $x \in X_1$. Then the following three statements are equivalent:*

(i) *f is continuous at x;*
(ii) *given any neighbourhood V of $f(x)$, there is a neighbourhood U of x such that $f(U) \subset V$;*
(iii) *$\lim_{n \to \infty} f(x_n) = f(x)$ if $x_n \to x$.*

Proof To prove that (i) implies (ii), let V be a neighbourhood of $f(x)$ and let $\varepsilon > 0$ be such that $B(f(x), \varepsilon) \subset V$. Since f is continuous at x, there exists $\delta > 0$ such that $f(y) \in B(f(x), \varepsilon)$ if $y \in B(x, \delta)$; thus $f(B(x, \delta)) \subset B(f(x), \varepsilon)$ and (ii) holds with $U = B(x, \delta)$. Next we show that (ii) implies (iii). Suppose that $x_n \to x$ in X_1 and let V be a neighbourhood of $f(x)$. As (ii) holds, there is a neighbourhood U of x such that $f(U) \subset V$; and there exists $N \in \mathbf{N}$ such that $x_n \in U$ if $n \geq N$. Hence $f(x_n) \in V$ for all $n \geq N$, which means that $f(x_n) \to f(x)$ as $n \to \infty$.

Finally, to prove that (iii) implies (i), suppose that (iii) holds but (i) is not true. Then there is an $\varepsilon > 0$ such that given any $n \in \mathbf{N}$, there exists $x_n \in X_1$ such that $d_1(x, x_n) < 1/n$ while $d_2(f(x), f(x_n)) \geq \varepsilon$; and so $x_n \to x$ but $f(x_n) \nrightarrow f(x)$, which contradicts (iii). $\qquad\square$

Lemma 2.1.29 *Let X_1, X_2 and X_3 be metric spaces, let $x \in X_1$, let $f : X_1 \to X_2$ be continuous at x and let $g : X_2 \to X_3$ be continuous at $f(x)$. Then $h := g \circ f$ is continuous at x. If f and g are continuous on X_1 and X_2 respectively, then h is continuous on X_1.*

Proof Suppose $x_n \to x$ as $n \to \infty$. Then as f is continuous at x, $f(x_n) \to f(x)$; and as g is continuous at $f(x)$, $g(f(x_n)) \to g(f(x))$. By Lemma 2.1.28, h is continuous at x. The rest is obvious. $\qquad\square$

Example 2.1.30

(i) Let $f : \mathbf{R}^m \to \mathbf{R}^n$, and suppose that $\mathbf{R}^m$ and $\mathbf{R}^n$ are endowed with the appropriate Euclidean metric. For each $x \in \mathbf{R}^m$ write $f(x) = (f_1(x), \ldots, f_n(x))$; we thus have defined functions $f_i : \mathbf{R}^m \to \mathbf{R}$ $(i = 1, \ldots, n)$, called the **coordinate functions of** f. It is now clear that f is continuous at $x_0 \in \mathbf{R}^m$ if, and only if, each f_i is continuous at x_0.
(ii) Just as in the case of maps from $\mathbf{R}$ to $\mathbf{R}$ it follows that if X is a metric space then sums and products of continuous maps from X to $\mathbf{R}$ are continuous; that is, if $x_0 \in X$ and $f_1, f_2 : X \to \mathbf{R}$ are continuous at x_0, then the maps $f_1 + f_2$ and $f_1 f_2$ (defined by $x \longmapsto f_1(x) + f_2(x)$ and $x \longmapsto f_1(x) f_2(x)$ respectively)

are continuous at x_0. Similarly, the map λf_1 (defined by $x \longmapsto \lambda f_1(x)$) is continuous at x_0, for all $\lambda \in \mathbf{R}$; and if $f_2(x_0) \neq 0$, then the map f_1/f_2 (defined by $x \longmapsto f_1(x)/f_2(x)$) is continuous at x_0. The proofs of all these assertions are identical to those of the corresponding assertions when $X = \mathbf{R}$ and which are familiar in elementary analysis.

It follows that every polynomial p on $\mathbf{R}^2$, where

$$p(x, y) = \sum_{m,n=0}^{N} a_{mn} x^m y^n (a_{mn} \in \mathbf{R}),$$

is continuous on $\mathbf{R}^2$; and any rational function f on $\mathbf{R}^2$, where $f(x, y) = p(x, y)/q(x, y)$ and p, q are polynomials with q never zero, is also continuous on $\mathbf{R}^2$.

(iii) Let $f : \mathbf{R}^2 \to \mathbf{R}$ be defined by

$$f(x, y) = \begin{cases} \frac{x^2 - y^2}{x^2 + y^2} & \text{if } (x, y) \neq (0, 0), \\ 0 & \text{if } (x, y) = (0, 0). \end{cases}$$

Reasoning as in (ii), f is continuous on $\mathbf{R}^2 \setminus \{0, 0\}$; it is not continuous at $(0, 0)$, for if $x \neq 0, f(x, 0) = 1 \nrightarrow 0 = f(0, 0)$ as $x \to 0$.

Functions between metric spaces commonly have points of discontinuity. As a tool for the investigation of discontinuity we introduce the concept of the oscillation of a function at a point.

Definition 2.1.31 Let X_1 and X_2 be metric spaces and let f be a map from X_1 to X_2. For each $x \in X_1$, let $\omega(x)$ be the extended real number defined by

$$\omega(x) = \inf\{\operatorname{diam}(f(U)) : U \text{ is a neighbourhood of } x\};$$

$\omega(x)$ is called the **oscillation of f at** x. The corresponding function ω is called the **oscillation function for** f.

Lemma 2.1.32 *Let X_1 and X_2 be metric spaces, f be a map from X_1 to X_2, and ω be the oscillation function for f. Then*

(a) *f is continuous at $x \in X_1$ if, and only if, $\omega(x) = 0$;*
(b) *for each real number α, the set $\{x \in X_1 : \omega(x) < \alpha\}$ is open in X_1.*

Proof (a) Let f be continuous at x and $\varepsilon > 0$. Then there exists $\delta > 0$ such that

$$f(B(x, \delta)) \subset B(f(x), \varepsilon).$$

Hence

$$\omega(x) \le \text{diam}(f(B(x,\delta))) \le 2\varepsilon,$$

and it follows that $\omega(x) = 0$.

Conversely, let $\omega(x) = 0$ and $\varepsilon > 0$. There is a neighbourhood U of x such that $\text{diam}(f(U)) < \varepsilon$; further, there exists $\delta > 0$ such that $B(x,\delta) \subset U$. Hence $f(B(x,\delta)) \subset B(f(x),\varepsilon)$ and f is continuous at x.

(b) Suppose $\alpha > 0$; the result is obvious otherwise. Let

$$E = \{x \in X_1 : \omega(x) < \alpha\}$$

and $y \in E$. Then there is a neighbourhood U of y such that $\text{diam}(f(U)) < \alpha$. Now U is a neighbourhood of each of its points and thus $U \subset E$. It follows that $y \in \overset{o}{E}$, that $E \subset \overset{o}{E}$ and so E is open. □

Lemma 2.1.33 *Let X_1 and X_2 be metric spaces and let $f : X_1 \to X_2$. The following three statements are equivalent:*

(i) *f is continuous (on X_1);*
(ii) *if V is an open subset of X_2, $f^{-1}(V)$ is open in X_1;*
(iii) *if F is a closed subset of X_2, $f^{-1}(F)$ is closed in X_1.*

Proof To prove that (i) implies (ii), assume that (i) holds, let V be open in X_2 and let $x \in f^{-1}(V)$. As f is continuous at x, there is a neighbourhood $U(x)$ of x such that $f(U(x)) \subset V$; that is, $U(x) \subset f^{-1}(V)$. Thus $f^{-1}(V)$ contains a neighbourhood of each of its points and hence is open.

Next suppose that (ii) holds, let $x \in X_1$ and let V be a neighbourhood of $f(x)$. Then by (ii), $f^{-1}(V)$ is open; and $x \in f^{-1}(V)$. Thus $f^{-1}(V)$ is a neighbourhood of x and $f(f^{-1}(V)) \subset V$, which by Lemma 2.1.28 means that f is continuous at x. Since x is an arbitrary point of X_1, f must be continuous on X_1. Hence (i) and (ii) are equivalent.

Finally, (ii) and (iii) are equivalent, in view of the identity $X_1 \setminus f^{-1}(F) = f^{-1}(X_2 \setminus F)$ for all $F \subset X_2$. □

Remark 2.1.34

(i) In view of Lemma 2.1.33 it is easy to see that $f : X_1 \to X_2$ is continuous if, and only if, $f^{-1}(B)$ is open for all open balls $B \subset X_2$.

(ii) Suppose that $f : X_1 \to X_2$ is continuous and that U is an open subset of X_1. It does not follow that $f(U)$ is open in X_2. To illustrate this important point, let $X_1 = X_2 = \mathbf{R}$, endowed with the usual metric, define $f : \mathbf{R} \to \mathbf{R}$ by $f(x) = (1 + x^2)^{-1}$ ($x \in \mathbf{R}$) and let $U = (-1, 1)$. Then f is continuous, U is open but $f(U) = \left(\frac{1}{2}, 1\right]$, which is not open. Similarly, it does not follow that the image of a closed set under a continuous map is closed.

Lemma 2.1.33 enables us to prove a simple and most useful result, often called the glueing lemma because it shows that under appropriate conditions, two continuous functions defined on subsets of a metric space may be 'glued together' to form a continuous function on the union of those subsets. Frequent reference to this lemma will be made in Sect. 2.5 and in our treatment of the Jordan curve theorem in Sect. 3.9.

Lemma 2.1.35 *Let X and Y be metric spaces and suppose that* $X = A \cup B$, *where A and B are non-empty and either both open or both closed. Let* $f : A \to Y$ *and* $g : B \to Y$ *be continuous (A and B are assumed equipped with the metric inherited from X), suppose that* $f(x) = g(x)$ *for all* $x \in A \cap B$, *and define* $h : X \to Y$ *by*

$$h(x) = \begin{cases} f(x), & x \in A, \\ g(x), & x \in B. \end{cases}$$

Then h is continuous.

Proof (i) Suppose A and B are both open in X. Let $\mathcal{O}$ be an open set in Y. By Lemma 2.1.33, $f^{-1}(\mathcal{O})$ and $g^{-1}(\mathcal{O})$ are open in A and B, respectively. Thus, by Lemma 2.1.5, there exist sets U, V open in X such that

$$f^{-1}(\mathcal{O}) = U \cap A, \ g^{-1}(\mathcal{O}) = V \cap B;$$

the sets $f^{-1}(\mathcal{O})$ and $g^{-1}(\mathcal{O})$ are open in X; and since $h^{-1}(\mathcal{O}) = f^{-1}(\mathcal{O}) \cup g^{-1}(\mathcal{O})$, $h^{-1}(\mathcal{O})$ is open in X. The continuity of h follows by further appeal to Lemma 2.1.33.

(ii) Suppose A and B are both closed in X and let F be a closed set in Y. By Lemma 2.1.33, $f^{-1}(F)$ and $g^{-1}(F)$ are closed in A and B, respectively. By Lemma 2.1.5, since $A \backslash f^{-1}(F)$ is open in A, there exists a set U open in X such that $A \backslash f^{-1}(F) = A \cap U$; also,

$$X \backslash f^{-1}(F) = (X \backslash A) \cup (A \backslash f^{-1}(F)) = (X \backslash A) \cup U,$$

a set open in X. Thus $f^{-1}(F)$ is closed in X as, by similar reasoning, is $g^{-1}(F)$. Now $h^{-1}(F) = f^{-1}(F) \cup g^{-1}(F)$ and so, by Lemma 2.1.11, $h^{-1}(F)$ is closed in X. Finally, by Lemma 2.1.33, the continuity of h is proved.

Note that use of Exercise 2.1.45/17 yields a simpler proof of (ii), one identical in form with that given in case (i).   □

2.1.2 Homeomorphisms

We now introduce the idea of homeomorphism, which enables a sensible classification of spaces to be made.

Definition 2.1.36 Let X_1, X_2 be metric spaces. A map $f : X_1 \to X_2$ is said to be a **homeomorphism** if it is a continuous bijection and f^{-1} is continuous. If such a map exists, X_1 and X_2 are said to be **homeomorphic**.

Remark 2.1.37

(i) Not every bijective continuous map is a homeomorphism; that is, the condition that f^{-1} be continuous is an essential part of the definition. For take $X_1 = \mathbf{R}$ endowed with the discrete metric, let $X_2 = \mathbf{R}$ endowed with the usual metric and let f be the identity map from X_1 to X_2; that is $f(x) = x$ for all $x \in \mathbf{R}$. Then f is not a homeomorphism: it is continuous, as $f^{-1}(O)$ is open in X_1 whenever O is open in X_2 (every subset of X_1 is open!); but f^{-1} is not continuous, for $(f^{-1})^{-1}([0, 1)) = [0, 1)$ is open in X_1 but not in X_2.

(ii) Let $f : X_1 \to X_2$ be a bijection. Then if f is a homeomorphism, a subset U of X_1 is open if, and only if, $f(U)$ is open in X_2: note that $U = (f^{-1})^{-1}(U)$. It follows that the open sets in a metric space may be put in one-to-one correspondence with the open sets in any metric space homeomorphic to it.

(iii) In general, homeomorphisms do not preserve distances. Thus let $X_1 = (0, 1)$, $X_2 = (1, \infty)$ and endow each set with the usual metric inherited from $\mathbf{R}$; let $f : X_1 \to X_2$ be defined by $f(x) = x^{-1}$ $(x \in X_1)$. Then f is plainly a homeomorphism, but if $x, y \in X_1$ and $x \neq y$, then $\left| x^{-1} - y^{-1} \right| \neq |x - y|$; that is, the distance between $f(x)$ and $f(y)$ differs from that between x and y. Homeomorphisms which do preserve distances are called **isometries**. We formalise this in the following definition.

Definition 2.1.38 Let (X_1, d_1) and (X_2, d_2) be metric spaces. A map $f : X_1 \to X_2$ is said to be an **isometry** if it is bijective and for all $x, y \in X_1$,

$$d_2(f(x), f(y)) = d_1(x, y).$$

If such a map exists, X_1 and X_2 are said to be **isometric**.

Example 2.1.39

(i) Consider $\mathbf{R}^n$, with the Euclidean metric, and the unit ball $B(0, 1)$ in $\mathbf{R}^n$, given the metric inherited from $\mathbf{R}^n$. Then $\mathbf{R}^n$ and $B(0, 1)$ are homeomorphic. To see this, let $f : \mathbf{R}^n \to B(0, 1)$ be defined by $f(x) = x/(1+|x|)$, where $|x| = \left(\sum_{i=1}^{n} x_i^2 \right)^{\frac{1}{2}}$. Since $|f(x)| = |x|/(1+|x|) < 1$ for all $x \in \mathbf{R}^n$, it follows that $f(\mathbf{R}^n) \subset B(0, 1)$. Moreover, given any $y \in B(0, 1)$, the point $x := y/(1 - |y|)$ is the unique point mapped by f to y: thus f is a bijection and $f^{-1}(y) = y/(1 - |y|)$. It is clear that f and f^{-1} are continuous, and hence f is a homeomorphism.

(ii) For any $n \in \mathbf{N}$, let S^n be the unit sphere in $\mathbf{R}^{n+1}$ (endowed with the Euclidean metric); that is,

$$S^n = \left\{ (x_i) \in \mathbf{R}^{n+1} : \sum_{i=1}^{n+1} x_i^2 = 1 \right\}.$$

Then $S^2 \setminus \{(0, 0, 1)\}$ is homeomorphic to $\mathbf{R}^2$, each set being given the Euclidean metric. This follows by consideration of the stereographic projection P, where

$$P(x_1, x_2, x_3) = \left(\frac{x_1}{1 - x_3}, \frac{x_2}{1 - x_3} \right).$$

First note that P is obviously a continuous map of $S^2 \setminus \{(0, 0, 1)\}$ to $\mathbf{R}^2$. Moreover, P is bijective, for given any $(y_1, y_2) \in \mathbf{R}^2$, the equation $Px = (y_1, y_2)$ has the unique solution $x_1 = 2y_1/(y_1^2 + y_2^2 + 1)$, $x_2 = 2y_2/(y_1^2 + y_2^2 + 1)$, $x_3 = (y_1^2 + y_2^2 - 1)/(y_1^2 + y_2^2 + 1)$, since any possible solution $x = (x_1, x_2, x_3)$ must satisfy

$$(x_1^2 + x_2^2)/(1 - x_3)^2 = y_1^2 + y_2^2, \ (1 - x_3^2)/(1 - x_3)^2 = y_1^2 + y_2^2,$$

so that $(1 + x_3)/(1 - x_3) = y_1^2 + y_2^2$ and hence

$$x_3 = (y_1^2 + y_2^2 - 1)/(y_1^2 + y_2^2 + 1), \ x_1 = y_1(1 - x_3) = 2y_1/(y_1^2 + y_2^2 + 1),$$

$$x_2 = y_2(1 - x_3) = 2y_2/(y_1^2 + y_2^2 + 1).$$

The continuity of P^{-1} is now clear, and so P is a homeomorphism.
(iii) Let $S, Q \subset \mathbf{R}^2$ be a circle and a square, respectively, each given the Euclidean metric inherited from $\mathbf{R}^2$. Then S and Q are homeomorphic. To prove this it is enough to consider the case in which $S = S^1$ (see (ii) above) and Q is the square with centre O, of side 2 and with sides parallel to the coordinate axes. Define $\phi : Q \to S$ by $\phi(x, y) = (x, y)/\sqrt{(x^2 + y^2)}$; ϕ is plainly continuous. It is bijective, for given $(u, v) \in S$, there is a unique (x, y) in Q such that $\phi(x, y) = (u, v)$. In fact, $x/\sqrt{(x^2 + y^2)} = u$, $y/\sqrt{(x^2 + y^2)} = v$, so that if we put $x = r \cos \theta$, $y = r \sin \theta$, then $u = \cos \theta$, $v = \sin \theta$; moreover, $(x, y) \in Q$ if, and only if, $\max\{|x|, |y|\} = 1$, and so $\max\{r|u|, r|v|\} = 1$, which gives $r = 1/\max\{|u|, |v|\}$. Hence

$$x = u/\max\{|u|, |v|\}, \ y = v/\max\{|u|, |v|\},$$

which shows that $\phi^{-1}(u, v) = (u, v)/\max\{|u|, |v|\}$. Since ϕ^{-1} is plainly continuous, it follows that ϕ is a homeomorphism.
(iv) Let $\mathbf{R}^n$ be given the Euclidean metric. Then a map $g : \mathbf{R}^n \to \mathbf{R}^n$ is an isometry if, and only if, it is of the form

$$g(t) = x_0 + f(t) \ (t \in \mathbf{R}^n)$$

where $x_0 \in \mathbf{R}^n$ and $f : \mathbf{R}^n \to \mathbf{R}^n$ is linear, and orthogonal in the sense that for all $s, t \in \mathbf{R}^n$, $\langle f(s), f(t) \rangle = \langle s, t \rangle$, where $\langle x, y \rangle = \sum_{i=1}^{n} x_i y_i$.

To prove this, first suppose that g is an isometry and let d be the Euclidean metric on $\mathbf{R}^n$. Put $f(t) = g(t) - g(0)$ $(t \in \mathbf{R}^n)$; then $f(0) = 0$, $d(f(s),$

$f(t)) = d(g(s), g(t)) = d(s, t)$ and $d(f(t), 0) = d(g(t), g(0)) = d(t, 0)$. For all $s, t \in \mathbf{R}^n$, it follows that since $d^2(f(s), f(t)) = d^2(s, t)$ we have

$$d^2(f(t), 0) - 2 \langle f(t), f(s) \rangle + d^2(f(s), 0) = d^2(s, 0) - 2 \langle s, t \rangle + d^2(t, 0),$$

and hence $\langle f(t), f(s) \rangle = \langle s, t \rangle$.

Thus f is orthogonal. To show that f is linear, let $s, t \in \mathbf{R}^n$ and $\alpha, \beta \in \mathbf{R}$. Then $d^2(f(\alpha s + \beta t), \alpha f(s) + \beta f(t))$ is given by

$$\langle f(\alpha s + \beta t) - \alpha f(s) - \beta f(t), f(\alpha s + \beta t) - \alpha f(s) - \beta f(t) \rangle$$

$$= \langle f(\alpha s + \beta t), f(\alpha s + \beta t) \rangle + \alpha^2 \langle f(s), f(s) \rangle + \beta^2 \langle f(t), f(t) \rangle$$

$$+ 2\alpha\beta \langle f(s), f(t) \rangle - 2\alpha \langle f(\alpha s + \beta t), f(s) \rangle - 2\beta \langle f(\alpha s + \beta t), f(t) \rangle$$

$$= \langle \alpha s + \beta t, \alpha s + \beta t \rangle + \alpha^2 \langle s, s \rangle + \beta^2 \langle t, t \rangle + 2\alpha\beta \langle s, t \rangle$$

$$- 2\alpha \langle \alpha s + \beta t, s \rangle - 2\beta \langle \alpha s + \beta t, t \rangle$$

$$= 2 \langle \alpha s + \beta t, \alpha s + \beta t \rangle - 2 \langle \alpha s + \beta t, \alpha s + \beta t \rangle = 0.$$

Hence f is linear.

Conversely, suppose that $g = x_0 + f$, where f is linear and orthogonal. Then for all $s, t \in \mathbf{R}^n$,

$$d^2(g(s), g(t)) = d^2(f(s), f(t)) = \langle f(s) - f(t), f(s) - f(t) \rangle$$

$$= \langle f(s - t), f(s - t) \rangle = \langle s - t, s - t \rangle = d^2(s, t),$$

which shows that g is an isometry.

2.1.3 An Extension Theorem

Let A be a subspace of a metric space X and let $f : A \to \mathbf{R}$ be continuous. A natural question to ask is whether or not f has a continuous real-valued extension defined on all of X. That is to say, does there exist a continuous map $g : X \to \mathbf{R}$ such that for all $x \in A$, $g(x) = f(x)$? In general, the answer is negative: the map $x \longmapsto 1/x : (0, 1) \to \mathbf{R}$ is continuous, but it has no continuous extension even to $[0, 1]$ because such an extension would have to be bounded. However, for an affirmative answer it turns out that a sufficient condition on A is that it is a closed subspace of X. We establish this below, beginning with a few preliminaries.

Lemma 2.1.40 *Let A and B be non-empty subsets of a metric space (X, d). For each $x \in X$, let the **distance from** x **to** A, denoted by $d(x, A)$ or by $\mathrm{dist}(x, A)$ when the metric d is understood, be defined by*

$$d(x, A) = \inf\{d(x, a) : a \in A\} \quad (x \in X),$$

*and the **distance from** A **to** B be*

$$d(A, B) = \inf\{d(a, b) : a \in A, b \in B\}.$$

Then

(i) $d(A, B) = \inf\{d(a, B) : a \in A\}$;
(ii) $d(x, A) = 0$ *if, and only if, $x \in \overline{A}$;*
(iii) *the map $x \longmapsto d(x, A) : X \to \mathbf{R}$ is continuous; in fact, for all $x, y \in X$,*

$$|d(x, A) - d(y, A)| \le d(x, y).$$

Proof (i) For all $a \in A$ and all $b \in B$, $d(A, B) \le d(a, b)$; hence, for all $a \in A$, $d(A, B) \le d(a, B)$ and so $d(A, B) \le \inf\{d(a, B) : a \in A\}$. Now let $\varepsilon > 0$. There exist $a \in A$ and $b \in B$ such that $d(a, B) \le d(a, b) < d(A, B) + \varepsilon$; thus

$$\inf\{d(x, B) : x \in A\} < d(A, B) + \varepsilon.$$

As this is true for all $\varepsilon > 0$, (i) follows.

(ii) $d(x, A) = 0$ if, and only if, for all $\varepsilon > 0$, $A \cap B(x, \varepsilon) \ne \emptyset$; and, by Lemma 2.1.16, this is true if, and only if, $x \in \overline{A}$.

(iii) Let $x, y \in X$. Then for all $a \in A$,

$$d(x, A) - d(x, y) \le d(x, a) - d(x, y) \le d(y, a).$$

Hence $d(x, A) - d(x, y) \le d(y, A)$. Interchange of x and y shows that $d(y, A) - d(y, x) \le d(x, A)$, and (iii) follows. $\qquad\square$

The notion of uniform convergence for sequences of real-valued functions, introduced in Sect. 1.7, may be developed further to include functions with range in a general metric space.

Definition 2.1.41 Let S be a non-empty set, let (X, d) be a metric space and, for each $n \in \mathbf{N}$, let $f_n : S \to X$. The sequence (f_n) is said to **converge pointwise on** S if there is a function $f : S \to X$ such that for each $s \in S$,

$$\lim_{n \to \infty} d(f_n(s), f(s)) = 0;$$

it is said to **converge uniformly on** S if there is a function $f : S \to X$ such that

$$\lim_{n \to \infty} \sup_S \, d(f_n(s), f(s)) = 0.$$

Evidently uniform convergence on S implies pointwise convergence on S; apart from special cases, the converse is false. In Sect. 1.7 it was observed that the limit of a uniformly convergent sequence of continuous real-valued functions defined on a subspace of $\mathbf{R}$ is itself a continuous function. This observation carries over to the setting of a general metric space: loosely speaking, uniform convergence preserves continuity. Henceforth, it will be convenient to use the symbol $C(X, Y)$ to denote the family of all continuous functions from a metric space X to a metric space Y; $C(X, \mathbf{R})$ ($\mathbf{R}$ being given the usual metric) may be abbreviated to $C(X)$.

Theorem 2.1.42 *Let X and Y be metric spaces and let $a \in X$. For each $n \in \mathbf{N}$ let $f_n : X \to Y$ be continuous at a, and suppose that the sequence (f_n) converges uniformly on X to $f : X \to Y$. Then f is continuous at a. In particular, if each $f_n \in C(X, Y)$, then $f \in C(X, Y)$.*

Proof This is an obvious modification of that of Theorem 1.7.7. □

Our goal in this subsection is to prove that a continuous real-valued function on a closed subspace of a metric space X has a continuous real-valued extension to all of X. The lemma which follows, usually referred to as Urysohn's lemma , is a special case of this result. Framed in the metric space context adequate for our purposes, it has an elementary proof: Urysohn established it in a more general setting.

Lemma 2.1.43 *Let A and B be disjoint closed subsets of a metric space (X, d). Then there exists a continuous map $f : X \to \mathbf{R}$ such that $f(x) = 1$ $(x \in A)$, $f(x) = -1$ $(x \in B)$ and $|f(x)| \leq 1$ on X.*

Proof If either A or B is empty, then a suitable constant map may be chosen for f. Suppose that neither A nor B is empty and, adopting the notation of Lemma 2.1.40, define $f : X \to \mathbf{R}$ by $f(x) = \{d(x, B) - d(x, A)\}/\{d(x, B) + d(x, A)\}$. Part (ii) of Lemma 2.1.40 shows that, for all $x \in X$, $d(x, B) + d(x, A) > 0$ and so f is well-defined; part (iii) shows that the maps $x \longmapsto d(x, A)$ and $x \longmapsto d(x, B)$ are continuous and therefore (see also Example 2.1.30 (ii)) f is continuous. That f has the remaining properties is clear; indeed, on $X \backslash (A \cup B)$, $|f| < 1$. □

With this lemma at our disposal we give the promised extension theorem, which is due to Tietze.

Theorem 2.1.44 (Tietze's extension theorem) *Let A be a non-empty closed subset of a metric space X and let $f : A \to \mathbf{R}$ be continuous. Then there exists a continuous map $g : X \to \mathbf{R}$ such that $g(x) = f(x)$ for all $x \in A$.*

Proof Let $\phi : \mathbf{R} \to (-1, 1)$ be defined by

$$\phi(x) = (1 + |x|)^{-1} x.$$

Let $h = \phi \circ f$; plainly, $|h| < 1$ on A. We begin by proving that the bounded map h has a continuous extension to X. Let E, F be the disjoint closed sets given by

$$E = \{x \in A : h(x) \leq -1/3\}, F = \{x \in A : h(x) \geq 1/3\}.$$

By the Urysohn Lemma, there exists a continuous map $u_1 : X \to \mathbf{R}$ such that

$$u_1(x) = -1/3 \ (x \in E), u_1(x) = 1/3 \ (x \in F)$$

and $|u_1(x)| \leq 1/3 \ (x \in X)$. Note that

$$|h(x) - u_1(x)| < 2/3 \ (x \in A).$$

With h replaced by $\frac{3}{2}(h - u_1)$, repetition of the above argument shows that there exists a continuous map $u_2 : X \to \mathbf{R}$ such that

$$|u_2(x)| \leq \frac{2}{3^2} \ (x \in X)$$

and

$$|h(x) - u_1(x) - u_2(x)| < (2/3)^2 \ (x \in A).$$

Inductively, it follows that there is a sequence (u_n) of continuous real-valued functions on X such that

$$|u_n(x)| \leq \frac{2^{n-1}}{3^n} \ (x \in X; n \in \mathbf{N}) \tag{2.1.1}$$

and

$$\left| h(x) - \sum_{k=1}^{n} u_k(x) \right| < (2/3)^n \ (x \in A; n \in \mathbf{N}). \tag{2.1.2}$$

Use of (2.1.1), Theorem 1.7.5 (the Weierstrass M−test) and Theorem 2.1.42 shows that $\sum u_n$ converges to a continuous function on X, say u. Then, for all $x \in X$,

$$|u(x)| \leq \sum_{n=1}^{\infty} |u_n(x)| \leq \sum_{n=1}^{\infty} \frac{2^{n-1}}{3^n} = 1;$$

also, from (2.1.2), $u |_A = h$.

The map u is a continuous extension of h to X, but this extension may assume the values -1 or 1. A continuous extension eliminating this possibility is constructed next. Let

$$B = \{x \in X : u(x) \in \{-1, 1\}\}.$$

Plainly, B is closed and $A \cap B = \emptyset$. By Urysohn's Lemma, there exists a continuous map $\psi : X \to \mathbf{R}$ such that

$$\psi(x) = 1 \ (x \in A), \ \psi(x) = 0 \ (x \in B)$$

and $0 \le \psi \le 1$ on $X \backslash (A \cup B)$. Let $v : X \to \mathbf{R}$ be defined by

$$v(x) = \psi(x)u(x) \ (x \in X).$$

Evidently, $v \mid_A = u \mid_A = h$, and v is a continuous extension of h with values in $(-1, 1)$.

Lastly, the conclusion of the theorem is immediate on setting $g = \phi^{-1} \circ v$. Incidentally, it can be shown (see [5], (4.5.1)) that there is an extension g with the property that

$$\sup_{x \in X} g(x) = \sup_{y \in A} f(y), \ \inf_{x \in X} g(x) = \inf_{y \in A} f(y).$$

$\square$

Exercise 2.1.45

1. Let $1 < p < \infty$ and define p' by $\frac{1}{p} + \frac{1}{p'} = 1$. Let $f : [0, \infty) \to \mathbf{R}$ be given by $f(t) = \frac{t^p}{p} + \frac{1}{p'} - t$. Show that the minimum of f on $[0, \infty)$ is attained only at $t = 1$ and that the minimum value is 0. Hence show that for all $a, b \in [0, \infty)$,

$$ab \le \frac{a^p}{p} + \frac{b^{p'}}{p'},$$

with equality if, and only if, $a = b^{1/(p-1)}$.

Deduce Hölder's inequality: for every $x = (x_1, \ldots x_n), y = (y_1, \ldots, y_n) \in \mathbf{R}^n$,

$$\sum_{k=1}^{n} |x_k y_k| \le \left(\sum_{k=1}^{n} |x_k|^p \right)^{1/p} \left(\sum_{k=1}^{n} |y_k|^{p'} \right)^{1/p'}.$$

(The case $p = 2$ is Schwarz's inequality.) Use this to prove Minkowski's inequality:

$$\left(\sum_{k=1}^{n} |x_k + y_k|^p \right)^{1/p} \le \left(\sum_{k=1}^{n} |x_k|^p \right)^{1/p} + \left(\sum_{k=1}^{n} |y_k|^p \right)^{1/p}.$$

Hence show that $(\mathbf{R}^n, d_p)$ is a metric space, where

$$d_p(x, y) = \left(\sum_{k=1}^{n} |x_k - y_k|^p \right)^{1/p}.$$

2. Let $a, b \in \mathbf{R}$, suppose that $a < b$ and put $I = [a, b]$; let $1 \le p < \infty$. Prove that $(C(I), d)$ is a metric space, where

$$d(f, g) = \left(\int_a^b |f(t) - g(t)|^p \, dt \right)^{1/p} \qquad (f, g \in C(I)).$$

3. Let $p \in [1, \infty)$ and set

$$\ell_p = \left\{ x = (x_i)_{i \in \mathbf{N}} : x_i \in \mathbf{R} \text{ for all } i \in \mathbf{N}, \sum_1^\infty |x_i|^p < \infty \right\},$$

$$d_p(x, y) = \left(\sum_{i=1}^\infty |x_i - y_i|^p \right)^{1/p} \qquad (x, y \in \ell_p).$$

Prove that (ℓ_p, d_p) is a metric space.

4. Let S be the set of all sequences of real numbers and define d by

$$d(x, y) = \sum_{n=1}^\infty \frac{|x_n - y_n|}{2^n [1 + |x_n - y_n|]} \qquad (x = (x_n), y = (y_n) \in S).$$

Show that (S, d) is a metric space.

$\left[\text{Hint: } t \longmapsto t/(1+t) \text{ is an increasing function on } [0, \infty). \right]$

5. Let p be a prime number. Given any distinct integers m, n, let $t = t(m, n)$ be the unique integer such that

$$m - n = kp^t$$

for some integer k not divisible by p. Define $d : \mathbf{Z} \times \mathbf{Z} \to \mathbf{R}$ by

$$d(m, n) = \begin{cases} 1/p^t & \text{if } m \neq n, \\ 0 & \text{if } m = n. \end{cases}$$

Prove that for all distinct $a, b, c \in \mathbf{Z}$,

$$t(a, c) \ge \min \{t(a, b), t(b, c)\},$$

and hence show that $(\mathbf{Z}, d)$ is a metric space.

6. Determine whether the following subsets of $\mathbf{R}$ (endowed with the usual metric) are open, closed or neither open nor closed:

(i) $\mathbf{N}$,   (ii) $\left\{ \dfrac{1}{n} : n \in \mathbf{N} \right\}$,   (iii) $\mathbf{Q}$,   (iv) $\left\{ (-1)^n \left(1 + \dfrac{1}{n} \right) : n \in \mathbf{N} \right\}$.

7. Show that each of the following sets is an open subset of $\mathbf{R}^2$, endowed with the Euclidean metric:

(i) $\{(x, y) : x^2 + y^2 < 1, \; x > 0, \; y > 0\}$,
(ii) $\{(x, y) : x + y \neq 0\}$,
(iii) $\{(x, y) : xy \neq 1\}$.

Is $\{(x, 0) : 0 < x < 1\}$ an open subset of $\mathbf{R}^2$?

8. Show that each subset of a discrete metric space is open and closed.

9. Let S be the set of all sequences of real numbers; given any $x = (x_n)$ and $y = (y_n)$ in S, with $x \neq y$, let $k(x, y)$ be the smallest integer n such that $x_n \neq y_n$. Define $d : S \times S \to \mathbf{R}$ by

$$d(x, y) = \begin{cases} 1/k(x, y) & \text{if } x \neq y, \\ 0 & \text{if } x = y. \end{cases}$$

Prove that for all $x, y, z \in S$,

$$d(x, y) \leq \max\{d(x, z), d(z, y)\},$$

and hence show that (S, d) is a metric space.

10. Let (X, d) be a metric space and suppose that for all $x, y, z \in X$,

$$d(x, y) \leq \max\{d(x, z), d(y, z)\}.$$

Prove that if $d(x, z) \neq d(y, z)$, then $d(x, y) = \max\{d(x, z), d(y, z)\}$. Show also that if $x \in X$ and $r > 0$, then $B(x, r) = B(y, r)$ for all $y \in B(x, r)$. Prove that if two open balls in (X, d) intersect, then one is contained in the other. Show that for all $x \in X$ and all $r > 0$, $B(x, r)$ is closed and $\{y \in X : d(x, y) \leq r\}$ is open.

11. Let X be a non-empty set and $d : X \times X \to \mathbf{R}$ a mapping such that

(i) $d(x, y) = 0$ if, and only if, $x = y$;
(ii) $d(x, z) \leq d(x, y) + d(z, y)$ for all $x, y, z \in X$.

Prove that (X, d) is a metric space.

12. Let d_1, d_2 be two metrics on a non-empty set X, and suppose that there are positive constants α, β such that for all $x, y \in X$,

$$\alpha d_1(x, y) \leq d_2(x, y) \leq \beta d_1(x, y).$$

Prove that the metric spaces (X, d_1) and (X, d_2) have the same open sets. Deduce that each of the metrics d_p of Example 2.1.2 (iii) generates the same family of open subsets of $\mathbf{R}^n$.

13. Show that for all subsets A and B of a metric space X,

$$\overset{\circ}{A} \cap \overset{\circ}{B} = \overset{\circ}{\overparen{A \cap B}}, \quad \overline{A} \cup \overline{B} = \overline{A \cup B}.$$

Show by means of examples that, in general,

$$\overset{\circ}{A} \cup \overset{\circ}{B} \neq \overset{\circ}{\overparen{A \cup B}}, \quad \overline{A} \cap \overline{B} \neq \overline{A \cap B}.$$

Find the closure and interior of the subset D of $\mathbf{R}^3$ (with the Euclidean metric) defined by

$$D = \left\{ (x, y, z) \in \mathbf{R}^3 : \cosh(x + yz) \geq 2 \right\}.$$

14. Determine the interiors and closures of the following subsets of $\mathbf{R}^2$ (with the Euclidean metric):
    (i) $\{(x, y) : 0 < x \leq y < 1\}$, (ii) $\{(x, 0) : 0 < x < 1\}$, (iii) $\{(x, y) : x, y \in \mathbf{Q}\}$.
15. Let S be the subset of $[0, 1]$ (with the usual metric) consisting of all those real numbers which have a decimal representation of the form

$$\sum_{n=1}^{\infty} \frac{a_n}{10^n},$$

where $a_n \in \{0, 1\}$ for all $n \in \mathbf{N}$. By consideration of any $y \in [0, 1] \backslash S$ and the first digit in the decimal representation of y which is not 0 or 1, find the closure of S.

16. By consideration of a discrete metric space, show that a closed ball in a metric space need not be the closure of the open ball with the same centre and the same radius.
17. Let (X, d) be a metric space, let Y be a non-empty subset of X and let d_Y be the restriction of d to $Y \times Y$.

    (i) Show that A is a closed subset of (Y, d_Y) if and only if there is a closed subset B of (X, d) such that $A = B \cap Y$.
    (ii) Let Y be a closed subset of (X, d) and $S \subset Y$; let $cl_Y(S)$ $(cl_X(S))$ denote the closure of S in (Y, d_Y) $((X, d))$. Show that $cl_Y(S) = cl_X(S)$.

18. Let $f : \mathbf{R}^2 \to \mathbf{R}$ be defined by

$$f(x, y) = \begin{cases} xy/(x^2 + y^2) & \text{if } (x, y) \neq (0, 0), \\ 0 & \text{if } (x, y) = (0, 0); \end{cases}$$

$\mathbf{R}^2$ and $\mathbf{R}$ are each supposed to be equipped with the appropriate Euclidean metric. Show that f is continuous at each point of $\mathbf{R}^2 \backslash \{(0, 0)\}$, and that it is not continuous at $(0, 0)$.

19. Let X be a metric space, let $f, g : X \to \mathbf{R}$ be continuous and define $h : X \to \mathbf{R}^2$ by $h(x) = (f(x), g(x))$ $(x \in X)$. Given that $\mathbf{R}^2$ is endowed with the Euclidean metric, show that h is continuous on X.

20. Discuss the continuity of the map $f : \mathbf{R}^2 \to \mathbf{R}^2$ ($\mathbf{R}^2$ is equipped with the Euclidean metric) defined by

$$f(x, y) = \begin{cases} \left(\dfrac{x^2 - y^2}{x^2 + y^2}, \dfrac{(x^2 - y^2)^2}{x^2 + y^2} \right) & \text{if } (x, y) \neq (0, 0), \\ (0, 0) & \text{if } (x, y) = (0, 0). \end{cases}$$

21. Show that $S := \{(x, y) : x^2 - y^2 + 2xy < 0\}$ is an open subset of $\mathbf{R}^2$ (equipped with the Euclidean metric).

22. Let A and B be non-empty subsets of a metric space (X, d). Prove that

   (i) A is bounded if, and only if, there exist $x \in X$ and $r > 0$ such that $A \subset B(x, r)$;

   (ii) $A \subset B$ implies that $\mathrm{diam}(A) \leq \mathrm{diam}(B)$;

   (iii) $\mathrm{diam}(A) = 0$ if, and only if, for some $x \in X$, $A = \{x\}$;

   (iv) if $a \in A$ and $b \in B$, then

$$\mathrm{diam}(A \cup B) \leq \mathrm{diam}(A) + \mathrm{diam}(B) + d(a, b);$$

   (v) if A and B are bounded, then $A \cup B$ is bounded; further, a finite union of bounded subsets of X is bounded.

23. Let A be a non-empty set of real numbers which is bounded above and let $a = \sup A$. Prove that $a \in \overline{A}$.

24. Let A and B be closed, disjoint subsets of a metric space X. Show that there are open, disjoint subsets U and V (of X) such that $A \subset U$ and $B \subset V$. [Hint: Urysohn's lemma.]

2.2 Complete Metric Spaces

An important property of real numbers is that every Cauchy sequence in $\mathbf{R}$ converges to a point of $\mathbf{R}$. We distinguish a class of metric spaces in which the same kind of property holds. These spaces, the *complete* spaces, are of the utmost theoretical and practical importance.

Definition 2.2.1 Let (X, d) be a metric space. A sequence (x_n) in X is called a **Cauchy sequence** if given any $\varepsilon > 0$, there exists $N \in \mathbf{N}$ such that $d(x_m, x_n) < \varepsilon$ whenever $m, n \geq N$; equivalently, $\mathrm{diam}\{x_m : m \geq n\} \to 0$ as $n \to \infty$. Loosely, these conditions may be written $d(x_m, x_n) \to 0$ as $m, n \to \infty$. The space X is said to be **complete** if given any Cauchy sequence (x_n) in X, there exists $x \in X$ such that $x_n \to x$ as $n \to \infty$.

Example 2.2.2

(i) **R**, with the usual metric, is complete: this was our prototype.

(ii) **R**^{n}, with the usual (Euclidean) metric d_2, is complete. To prove this, let $(x^{(m)})$ be a Cauchy sequence in **R**^{n}, with $x^{(m)} = (x_1^{(m)}, \ldots, x_n^{(m)})$. For each $j \in \{1, \ldots, n\}$,

$$\left| x_j^{(m)} - x_j^{(p)} \right| \le \left(\sum_{k=1}^{n} \left| x_k^{(m)} - x_k^{(p)} \right|^2 \right)^{1/2} = d_2(x^{(m)}, x^{(p)}) \to 0$$

as $m, p \to \infty$; that is, $(x_j^{(m)})_{m \in \mathbf{N}}$ is a Cauchy sequence in **R** and hence converges, to $x_j \in \mathbf{R}$, say. Put $x = (x_1, \ldots, x_n) \in \mathbf{R}^n$. Then $d_2(x^{(m)}, x) = \left(\sum_{k=1}^{n} \left| x_k^{(m)} - x_k \right|^2 \right)^{1/2} \to 0$ as $m \to \infty$: **R**^{n} is complete.

(iii) **Q**, the set of all rationals, with the usual metric inherited from **R**, is not complete: $\left(\left(1 + \frac{1}{n} \right)^n \right)_{n \in \mathbf{N}}$ is a Cauchy sequence in **Q** which does not converge to an element of **Q**.

(iv) The open interval $(0, 2)$, with the usual metric inherited from **R**, is not complete: $\left(\frac{1}{n} \right)_{n \in \mathbf{N}}$ is a Cauchy sequence in $(0, 2)$ which fails to converge to an element of $(0, 2)$.

(v) Let $I = [0, 1]$, take $X = C(I)$ (the set of all continuous, real-valued functions on I) and define a metric d on $C(I)$ by

$$d(f, g) = \int_0^1 |f(t) - g(t)| \, dt.$$

Then $(C(I), d)$ is not complete. To establish this, consider the sequence $(f_n)_{n \ge 2}$, where

$$f_n(t) = \begin{cases} 0, & \text{if } 0 \le t \le \frac{1}{2}, \\ n(t - \frac{1}{2}), & \text{if } \frac{1}{2} < t \le \frac{1}{2} + \frac{1}{n}, \\ 1, & \text{if } \frac{1}{2} + \frac{1}{n} \le t \le 1. \end{cases}$$

Since $d(f_n, f_m) = \frac{1}{2} |m^{-1} - n^{-1}| \to 0$ as $m, n \to \infty$, (f_n) is a Cauchy sequence. Suppose that there is a function $f \in C(I)$ such that $d(f_n, f) \to 0$ as $n \to \infty$. Then $\int_0^{1/2} |f_n(t) - f(t)| \, dt \le d(f_n, f) \to 0$ as $n \to \infty$; thus $\int_0^{1/2} |f(t)| \, dt = 0$, and hence $f(t) = 0$ for all $t \in [0, \frac{1}{2}]$. Now let $\varepsilon \in (0, \frac{1}{2})$. Since

$$\int_{\frac{1}{2} + \varepsilon}^{1} |f_n(t) - f(t)| \, dt \le d(f, f_n) \to 0 \text{ as } n \to \infty,$$

and for all large enough n, $f_n(t) = 1$ on $[\frac{1}{2} + \varepsilon, 1]$, we see that

$$\int_{\frac{1}{2}+\varepsilon}^{1} |f(t) - 1| \, dt = 0,$$

so that $f(t) = 1$ for all $t \in \left[\frac{1}{2} + \varepsilon, 1\right]$. As this holds for all $\varepsilon \in (0, \frac{1}{2})$, it follows that $f(t) = 1$ for all $t \in \left(\frac{1}{2}, 1\right]$, which implies that f is discontinuous at $t = \frac{1}{2}$. This contradiction shows that $(C(I), d)$ is not complete.

(vi) Let $p \in [1, \infty)$, let

$$\ell_p = \left\{ x = (x_i)_{i \in \mathbf{N}} : x_i \in \mathbf{R} \text{ for all } i \in \mathbf{N}, \sum_{1}^{\infty} |x_i|^p < \infty \right\},$$

and let

$$d_p(x, y) = \left(\sum_{1}^{\infty} |x_i - y_i|^p \right)^{1/p},$$

where $x = (x_i), y = (y_i) \in \ell_p$. Then (ℓ_p, d_p) is complete. To prove this, let $(x^n)_{n \in \mathbf{N}}$ be a Cauchy sequence in ℓ_p, where $x^n = \left(x_i^n\right)_{i \in \mathbf{N}}$. For each $i \in \mathbf{N}$,

$$\left| x_i^m - x_i^n \right| \le d_p(x^m, x^n)$$

and hence $\left(x_i^n\right)_{n \in \mathbf{N}}$ is a Cauchy sequence in $\mathbf{R}$. Using the completeness of $\mathbf{R}$, let $x_i = \lim_{n \to \infty} x_i^n$ and put $x = (x_i)_{i \in \mathbf{N}}$. It remains to show that $x \in \ell_p$ and that $x^n \to x$ in ℓ_p.

Let $\varepsilon > 0$. There exists $N \in \mathbf{N}$ such that $d_p(x^m, x^n) < \varepsilon$ if $m, n \ge N$. Thus for each $k \in \mathbf{N}$,

$$\sum_{i=1}^{k} |x_i^m - x_i^n|^p < \varepsilon^p \text{ if } m, n \ge N;$$

thus (letting $m \to \infty$),

$$\sum_{i=1}^{k} |x_i - x_i^n|^p \le \varepsilon^p \text{ if } n \ge N. \tag{2.2.1}$$

Use of (2.2.1) in conjunction with Minkowski's inequality shows that, for each $k \in \mathbf{N}$,

$$\left(\sum_{i=1}^{k}|x_i|^p\right)^{1/p} \le \left(\sum_{i=1}^{k}\left|x_i - x_i^N\right|^p\right)^{1/p} + \left(\sum_{i=1}^{k}\left|x_i^N\right|^p\right)^{1/p}$$

$$\le \varepsilon + \left(\sum_{i=1}^{\infty}\left|x_i^N\right|^p\right)^{1/p}.$$

Hence $x \in \ell_p$. Further use of (2.2.1) shows that

$$d_p(x^n, x) = \left(\sum_{i=1}^{\infty}\left|x_i - x_i^n\right|^p\right)^{1/p} \le \varepsilon.$$

Hence $x^n \to x$.

To make the interval $(0, 2)$ of Example 2.2.2 (iv) above into a complete space all we have to do is to adjoin the two points 0 and 2; the space $\mathbf{Q}$ of Example 2.2.2 (iii) may be 'completed' by adjoining all irrationals. These two examples illustrate the general principle, examined later, that any incomplete space may be enlarged so as to make it into a complete space.

The following result gives a useful characterisation of complete spaces; it uses the so-called *Cantor intersection property*: a metric space (X, d) is said to have this property if whenever (A_n) is a sequence of non-empty, closed, bounded subsets of X such that $A_{n+1} \subset A_n$ for all $n \in \mathbf{N}$ and $\lim_{n\to\infty} \operatorname{diam} A_n = 0$, then $\bigcap_{n=1}^{\infty} A_n$ has exactly one point.

Theorem 2.2.3 (Cantor's characterisation of completeness) *A metric space (X, d) is complete if, and only if, X has the Cantor intersection property.*

Proof First suppose that X is complete, and let (A_n) be a sequence of non-empty, closed bounded subsets of X such that $A_{n+1} \subset A_n$ for all $n \in \mathbf{N}$ and $\lim_{n\to\infty} \operatorname{diam} A_n = 0$; let (x_n) be a sequence such that for all $n \in \mathbf{N}$, $x_n \in A_n$. If $m \ge n$, then $x_m \in A_n$ and $\operatorname{diam}\{x_m : m \ge n\} \le \operatorname{diam} A_n \to 0$ as $n \to \infty$. Hence (x_n) is a Cauchy sequence and $x := \lim_{n\to\infty} x_n \in \overline{A}_k = A_k$ for all $k \in \mathbf{N}$; so $x \in \bigcap_1^{\infty} A_k$. If $y \in \bigcap_1^{\infty} A_k$, then $d(x, y) \le \operatorname{diam} A_n \to 0$ as $n \to \infty$; hence $y = x$. It follows that $\bigcap_1^{\infty} A_n = \{x\}$.

Conversely, suppose that X has the Cantor intersection property and let (x_n) be a Cauchy sequence in X. Let A_n be the closure of $\{x_m : m \ge n\}$ $(n \in \mathbf{N})$; then $A_{n+1} \subset A_n$ and $\operatorname{diam} A_n \to 0$ as $n \to \infty$. Thus there exists a unique $x \in X$ such that $x \in \bigcap_1^{\infty} A_n$; and $d(x, x_m) \le \operatorname{diam} A_m \to 0$ as $m \to \infty$, that is, $x_m \to x$. Hence X is complete. $\square$

Augmenting the complete metric spaces already described, we now introduce further examples each of which provides a suitable context for specific problems.

Definition 2.2.4 Let S be a non-empty set and let $\mathscr{B}(S)$ be the family of all bounded, real-valued functions on S. The **uniform metric** d_∞ on $\mathscr{B}(S)$ is given by

$$d_\infty(f, g) = \sup\{|f(s) - g(s)| : s \in S\} \quad (f, g \in \mathscr{B}(S)).$$

If S is a metric space, $\mathscr{C}(S)$ stands for the family of all continuous, bounded, real-valued functions on S; the restriction of d_∞ to $\mathscr{C}(S)$ is again denoted by d_∞.

Note that $d_\infty(f_n, f) \to 0$ if, and only if, (f_n) converges to f uniformly on S.

The arguments needed for the proofs of the next two theorems are essentially those given in Sect. 1.7, but we give the details for the convenience of the reader.

Theorem 2.2.5 *The metric space* $(\mathscr{B}(S), d_\infty)$ *is complete.*

Proof Let (f_n) be a Cauchy sequence in $\mathscr{B}(S)$. Then given any $\varepsilon > 0$, there exists $N \in \mathbf{N}$ such that $d_\infty(f_n, f_m) < \varepsilon$ if $m, n \geq N$; and so for each $s \in S$, $|f_n(s) - f_m(s)| < \varepsilon$ if $m, n \geq N$. Thus for each $s \in S$, $(f_n(s))$ is a Cauchy sequence in $\mathbf{R}$ and hence converges, to $f(s)$, say. We thus have a map $f : S \to \mathbf{R}$, where $f(s) = \lim_{n \to \infty} f_n(s)$ for all $s \in S$. To complete the proof we must show that $f \in \mathscr{B}(S)$ and $d(f_n, f) \to 0$ as $n \to \infty$. As above, we see that for all $s \in S$, $|f_n(s) - f_m(s)| < \varepsilon$ if $m, n \geq N$. Let $m \to \infty$: then for all $s \in S$, $|f_n(s) - f(s)| \leq \varepsilon$ if $n \geq N$. Since $|f(s)| \leq |f_N(s)| + \varepsilon$, it follows that $f \in \mathscr{B}(S)$; also we have $d_\infty(f, f_n) \leq \varepsilon$ if $n \geq N$. Hence $f_n \to f$ in $\mathscr{B}(S)$. $\square$

Theorem 2.2.6 *Let* (S, d) *be a metric space. Then* $(\mathscr{C}(S), d_\infty)$ *is complete.*

Proof Let (f_n) be a Cauchy sequence in $\mathscr{C}(S)$. Then (f_n) is a Cauchy sequence in $\mathscr{B}(S)$, and so by Theorem 2.2.5, there exists $f \in \mathscr{B}(S)$ such that $d_\infty(f, f_n) \to 0$ as $n \to \infty$. Let $\varepsilon > 0$. Then there exists $N \in \mathbf{N}$ such that for all $n \geq N$ and all $s \in S$, $|f_n(s) - f(s)| < \varepsilon/3$. Let $s_0 \in S$. Since f_N is continuous at s_0, there exists $\delta > 0$ such that $|f_N(s) - f_N(s_0)| < \varepsilon/3$ if $d(s, s_0) < \delta$. Thus if $d(s, s_0) < \delta$, then

$$|f(s) - f(s_0)| \leq |f(s) - f_N(s)| + |f_N(s) - f_N(s_0)| + |f_N(s_0) - f(s_0)| < \varepsilon.$$

Hence $f \in \mathscr{C}(S)$, and the theorem follows. $\square$

Corollary 2.2.7 *Let* $I = [a, b] \subset \mathbf{R}$. *Then* $C(I) = \mathscr{C}(I)$ *and* $(C(I), d_\infty)$ *is complete.*

Proof That $C(I) = \mathscr{C}(I)$ follows immediately from the fact that every continuous real-valued function on the closed, bounded interval I is bounded. The rest is now clear from Theorem 2.2.6. $\square$

We take up in the next section the question of under what conditions on a metric space S can it be shown that $C(S) = \mathscr{C}(S)$.

Theorem 2.2.8 *Let* $a, b \in \mathbf{R}$ *and* $a < b$. *Then* $(\mathscr{R}[a, b], d_\infty)$ *is complete.*

Proof Let (f_n) be a Cauchy sequence in $\mathscr{R}[a, b]$; it is also a Cauchy sequence in $\mathscr{B}[a, b]$ and so there is an $f \in \mathscr{B}[a, b]$ such that $d_\infty(f, f_n) \to 0$ *as* $n \to \infty$. Evidently (f_n) converges uniformly to f on $[a, b]$. By Theorem 1.7.12 it follows that $f \in \mathscr{R}[a, b]$. $\qquad\square$

Theorem 2.2.9 *Let* $a, b \in \mathbf{R}$, $a < b$, *let* $I = [a, b]$ *and let* $C^1(I)$ *denote the family of all continuously differentiable real-valued functions on* I. *Let* $v : C(I) \times C(I) \to \mathbf{R}$ *be defined by*

$$v(f, g) = \sup\{|f(x) - g(x)| : x \in I\} + \sup\{|f'(x) - g'(x)| : x \in I\}$$

(that is, $v(f, g) = d_\infty(f, g) + d_\infty(f', g')$*). Then* $(C^1(I), v)$ *is a complete metric space.*

Proof Routine arguments show that v is a metric on $C^1(I)$. To prove completeness, let (f_n) be a Cauchy sequence in $C^1(I)$. Then (f_n) and (f_n') are Cauchy sequences in the complete space $(C(I), d_\infty)$, and so there exist $f, g \in C(I)$ such that $d_\infty(f_n, f) \to 0$ and $d_\infty(f_n', g) \to 0$ as $n \to \infty$. The result is immediate if we can prove that $f' = g$. However, by Theorem 1.4.4,

$$f_n(x) - f_n(a) = \int\limits_a^x f_n' \ (x \in I, n \in \mathbf{N});$$

and since $\left| \int_a^x (f_n' - g) \right| \leq (x - a) d_\infty(f_n', g) \to 0$ as $n \to \infty$, we have that

$$f(x) - f(a) = \lim_{n \to \infty} (f_n(x) - f_n(a)) = \lim_{n \to \infty} \int\limits_a^x f_n' = \int\limits_a^x g \ (x \in I).$$

Thus by Theorem 1.4.9, f is differentiable and $f' = g$. $\qquad\square$

Corollary 2.2.10 *Let* (f_n) *be a sequence in* $C^1(I)$ $(I = [a, b])$ *such that* (f_n') *converges uniformly on* I *and for some* $x_0 \in I$, $(f_n(x_0))$ *is convergent. Then there exists* $f \in C^1(I)$ *such that* (f_n) *converges uniformly on* I *to* f *and*

$$f'(x) = \lim_{n \to \infty} f_n'(x) \ (x \in I).$$

Proof It is enough to show that (f_n) is a Cauchy sequence in $(C(I), d_\infty)$, for then (f_n) will be a Cauchy sequence in $(C^1(I), v)$ and the result will follow immediately from Theorem 2.2.9. To do this, let $\varepsilon > 0$ and let $N \in \mathbf{N}$ be such that

$$|f_m(x_0) - f_n(x_0)| < \varepsilon/2 \text{ and } d_\infty(f_m', f_n') < \varepsilon/2(b - a) \text{ if } m, n > N.$$

Since for all $x \in I$ we have, by Theorem 1.4.4,

$$|f_m(x) - f_n(x)| \leq |f_m(x_0) - f_n(x_0)| + \left| \int_{x_0}^{x} (f_m' - f_n') \right|$$

$$\leq |f_m(x_0) - f_n(x_0)| + (b-a)d_\infty(f_m', f_n') < \varepsilon$$

if $m, n > N$. The result follows. $\qquad\square$

Returning to the observations concerning the completion of an incomplete metric space following Example 2.2.2, we see that Theorem 2.2.6 leads to the following result.

Theorem 2.2.11 *Let (S, d) be a metric space. Then there is a complete metric space $(\widehat{S}, \widehat{d})$ such that S is isometric to a dense subset S_0 of $\widehat{S}$; that is, a subset S_0 such that $\overline{S}_0 = \widehat{S}$.*

Proof Fix $a \in S$ and for every $p \in S$, define $f_p : S \to \mathbf{R}$ by

$$f_p(x) = d(x, p) - d(x, a).$$

Use of the triangle inequality shows that, for all $x, y \in S$,

$$|f_p(x) - f_p(y)| \leq 2d(x, y), \quad |f_p(x)| \leq d(a, p).$$

Hence $f_p \in \mathscr{C}(S)$. Let $S_0 = \{f_p : p \in S\}$. Since, for all $p, q \in S$,

$$d_\infty(f_p, f_q) = \sup_{x \in S} |d(x, p) - d(x, q)| = d(p, q),$$

the map $p \longmapsto f_p : S \to S_0 \subset \mathscr{C}(S)$ is an isometry of S onto S_0. Let $\widehat{S}$ be the closure of S_0 in $(\mathscr{C}(S), d_\infty)$ and $\widehat{d}$ be the restriction of d_∞ to $\widehat{S} \times \widehat{S}$. Since $\widehat{S}$ is closed in $(\mathscr{C}(S), d_\infty)$, use of Lemma 2.1.24 shows that $(\widehat{S}, \widehat{d})$ is complete. Let $\overline{S}_0$ denote the closure of S_0 in $(\widehat{S}, \widehat{d})$. As $\overline{S}_0 = \widehat{S}$, S_0 is dense in $(\widehat{S}, \widehat{d})$ and S is isometric to it. $\qquad\square$

Even though $\widehat{X}$ is not unique in having the property of being complete and having X isometric to a dense subset of $\widehat{X}$—it is only unique up to an isometry—we shall refer to it as *the completion* of X.

Two of the most celebrated results associated with complete metric spaces are the Contraction Mapping Theorem and the Baire Category Theorem. The rest of this section is devoted to establishing these and illustrating their application.

2.2.1 The Contraction Mapping Theorem

This is one of the most useful, yet simple, theorems in mathematics.

Definition 2.2.12 Let (X, d) be a metric space. A map $f : X \to X$ is called a **contraction** if there is a number $\lambda \in [0, 1)$ such that for all $x, y \in X$,

$$d(f(x), f(y)) \le \lambda d(x, y).$$

Theorem 2.2.13 (Banach's contraction mapping theorem) *Let (X, d) be a complete metric space and let $f : X \to X$ be a contraction. Then there is exactly one point $x \in X$ such that $f(x) = x$; that is, f has exactly one fixed point.*

Proof As f is a contraction, there exists $\lambda \in [0, 1)$ such that $d(f(x), f(y)) \le \lambda d(x, y)$ for all $x, y \in X$. Let $x_0 \in X$ and define a sequence (x_n) by $x_n = f(x_{n-1})$ $(n \in \mathbf{N})$. Then for each $n \in \mathbf{N}$,

$$d(x_{n+1}, x_n) = d(f(x_n), f(x_{n-1})) \le \lambda d(x_n, x_{n-1}) \le \lambda^n d(x_1, x_0).$$

If $m > n$,

$$d(x_m, x_n) \le d(x_m, x_{m-1}) + d(x_{m-1}, x_{m-2}) + \ldots + d(x_{n+1}, x_n)$$

$$\le (\lambda^{m-1} + \lambda^{m-2} + \ldots + \lambda^n) d(x_1, x_0)$$

$$= \frac{(\lambda^n - \lambda^m)}{1 - \lambda} d(x_1, x_0).$$

It follows that (x_n) is a Cauchy sequence in X and, as X is complete, there exists $x \in X$ such that $x_n \to x$. Thus

$$d(x, f(x)) \le d(x, x_{n+1}) + d(x_{n+1}, f(x)) \le d(x, x_{n+1}) + \lambda d(x_n, x)$$

$$\to 0 \text{ as } n \to \infty.$$

Hence $f(x) = x$; that is, x is a fixed point of f.

If there exists $y \in X$ such that $f(y) = y$, then

$$d(x, y) = d(f(x), f(y)) \le \lambda d(x, y);$$

and as $\lambda < 1$, $d(x, y) = 0$. Hence $x = y$, and the proof is complete. $\square$

Note the constructive nature of this proof: no matter what point x_0 of X is chosen, the fixed point x of f is given by the formula

$$x = \lim_{n \to \infty} f^n(x_0).$$

In practical circumstances, approximations to the fixed point may be derived by choosing a convenient point x_0 and determining $f^n(x_0)$ for various values of n.

Corollary 2.2.14 *Let X be a complete metric space and let $f : X \to X$ be such that, for some $k \in \mathbf{N}$, f^k is a contraction. Then f has a unique fixed point.*

Proof By Theorem 2.2.13, there is a unique $x \in X$ such that $f^k(x) = x$. But $f^k(f(x)) = f(f^k(x)) = f(x)$, and so $f(x)$ is a fixed point of f^k. Hence $f(x) = x$. That f has a unique fixed point now follows since evidently any fixed point of f must be a fixed point of f^k, and so must coincide with x. $\qquad\square$

We can now give an application of the contraction mapping theorem to the theory of ordinary differential equations.

Theorem 2.2.15 *Let $a, b \in \mathbf{R}$, $a < b$, put $I = [a, b]$, let $f : I \times \mathbf{R} \to \mathbf{R}$ be continuous and suppose there exists $M > 0$ such that for all $x \in I$ and all $y_1, y_2 \in \mathbf{R}$,*

$$|f(x, y_1) - f(x, y_2)| \le M |y_1 - y_2|.$$

Let $c \in \mathbf{R}$. Then there is a unique function $u \in C^1(I)$ such that

$$u'(x) = f(x, u(x)) \ (x \in I), u(a) = c. \tag{2.2.2}$$

Proof First observe that if $u \in C(I)$, then $t \longmapsto f(t, u(t)): I \to \mathbf{R}$ is continuous, for if $t \in I$ and (t_n) is a sequence in I with $t_n \to t$, then $(t_n, u(t_n)) \to (t, u(t))$ in $I \times \mathbf{R}$ and so $f(t_n, u(t_n)) \to f(t, u(t))$. Further, by the Fundamental Theorem of Integral Calculus, there is a unique $u \in C^1(I)$ such that (2.2.2) holds if, and only if, the integral equation

$$u(x) = c + \int_a^x f(t, u(t))dt \ (x \in I) \tag{2.2.3}$$

has a unique solution $u \in C(I)$. Define a map $T : C(I) \to C(I)$ by

$$(Tu)(x) = c + \int_a^x f(t, u(t))dt \ (x \in I; \ u \in C(I)).$$

For each $n \in \mathbf{N}_0$, let $P(n)$ be the proposition

$$|(T^n u)(x) - (T^n v)(x)| \le (M |x - a|)^n d_\infty(u, v)/n!$$

for all $u, v \in C(I)$ and all $x \in I$. (Here T^0 is the identity map of $C(I)$ to itself.) Plainly $P(0)$ is true; moreover, if $P(n)$ holds for some $n \in \mathbf{N}_0$, then

$$\left|(T^{n+1}u)(x) - (T^{n+1}v)(x)\right| = \left|\int_a^x \left\{f(t, (T^n u)(t)) - f(t, (T^n v)(t))\right\} dt\right|$$

$$\leq M \left|\int_a^x \{(M\,|t-a|)^n d_\infty(u,v)/n!\} dt\right|$$

$$\leq (M\,|x-a|)^{n+1} d_\infty(u,v)/(n+1)!$$

for all $u, v \in C(I)$ and all $x \in I$, and so $P(n+1)$ is true. Hence $P(n)$ is true for all $n \in \mathbf{N}$. It follows that

$$d_\infty(T^n u, T^n v) \leq (M(b-a))^n d_\infty(u,v)/n!$$

for all $u, v \in C(I)$ and all $n \in \mathbf{N}$. Choose $k \in \mathbf{N}$ so large that $(M(b-a))^k/k! < 1$: then T^k is a contraction on the complete space $(C(I), d_\infty)$. By Corollary 2.2.14, there is a unique $u \in C(I)$ such that $Tu = u$; that is, such that

$$u(x) = c + \int_a^x f(t, u(t)) dt \ (x \in I).$$

The result follows.   □

2.2.2 The Baire Category Theorem

Several formulations of this theorem exist. One of the most accessible is as follows.

Theorem 2.2.16 *Let X be a complete metric space and let $(\mathscr{O}_n)$ be a sequence of dense open subsets of X. Then $\cap_{n=1}^\infty \mathscr{O}_n$ is dense in X; that is, $\overline{\cap_{n=1}^\infty \mathscr{O}_n} = X$.*

Proof Suppose that the conclusion is false. Let $U := X \setminus \overline{\cap_{n=1}^\infty \mathscr{O}_n}$: U is open and non-empty. Since $\overline{\mathscr{O}_1} = X, U \cap \mathscr{O}_1 \neq \emptyset$. Hence there exists a non-empty open set U_1 such that

$$U_1 \subset \overline{U_1} \subset U \cap \mathscr{O}_1 \text{ and } \operatorname{diam} \overline{U_1} < 1.$$

(U_1 may be taken to be an open ball of suitable radius). Since $\overline{\mathscr{O}_2} = X, U_1 \cap \mathscr{O}_2 \neq \emptyset$ and so there is a non-empty open set U_2 such that

$$U_2 \subset \overline{U_2} \subset U_1 \cap \mathscr{O}_2 \text{ and } \operatorname{diam} \overline{U_2} < 2^{-1}.$$

Continuing in this way we see that there exists a sequence (U_n) of non-empty open subsets of X such that, for all $n \in \mathbf{N}$,

$$U_n \subset \overline{U_n} \subset U_{n-1} \cap \mathscr{O}_n \text{ and } \operatorname{diam} \overline{U_n} < n^{-1}.$$

(Here $U_0 := U$.) Thus Cantor's characterisation of completeness (Theorem 2.2.3) shows that for some $x \in X$,

$$\{x\} = \cap_{n=1}^{\infty} \overline{U_n}.$$

Since $\overline{U_1} \subset U$ and, for all $n \in \mathbf{N}$, $\overline{U_n} \subset \mathscr{O}_n$, it follows that $x \in U \cap (\cap_{n=1}^{\infty} \mathscr{O}_n)$, a contradiction. $\qquad\square$

Taking complements, and recalling that $^c(\overset{o}{A}) = \overline{^cA}$ whenever $A \subset X$, we immediately obtain the equivalent result:

Theorem 2.2.17 *Let X be a complete metric space and let (F_n) be a sequence of closed subsets of X, each with empty interior. Then $\cup_{n=1}^{\infty} F_n$ has empty interior.*

Breaking the theoretical development for a moment, we use this last result to give a striking demonstration of the existence of a continuous nowhere-differentiable function.

Theorem 2.2.18 *Let I be the closed interval $[0, 1]$. Then there exists an element of $C(I)$ which is not differentiable at any point of I.*

Proof For each $n \in \mathbf{N}$ put

$$M_n = \left\{ f \in C([0, 2]) : \text{ for some } x_0 \in I, \ \sup_{0<h<1} \frac{|f(x_0 + h) - f(x_0)|}{h} \leq n \right\}.$$

We claim that each M_n is closed in $C([0, 2])$. To prove this, let $f \in \overline{M_n}$ and let (f_k) be a sequence in M_n that converges to f. For each $k \in \mathbf{N}$, there exists $x_k \in I$ with

$$|f_k(x_k + h) - f_k(x_k)| \leq nh \text{ if } 0 < h < 1.$$

As the bounded sequence (x_k) contains a convergent subsequence, we may and shall assume, without loss of generality, that $x_k \to x_0 \in I$. For all $k \in \mathbf{N}$,

$$|f(x_0 + h) - f(x_0)| \leq |f(x_0 + h) - f(x_k + h)| + |f(x_k + h) - f_k(x_k + h)|$$

$$+ |f_k(x_k + h) - f_k(x_k)| + |f_k(x_k) - f(x_k)|$$

$$+ |f(x_k) - f(x_0)|,$$

and using the fact that $|f_k(x_k + h) - f_k(x_k)| \leq nh$ we see, on letting $k \to \infty$, that

$$|f(x_0 + h) - f(x_0)| \leq nh \text{ if } 0 < h < 1.$$

It follows that $f \in M_n$ and our claim is justified.

Next we claim that each M_n has empty interior. For let g be a piecewise-linear continuous function on $[0, 2]$, so that the graph of g consists of a finite number of straight-line segments; let M be the maximum absolute value of the gradients of these segments. Given $\varepsilon > 0$, choose $m \in \mathbf{N}$ so that $m\varepsilon > n + M$, define $\phi : \mathbf{R} \to I$ by

$$\phi(x) = \min\{x - [x], [x] + 1 - x\}, \ x \in \mathbf{R}$$

(here $[x]$ denotes the integer part of x; $\phi(x)$ is simply the distance of x from the nearest integer), and put

$$F(x) = g(x) + \varepsilon\phi(mx), \ x \in I.$$

Then if $x \in I$ and $0 < h < 1$,

$$|F(x + h) - F(x)| = |g(x + h) - g(x) + \varepsilon\{\phi(m(x + h)) - \phi(mx)\}|$$

$$\geq \varepsilon mh - |g(x + h) - g(x)|$$

$$\geq \varepsilon mh - Mh > nh.$$

Hence $F \in C([0, 2])\backslash M_n$. Moreover, $d_\infty(g, F) = \varepsilon$. Assuming for the moment that the set of all piecewise-linear continuous functions is dense in $C([0, 2])$, our analysis shows that any $f \in C([0, 2])$ may be approximated arbitrarily closely in $C([0, 2])$ by an element of $C([0, 2])\backslash M_n$, so that the interior of M_n must be empty, as claimed. Since $C([0, 2])$ is complete, it follows from Theorem 2.2.17 that $C([0, 2])\backslash \bigcup_{n=1}^{\infty} M_n$ is non-empty; and as every function in $C([0, 2])$ that is differentiable at some point of I must lie in some M_n, the theorem follows.

All that remains is to establish the density of the piecewise-linear continuous functions in $C([0, 2])$. Let $f \in C([0, 2])$ and let $\varepsilon > 0$. Since f is uniformly continuous on $[0, 2]$, there is a partition $P = \{0, 2/m, 4/m, \ldots, 2\}$ of $[0, 2]$ such that for $j = 1, \ldots, m$ we have $\mathrm{osc}(f, [(2j - 1)/m, 2j/m]) < \varepsilon$ (see Exercise 1.1.10 /2). Define ψ on $[0, 2]$ by

$$\psi(x) = \frac{m}{2}\left\{\left(x - \frac{2(j-1)}{m}\right)f(2j/m) + \left(\frac{2j}{m} - x\right)f(2(j-1)/m)\right\}$$

if $2(j - 1)/m \leq x \leq 2j/m, \ j = 1, \ldots, m$,
so that ψ is piecewise linear and coincides with f at the points of the partition. Evidently $d_\infty(f, \psi) < \varepsilon$, and the density follows. $\square$

Alternative formulations of the Baire theorem demand a little preparation.

Definition 2.2.19 Let A and B be subsets of a metric space X. Then A is said to be **dense in B** if $B \subset \overline{A}$; it is **everywhere dense** (or simply **dense**) if it is dense in X; and it is **nowhere dense** (or **rare**) if it is not dense in any non-empty open subset of X, or equivalently, if its closure contains no interior points.

Remark 2.2.20

(i) Plainly, a subset of a nowhere dense set is nowhere dense; also, the closure of a nowhere dense set is itself nowhere dense.

(ii) A closed set is nowhere dense if, and only if, it coincides with its own boundary. That is, if $A \subset X$ and $A = \overline{A}$, then $\overset{o}{A} = \emptyset$ if, and only if, $\partial A = A \backslash \overset{o}{A} = A$.

(iii) To say that a set is everywhere dense is not the antithesis of saying that it is nowhere dense.

Example 2.2.21

1. Let $x \in X$. Then $\{x\}$ is nowhere dense if, and only if, x is not an isolated point of X : an isolated point y is one having a neighbourhood containing no point of X except y. Note that $\{x\}$ is closed. Thus $\overset{o}{\widehat{\{x\}}} = \emptyset$ if, and only if, every neighbourhood U of x is such that $U \cap {}^c \{x\} \neq \emptyset$; and this is so if, and only if, x is not an isolated point of X.

2. The boundary of an open (or closed) set in X is always nowhere dense. For let U be an open set in X and let V be an open set in X such that $V \subset \partial U = \overline{U} \cap \overline{{}^c U} = \overline{U} \cap {}^c U$. Then ${}^c V$ is closed and contains U, so that ${}^c V \supset \overline{U} \supset V$, which is possible only if $V = \emptyset$. Note that the boundary of an arbitrary set A in X need not be nowhere dense: for example, A and ${}^c A$ might both be dense, in which case $\partial A = X$.

Lemma 2.2.22 *Let A be a subset of a metric space X. The following three statements are equivalent:*

(i) *A is nowhere dense in X.*

(ii) *${}^c A$ contains an everywhere dense open subset of X.*

(iii) *Each non-empty open set U in X contains a non-empty open set V such that $V \cap A = \emptyset$.*

Proof (i)$\Rightarrow$(ii) Suppose $\overset{o}{\overline{A}} = \emptyset$. Then ${}^c(\overline{\overline{A}}) = {}^c(\overset{o}{\overline{A}}) = X$ and so the open set ${}^c(\overline{A})$, which is contained in ${}^c A$, is everywhere dense.

(ii)$\Rightarrow$(iii) Let $\mathcal{O}$ be a dense open subset of X contained in ${}^c A$ and let U be a non-empty open subset of X. Note that $U \cap \mathcal{O} \neq \emptyset$: otherwise, $\mathcal{O}$ is contained in the closed set ${}^c U$, so that $X = \overline{\mathcal{O}} \subset {}^c U$ which implies that $U = \emptyset$. Let $V = U \cap \mathcal{O}$. Then V is non-empty and open, $V \subset U$ and $V \cap A = \emptyset$, since $V \subset \mathcal{O} \subset {}^c A$.

(iii)$\Rightarrow$ (i) To obtain a contradiction, suppose $U := \overset{o}{\overline{A}} \neq \emptyset$. Then there is a non-empty open set $V \subset U$ such that $A \cap V = \emptyset$. Since ${}^c V$ is closed and $A \subset {}^c V$, it follows that $U \subset \overline{A} \subset {}^c V$ and so $\emptyset = U \cap V = V \neq \emptyset$, a contradiction. $\qquad \square$

Lemma 2.2.23 *Let X be a metric space.*

(i) *If U and V are each dense open subsets of X, then $U \cap V$ is a dense open subset of X.*

(ii) *If A and B are each nowhere dense subsets of X, then $A \cup B$ is nowhere dense.*

Proof (i) To obtain a contradiction, suppose $\overline{U \cap V} \neq X$. Then $G :=^c \left(\overline{U \cap V} \right)$ is open and non-empty. Since $\overline{U} = X$, $G \cap U$ is open and non-empty: otherwise, $^cG \supset U$ and, since ^cG is closed, $^cG \supset \overline{U} = X$, implying that $G = \emptyset$. Since $\overline{V} = X$, similar reasoning shows that $G \cap U \cap V$ is open and non-empty. But this contradicts the fact that $U \cap V \subset {}^cG$. Hence $U \cap V$ is dense in X.

(ii) Since A and B are nowhere dense, each of the sets $^c(\overline{A})$, $^c(\overline{B})$ is open and dense in X. Thus, using (i), it follows that $^c(\overline{A}) \cap {}^c(\overline{B}) = {}^c(\overline{A} \cup \overline{B}) = {}^c(\overline{A \cup B})$ is dense in X. Since

$$^c\left(\overset{0}{\overline{A \cup B}} \right) = \overline{^c(\overline{A \cup B})} = X,$$

the set $A \cup B$ is nowhere dense.   □

Note that (i) and (ii) can obviously be extended to arbitrary **finite** intersections and **finite** unions.

Taken together, Theorems 2.2.16 and 2.2.17 extend the last Lemma to countably infinite families of sets. But the extension comes at a price. Recall that a countable intersection of open sets need not be open, and a countable union of closed sets need not be closed. The theorems demand a stronger hypothesis, namely the completeness of X, and support a weaker conclusion than that of the Lemma. To illustrate by example, let $X = \mathbb{R}$ and $A_n = \{x_n\}$, where the sequence $(x_n)_{n \in \mathbb{N}}$ is an enumeration of the rationals. Then $\mathbb{Q} = \cup_{n=1}^{\infty} A_n$ is a countably infinite union of nowhere dense sets and its interior is empty. However, it is not nowhere dense; indeed, it is everywhere dense.

Definition 2.2.24 A subset A of a metric space X is said to be **of first category** (or **meagre**) **in** X if it can be represented as a countable union of nowhere dense subsets of X. Otherwise, it is said to be **of second category** (or **nonmeagre**) **in** X. A set $B \subset X$ is said to be **residual in** X if ^cB is of first category in X.

To give an example of a set of first category arising naturally in a non-trivial context, we establish the following result.

Theorem 2.2.25 *Let X be a metric space, let (f_n) be a sequence of continuous real-valued functions on X which is pointwise convergent, and let the function $f : X \to \mathbf{R}$ be defined by*

$$f(x) = \lim_{n \to \infty} f_n(x) \ (x \in X).$$

Then

$$\mathscr{D} := \{x \in X : f \text{ is not continuous at } x\}$$

is of first category in X.

Proof Let ω be the oscillation function of f (see Definition 2.1.31). The identity

$$\mathscr{D} = \cup_{n=1}^{\infty}\{x \in X : \omega(x) \geq n^{-1}\}$$

exhibits $\mathscr{D}$ as a countable union of closed sets each of which will be shown to be of first category (in X). Plainly, a countable union of sets of first category is itself of first category. Thus $\mathscr{D}$ is of first category (in X).

Let $\varepsilon > 0$ and let

$$F = \{x \in X : \omega(x) \geq \varepsilon\}.$$

It is enough to establish that F is of first category. To do this, for $n \in \mathbf{N}$ let

$$E_n := \cap_{i,j \geq n}\{x \in X : |f_i(x) - f_j(x)| \leq \varepsilon/8\} :$$

each E_n is closed, $E_n \subset E_{n+1}$ and $X = \cup_{n=1}^{\infty} E_n$. Evidently

$$F = \cup_{n=1}^{\infty}(F \cap E_n),$$

and the matter of category is settled provided that, for each n, $\overline{F \cap E_n}^{\,o} = \emptyset$. To obtain a contradiction, suppose that for some n, $F \cap E_n$ is not nowhere dense. Then there exists an open set U such that

$$U \neq \emptyset, U \subset \overline{F \cap E_n} = F \cap E_n;$$

moreover, for each $x \in U$,

$$|f_i(x) - f_j(x)| \leq \varepsilon/8 \text{ if } i, j \geq n.$$

Setting $i = n$ and letting $j \to \infty$, it follows that

$$|f_n(x) - f(x)| \leq \varepsilon/8 \ (x \in U).$$

Let $y \in U$. Since f_n is continuous at y, there is a neighbourhood U_y of y such that $U_y \subset U$ and

$$|f_n(x) - f_n(y)| \leq \varepsilon/8 \ (x \in U_y).$$

It follows that

$$|f(x) - f_n(y)| \leq \varepsilon/4 \ (x \in U_y);$$

that

$$|f(x) - f(x')| \leq \varepsilon/2 \ (x, x' \in U_y);$$

and that

$$\omega(y) \leq \varepsilon/2.$$

But the last inequality, valid for all $y \in U$, implies that $U \cap F = \emptyset$, a conclusion incompatible with $U \neq \emptyset$, $U \subset F$. Thus, for all $n \in \mathbf{N}$, $F \cap E_n$ is nowhere dense, as required. □

Paraphrasing, Theorem 2.2.25 shows that, for an arbitrary metric space X, the set $\mathscr{D}$ of points of discontinuity of a function f generated as a pointwise limit of a sequence of continuous real-valued functions is of first category. Naturally, isolation of those metric spaces in which more can be said about $\mathscr{D}$ is of interest. If as an ideal one might wish to have $\mathscr{D} = \emptyset$, $^c\mathscr{D} = X$, then as a step towards this, generally, for complete metric spaces it turns out that $\overset{o}{\mathscr{D}} = \emptyset$, $\overline{^c\mathscr{D}} = X$. This is a consequence of the following theorem.

Theorem 2.2.26 *Let X be a metric space. If X has one of the following properties, then it has all of them.*

(i) *Every countable intersection of dense open subsets of X is dense in X.*
(ii) *The complement of every set of first category in X is dense in X.*
(iii) *Every set of first category in X has empty interior in X.*
(iv) *Every non-empty open set in X is of second category in X.*

*A metric space with one, and hence all, of the above properties is said to be a **Baire space**.*

Proof (i) $\Longrightarrow$ (ii) Let A be a set of first category in $X : A = \cup_{n=1}^{\infty} H_n$, where $\overset{o}{\overline{H_n}} = \emptyset$ ($n \in \mathbf{N}$). Let $B = \cup_{n=1}^{\infty} \overline{H_n}$. Then B is of first category in X and $A \subset B$. Now

$$^cB = \cap_{n=1}^{\infty} {}^c(\overline{H_n}) \text{ and } \overline{^c(\overline{H_n})} = {}^c\left(\overset{o}{\overline{H_n}}\right) = X \ (n \in \mathbf{N}).$$

Hence, given that (i) holds and that $^cB \subset {}^cA$, it follows that $X = \overline{^cB} \subset \overline{^cA} \subset X$ and that ^cA is dense in X.

(ii) $\Longrightarrow$ (iii) Let A be of first category in X. Using (ii) we see that $\overline{^cA} = X$. Thus $\emptyset = {}^c(\overline{^cA}) = \overset{o}{A}$.

(iii) $\Longrightarrow$ (iv) To obtain a contradiction, suppose that U is a non-empty open subset of X which is of first category in X. Then, since (iii) holds, $\emptyset = \overset{o}{U} = U$.

(iv) $\Longrightarrow$ (i) Let $(\mathscr{O}_n)$ be a sequence of dense open subsets of X and let $E = \cap_{n=1}^{\infty} \mathscr{O}_n$. Then $^cE = \cup_{n=1}^{\infty} {}^c\mathscr{O}_n$ and

$$\overset{o}{\overline{^c\mathscr{O}_n}} = \overset{o}{^c\mathscr{O}_n} = {}^c(\overline{\mathscr{O}_n}) = \emptyset \ (n \in \mathbf{N}).$$

Hence ^cE is of first category in X and consequently so is $\overset{o}{^cE}$. Since (iv) holds, it follows that $\emptyset = \overset{o}{^cE}$ and $X = \overline{E}$. □

In view of the above result it is obvious that our first version of Baire's theorem may now be recast in a final one as follows.

Theorem 2.2.27 *Every complete metric space is a Baire space.*

The reader should note that there exist incomplete metric spaces which are Baire spaces (see [2], Sect. 5, Exercise 14).

We conclude this section with the observation that, in the context of complete metric spaces, Baire's theorem immediately permits the following strengthened version of Theorem 2.2.25.

Theorem 2.2.28 *Let X be a complete metric space and let $f : X \to \mathbb{R}$ be the pointwise limit of a sequence of continuous real-valued functions on X. Then the set of points of continuity of f is residual and dense in X.*

Exercise 2.2.29

1. Let (X_1, d_1) and (X_2, d_2) be complete metric spaces, let $X = X_1 \times X_2$ and define $d : X \times X \to \mathbb{R}$ by

$$d(x, y) = \left\{ d_1^2(x_1, y_1) + d_2^2(x_2, y_2) \right\}^{1/2} \quad (x = (x_1, x_2), y = (y_1, y_2) \in X).$$

   Prove that (X, d) is complete.

2. Let (X, d) be a metric space and let F be a non-empty subset of X. Prove that
   (i) if (X, d) is complete and F is closed relative to (X, d), then (F, d) is complete;
   (ii) if (F, d) is complete, then F is closed relative to (X, d).
   [By convention, (F, d) stands in place of $(F, d |_{F \times F})$.]

3. Let $I = [0, 1]$ and define $T : C(I) \to C(I)$ by

$$(Tf)(x) = x + \int_0^x (x - t)f(t)dt \quad (x \in I, f \in C(I)).$$

   Show that T is a contraction on $C(I)$ (assumed to be endowed with the uniform metric) and deduce that the only element f of $C(I)$ such that

$$f(x) = x + \int_0^x (x - t)f(t)dt \quad (x \in I)$$

   is the restriction to $[0, 1]$ of the hyperbolic sine function.

4. Use the contraction mapping theorem to show that for each $k \in (0, 1)$ the equation

$$f(x) = 1 + \int_0^x f(t^2)dt \quad (0 \le x \le k)$$

has exactly one solution $f \in C([0, k])$. Hence show that this result is also true when $k = 1$.

5. (i) Give an example of a contraction mapping of an incomplete metric space into itself which has no fixed point.

   (ii) Give an example of a mapping T of a complete metric space (X, d) into itself with the property

$$d(Tx, Ty) < d(x, y) \text{ for all } x, y \in X, x \neq y,$$

   but which has no fixed point.

   (iii) Give an example of a mapping T of a complete metric space into itself such that T^m is a contraction mapping for some $m \in \mathbf{N}$, but T is not a contraction.

6. Let $X = \{x \in \mathbf{R} : 0 < x \leq 1\}$ and let d_1 and d_2 be metrics on X defined by

$$d_1(x, y) = |x - y| , d_2(x, y) = \left| \frac{1}{x} - \frac{1}{y} \right| \ (x, y \in X).$$

Prove that the two metric spaces (X, d_1) and (X, d_2) have the same convergent sequences, but that (X, d_2) is complete while (X, d_1) is not complete.

7. Let S be the set of all real sequences $x = (x_n)$ and let $d : S \times S \to \mathbf{R}$ be defined by

$$d(x, y) = \sum_{n=1}^{\infty} \frac{|x_n - y_n|}{2^n \left[1 + |x_n - y_n| \right]} \ (x = (x_n), y = (y_n) \in S).$$

Prove that (S, d) is a complete metric space.

8. Let (X, d) be a metric space.

   (i) Show that if (x_n) and (y_n) are Cauchy sequences in X, then $(d(x_n, y_n))$ is a Cauchy sequence in $\mathbf{R}$ and is therefore convergent.

   (ii) Let $\mathscr{X}$ be the set of all Cauchy sequences in X. Call elements (x_n), (y_n) of $\mathscr{X}$ equivalent, and write $(x_n) \sim (y_n)$, if $\lim_{n\to\infty} d(x_n, y_n) = 0$. Show that $\sim$ is an equivalence relation on $\mathscr{X}$.

   (iii) Let (x_n), (x'_n), (y_n) and $(y'_n) \in \mathscr{X}$ and suppose that $(x_n) \sim (x'_n)$ and $(y_n) \sim (y'_n)$. Show that

$$\lim_{n\to\infty} d(x_n, y_n) = \lim_{n\to\infty} d(x'_n, y'_n).$$

   (iv) For $(x_n) \in \mathscr{X}$, let $[(x_n)]$ denote the equivalence class of which it is a member:

$$[(x_n)] = \{(y_n) \in \mathscr{X} : (y_n) \sim (x_n)\}.$$

Let $\widehat{X}$ be the set of all equivalence classes and define $\widehat{d} : \widehat{X} \times \widehat{X} \to \mathbf{R}$ by

$$\widehat{d} \left([(x_n)], [(y_n)]\right) = \lim_{n\to\infty} d(x_n, y_n).$$

Show that $\widehat{d}$ is a metric on $\widehat{X}$ (it is well-defined by virtue of (iii)),

(v)   For each $x \in X$, let $\phi(x) = [(x_n)]$, where $x_n = x$ for all $n \in \mathbf{N}$. Let $X_0 = \{\phi(x) : x \in X\}$. Show that, if X_0 is equipped with the metric inherited from $\widehat{X}$, then $x \mapsto \phi(x) : X \to X_0$ is an isometry.

(vi) Prove that X_0 is dense in $(\widehat{X}, \widehat{d})$, i.e., $\overline{X_0} = \widehat{X}$.

(vii) Prove that $(\widehat{X}, \widehat{d})$ is a complete metric space.

2.3 Compactness

We focus here on those metric spaces X with the following property: if a map $f : X \to \mathbf{R}$ is continuous then it is bounded, that is, its range, $f(X)$, is bounded. Spaces with this property are precisely those for which the sets $C(X)$ and $\mathscr{C}(X)$ coincide, a coincidence already noted in the case of each non-degenerate, closed, bounded interval in $\mathbf{R}$. In seeking to ensure the property three main strategies have emerged. These we now examine in turn.

Strategy I Let $f : X \to \mathbf{R}$ be continuous. Then each $x \in X$ has a neighbourhood U_x such that, for all $u \in U_x$,

$$|f(u)| < 1 + |f(x)|.$$

Evidently $X = \bigcup_{x \in X} U_x$. Hence if a **finite** set $\{x_1, x_2, \ldots, x_m\} \subset X$ exists such that $X = \bigcup_{k=1}^{m} U_{x_k}$, then $f(X)$ is bounded, since for all $u \in X$,

$$|f(u)| < 1 + \max_{1 \le k \le m} |f(x_k)|.$$

This observation motivates the next definition and establishes the theorem that follows.

Definition 2.3.1 A metric space X is said to be **compact** if every family $\mathscr{U}$ of open subsets of X such that $X = \cup \mathscr{U}$ contains a finite subfamily $\mathscr{V}$ such that $X = \cup \mathscr{V}$.

Theorem 2.3.2 *If X is a compact metric space, then every continuous map $f : X \to \mathbf{R}$ is bounded.*

By way of illustration, let $a, b \in \mathbf{R}$ and $a < b$. We claim that, viewed as a subspace of $\mathbf{R}$, the interval $[a, b]$ is compact. For if this were not so, then there would be a family $\mathscr{U}$ of open subsets of $[a, b]$, with union $[a, b]$, such that no finite collection of sets in $\mathscr{U}$ has union $[a, b]$. Bisect $[a, b]$: then at least one of the sub-intervals $[a, \frac{1}{2}(a + b)], [\frac{1}{2}(a + b), b]$ is not contained in the union of any finite collection of members of $\mathscr{U}$. Repetition of this process gives a sequence of nested, closed sub-intervals of $[a, b]$, (I_n) say, with the length of I_n equal to $2^{-n}(b - a)$. By Cantor's intersection theorem (Theorem 2.2.3) these intervals I_n have intersection consisting of a single-point set in $[a, b]$, $\{x\}$ say. Obviously, there exists $U \in \mathscr{U}$ such that $x \in U$;

since U is open, $I_n \subset U$ for all large enough n. This contradicts the fact that no I_n is contained in the union of a finite number of members of $\mathscr{U}$, and our claim is justified.

Strategy II Suppose that X does not have the required property and that $f : X \to \mathbf{R}$ is a continuous but unbounded map. Then, for each $n \in \mathbf{N}$, there exists $x_n \in X$ such that $|f(x_n)| \geq n$. The sequence (x_n) does not have a convergent subsequence. To see this, suppose that (x_n) has a subsequence $(x_{m(n)})$ which converges to an element $x \in X$. Since f is continuous at x,

$$\left|f(x_{m(n)})\right| \to |f(x)| ;$$

however, for all $n \in \mathbf{N}$, $\left|f(x_{m(n)})\right| \geq m(n) \geq n$, and so $\left|f(x_{m(n)})\right| \to \infty$.

We have shown that any metric space X without the required property has a sequence with no convergent subsequence. Put equivalently, if each sequence in X has a convergent subsequence, then each continuous, real-valued function on X is bounded. These matters are summarised below.

Definition 2.3.3 A metric space X is said to be **sequentially compact** if each sequence in X has a subsequence which converges to a point of X.

Theorem 2.3.4 *If X is a sequentially compact metric space, then every continuous map $f : X \to \mathbf{R}$ is bounded.*

That each closed, bounded interval in $\mathbf{R}$ is sequentially compact is immediate from the Bolzano-Weierstrass theorem.

We preface the final strategy with a definition.

Definition 2.3.5 A metric space X is said to be **totally bounded** if to each $\varepsilon > 0$ there corresponds a finite family $\mathscr{F}$ of subsets of X such that $X = \cup \mathscr{F}$ and, for each $F \in \mathscr{F}$, diam $F < \varepsilon$.

Plainly, the interval $[0, 1]$, inheriting the usual metric from $\mathbf{R}$, is totally bounded. Indeed, every bounded subspace of $\mathbf{R}$ is totally bounded.

Strategy III Let X be complete and totally bounded. To obtain a contradiction, suppose that it carries a continuous but unbounded map $f : X \to \mathbf{R}$. Since X is totally bounded it is a union of finitely many closed sets each with diameter ≤ 1. (Observe that, if $A \subset X$, then diam $A = $ diam $\overline{A}$.) The restriction of f to one of these, X_1 say, is unbounded. A further appeal to the total boundedness of X shows that it, and therefore X_1, is a union of finitely many sets each of which is closed in X and of diameter $\leq 1/2$. The restriction of f to one of these subsets of X_1, X_2 say, is unbounded. Proceeding in this way, the result is a sequence (X_n) of sets closed in X such that, for all $n \in \mathbf{N}$, (i) $X_{n+1} \subset X_n$, (ii) diam $X_n \leq 1/n$, (iii) $f(X_n)$ is unbounded. By the Cantor intersection theorem, there exists $x \in X$ such that $\{x\} = \cap_{n=1}^{\infty} X_n$. Since f is continuous at x, there is a neighbourhood U_x of x on which f is bounded. But, for sufficiently large n, $X_n \subset U_x$ and therefore $f(X_n)$ is bounded. This contradicts (iii), and we have proved the following theorem.

Theorem 2.3.6 *If X is a complete and totally bounded metric space, then every continuous map $f : X \to \mathbf{R}$ is bounded.*

Conditions sufficient to ensure our property of interest are offered by each of Theorems 2.3.2, 2.3.4 and 2.3.6. Remarkably, they are also necessary conditions and hence equivalent. The position is formalised in our next result, the definition preceding which concerns terminology useful in its proof.

Definition 2.3.7 Let S be a subset of a set X and let $\mathscr{F}$ be a family of subsets of X such that $S \subset \cup \mathscr{F}$. Then $\mathscr{F}$ is called a **covering of** S : if $\mathscr{F}$ has only a finite number of members then it is called a **finite covering of** S. If X is a metric space and the members of $\mathscr{F}$ are open sets, $\mathscr{F}$ is called an **open covering of** S.

Theorem 2.3.8 *Let X be a metric space. The following are equivalent statements:*

(a) *X is compact.*
(b) $C(X) = \mathscr{C}(X)$, *i.e. each continuous map $f : X \to \mathbf{R}$ is bounded.*
(c) *X has the **Bolzano-Weierstrass property** : every infinite subset of X has a limit point in X.*
(d) *X is sequentially compact.*
(e) *X is complete and totally bounded.*

Proof (a)$\Longrightarrow$(b): this has already been established in Theorem 2.3.2.

(b)$\Longrightarrow$(c): let (b) hold and suppose that X has an infinite subset with no limit point in X. Then there is a sequence (x_n) of distinct points of X such that $S := \{x_n : n \in \mathbf{N}\}$ also has no limit point in X (the countable axiom of choice A.5.2 is used here). Thus, for each $n \in \mathbf{N}$, there exists $r_n \in (0, 1/n)$ such that $B(x_n, r_n) \cap S = \{x_n\}$. Letting d denote the metric on X, for each $n \in \mathbf{N}$ define $f_n : X \to \mathbf{R}$ by

$$f_n(x) = \max \left\{ n \left(1 - 2r_n^{-1}d(x, x_n) \right), 0 \right\}.$$

Evidently, each f_n is continuous.

Now, each element of X has a neighbourhood of itself restricted to which all save finitely many f_n are identically zero. To see this, let $x \in X$ and let $\rho > 0$ be such that $B(x, \rho) \cap S \subset \{x\}$. Further, let $N \in \mathbf{N}$ be such that $N > \rho^{-1}$ and, for all $n \geq N$, $x_n \neq x$. Then

$$B(x, \rho/2) \cap \bigcup_{n=N}^{\infty} B(x_n, r_n/2) = \emptyset;$$

for otherwise an $n \geq N$ would exist such that

$$\rho \leq d(x, x_n) < (\rho + r_n)/2 < (\rho + 1/n)/2 < \rho,$$

which is impossible. Hence, for all $n \geq N$ and all $y \in B(x, \rho/2), f_n(y) = 0$.

Define $f : X \to \mathbf{R}$ by

$$f(x) = \sum_{n=1}^{\infty} f_n(x).$$

The existence for each $x \in X$ of an N and a ρ, both depending on x, such that

$$f(y) = \sum_{n=1}^{N} f_n(y) \quad (y \in B(x, \rho/2)),$$

shows that f is continuous: the fact that, for each $n \in \mathbf{N}, f(x_n) = n$, shows that it is unbounded. Such an f is incompatible with (b) and so X has the Bolzano-Weierstrass property.

(c)$\Longrightarrow$(d): Suppose that (c) holds and let (x_n) be a sequence in X. If there is a point $x \in X$ such that $x_n = x$ for infinitely many values of n, then evidently there is a subsequence of (x_n) which is constant (has all its terms equal to x) and converges to x. If no such x exists then $S := \{x_n : n \in \mathbf{N}\}$ is infinite and has a limit point $y \in X$. We now choose $m(1)$ to be the least positive integer n such that $0 < d(y, x_n) < 1$, and define inductively a subsequence $(x_{m(n)})$ of (x_n) which converges to y. Suppose that $m(1) < m(2) < \ldots < m(n)$ have been chosen so that $0 < d(y, x_{m(j)}) < 1/j$ for $j = 1, 2, \ldots, n$. Choose $m(n + 1)$ to be the least integer exceeding $m(n)$ such that $0 < d(y, x_{m(n+1)}) < 1/(n+1)$. This establishes (d).

(d)$\Longrightarrow$(e): Suppose that (d) holds, let $\varepsilon > 0$ and select $x_1 \in X$. Suppose that $x_1, \ldots, x_n$ have been chosen in X so that $d(x_i, x_j) \geq \varepsilon/3$ if $i \neq j$. If possible, choose $x_{n+1} \in X$ such that $d(x_i, x_{n+1}) \geq \varepsilon/3$ for all $i, 1 \leq i \leq n$. This process must stop after a finite number of steps because of our assumption that (d) holds. It follows that $X = \bigcup_{j=1}^{N} B(x_j, \varepsilon/3)$ for some $N \in \mathbf{N}$. Since $B(x_j, \varepsilon/3)$ has diameter $< \varepsilon$, X is totally bounded.

To prove X complete, let (x_n) be a Cauchy sequence in X. By (d), there is a subsequence $(x_{k(n)})$ of (x_n) which converges to a point in X. Suppose $\lim_{n\to\infty} x_{k(n)} = x$ and let $\eta > 0$. There exists $N \in \mathbf{N}$ such that $d(x_m, x_n) < \eta/2$ and $d(x, x_{k(n)}) < \eta/2$ if $m, n \geq N$. Thus, for all $n \geq N$,

$$d(x, x_n) \leq d(x, x_{k(n)}) + d(x_{k(n)}, x_n) < \eta;$$

hence $\lim_{n\to\infty} x_n = x$; X is complete; and (e) holds.

(e)$\Longrightarrow$(a): Let $\mathscr{U}$ be an open covering of X and suppose that no finite subfamily of $\mathscr{U}$ is a covering of X. By (e), X is a union of finitely many closed sets each with diameter ≤ 1. One of these, say X_1, cannot be covered by finitely many members of $\mathscr{U}$. Repeat this argument with X_1 in place of X and continue indefinitely: there is a sequence (X_n) of closed sets such that, for all $n \in \mathbf{N}$, (i) $X_{n+1} \subset X_n$, (ii) diam $(X_n) < 1/n$, (iii) X_n is not covered by a finite subfamily of $\mathscr{U}$. By the Cantor intersection theorem there is a point $x \in X$ such that $\{x\} = \cap_{n=1}^{\infty} X_n$. Hence $x \in U$ for some $U \in \mathscr{U}$. By (ii), $X_n \subset U$ for all large enough n and this contradicts (iii). Hence X is compact and the proof of the theorem is complete. $\square$

Corollary 2.3.9 *If X is compact then C(X), equipped with the uniform metric*

$$d_\infty(f, g) = \sup_{x \in X} |f(x) - g(x)|,$$

is complete.

Proof The result is immediate from Theorems 2.2.6 and 2.3.8. ☐

A standard method generates new compact spaces from old. Proceeding through a reformulation of total boundedness we show that a finite Cartesian product of compact spaces equipped with the standard metric is compact.

Lemma 2.3.10 *Let (X, d) be a metric space. The following three statements are equivalent.*

(a) *X is totally bounded.*
(b) *Given any $\varepsilon > 0$, there exists a finite set $F \subset X$ such that, for each $x \in X$,*

$$d(x, F) := \inf \{d(x, f) : f \in F\} < \varepsilon.$$

(c) *Given any $\varepsilon > 0$, there exists a finite set $F \subset X$ such that $X \subset \bigcup_{f \in F} B(f, \varepsilon)$.*

Proof Suppose that (a) holds, that $\varepsilon > 0$ and that $x \in X$. There exist finitely many non-empty sets $A_1, A_2, \ldots, A_k \subset X$, whose union covers X, and each of which has diameter less than ε. Select $a_j \in A_j$ for $1 \le j \le k$; put $F = \{a_1, a_2, \ldots, a_k\}$. Then, for some j, $x \in A_j$ and $d(x, a_j) < \varepsilon$. Hence $d(x, F) < \varepsilon$ and (b) holds.

Next, suppose that (b) holds and that $\varepsilon > 0$. Let F be a finite set in X such that, for each $x \in X$, $d(x, F) < \varepsilon$. Plainly $X = \bigcup_{f \in F} B(f, \varepsilon)$, since, for each $x \in X$ there is an $f \in F$ with $d(x, f) < \varepsilon$. Thus (b) implies (c).

Finally, suppose that (c) holds and that $\varepsilon > 0$. Evidently $X = \bigcup_{f \in F} B(f, \varepsilon/3)$ for some finite set $F \subset X$, and so X is covered by finitely many sets each with diameter $< \varepsilon$. Hence (a) holds. ☐

Corollary 2.3.11 *If X is totally bounded, then it is bounded.*

Proof Let X be totally bounded. There exists a finite set $F \subset X$ such that for all $x \in X$, $d(x, F) < 1$. Hence diam $X \le 2 + $ diam F. ☐

Theorem 2.3.12 *Let $(X_1, d_1), \ldots, (X_n, d_n)$ be compact metric spaces. Let $X = \prod_{k=1}^{n} X_k$ and let d be defined for all $x = (x_1, \ldots, x_n), y = (y_1, \ldots, y_n) \in X$ by*

$$d(x, y) = \left\{ \sum_{k=1}^{n} d_k^2(x_k, y_k) \right\}^{1/2}.$$

Then (X, d) is a compact metric space.

Proof The obvious extension of Exercise 2.2.29 /1 shows that (X, d) is complete. It remains to prove that it is totally bounded.

Let $\varepsilon > 0$. Use of Lemma 2.3.10 shows that for each $k \in \{1, \ldots, n\}$, there is a finite set $F_k \subset X_k$ such that, for all $u \in X_k, d_k(u, F_k) < \varepsilon/\sqrt{n}$. Let $F = \prod_{k=1}^{n} F_k : F$ is finite. Let $x = (x_1, \ldots, x_n) \in X$. For each k, there exists $f_k \in F_k$ such that $d_k(x_k, f_k) < \varepsilon/\sqrt{n}$. Let $f = (f_1, \ldots, f_n) : f \in F$ and $d(x, f) < \varepsilon$. Hence (X, d) is totally bounded.   □

Metric spaces are often encountered as subspaces of others. Language for the situation in which the embedded space is compact is introduced in the next definition. The two lemmas which follow it describe attributes of a compact subspace relative to its host.

Definition 2.3.13 A subset of a metric space X is said to be a **compact set** in X if either it is empty or it is compact as a subspace of X. The obvious substitutions respectively define (a) a **sequentially compact set** in X, (b) a **totally bounded set** in X.

Lemma 2.3.14 *Let E be a subset of a metric space X. Then E is a compact set in X if, and only if, every covering of E by sets open in X contains a finite covering of E.*

Proof If $E = \emptyset$ the result is obvious.

Let E be a non-empty compact set in X and let $\mathscr{U}$ be a covering of E by sets open in X. The family $\widetilde{\mathscr{U}} = \{E \cap U : U \in \mathscr{U}\}$ is a covering of E by sets open in the subspace E and, therefore, there exist sets $U_1, \ldots, U_n \in \mathscr{U}$ such that $E = \cup_{j=1}^{n} E \cap U_j \subset \cup_{j=1}^{n} U_j$. Thus $\mathscr{U}$ contains a finite covering of E.

Conversely, let E be non-empty and let $\mathscr{V}$ be a covering of E by sets open in the subspace E. Then there exists a covering $\mathscr{U}$ of E, whose elements are open in the metric space X, such that $\mathscr{V} = \{E \cap U : U \in \mathscr{U}\}$. Since there are sets $U_1, \ldots, U_n \in \mathscr{U}$ such that $E \subset \cup_{j=1}^{n} U_j$, $\mathscr{V}$ contains a finite covering of E. Hence E is a compact set in X.   □

Lemma 2.3.15 *Let X be a metric space.*

(i) *If E is a compact set in X then it is closed and bounded in X.*
(ii) *If X is a compact space and E is a closed set in X, then E is a compact set in X.*

Proof (i) If $E = \emptyset$ then the matter is clear.

Suppose $E \neq \emptyset$. The subspace E is complete and totally bounded. Hence the set E is closed in the space X (Exercise 2.2.29/2(ii)) and bounded in the space X, since it is bounded in the subspace E (Corollary 2.3.11).

(ii) Suppose $E \neq \emptyset$; otherwise the result holds trivially. Since E is closed in X, E is a complete subspace of X (Exercise 2.2.29/2(i)). Since X is a totally bounded space, so also is the subspace E. Hence E is a compact set in X.   □

For general spaces X, the converse of Lemma 2.3.15 (i) is false. However, in the important special case when $X = \mathbf{R}^n$, it is true.

Theorem 2.3.16 (Heine-Borel) *Let K be a subset of* $\mathbf{R}^n$. *Then K is a compact set in* $\mathbf{R}^n$ *if, and only if, it is closed and bounded.*

Proof If K is compact in $\mathbf{R}^n$ then by Lemma 2.3.15 (i), it is a closed and bounded set in $\mathbf{R}^n$.

Conversely, suppose that K is a non-empty, closed and bounded set in $\mathbf{R}^n$; if $K = \emptyset$ then the result holds trivially. Observe that K is contained in a cube $I^n \subset \mathbf{R}^n$, where I is a closed and bounded interval in $\mathbf{R}$. Since K is closed in $\mathbf{R}^n$ it is closed in the subspace I^n. Hence, by Lemma 2.3.15 (ii), if I^n is compact then K is a compact set in I^n and also in $\mathbf{R}^n$. It remains to prove that I^n is compact.

Note that I regarded as a subspace of $\mathbf{R}$ is complete and totally bounded and therefore compact. By Theorem 2.3.12, I^n is compact, as required. $\qquad\square$

The example to follow reinforces the fact that the converse of Lemma 2.3.15 (i) is false. It is sited in ℓ_2, and is complemented by a characterisation of the compact sets therein.

Example 2.3.17 From Examples 2.1.2 (vi) and 2.2.2 (vi),

$$\ell_2 = \left\{ x = (x_n)_{n \in \mathbf{N}} : x_n \in \mathbf{R} \text{ for all } n \in \mathbf{N}, \ \sum_{n=1}^{\infty} x_n^2 < \infty \right\}$$

is a complete metric space when equipped with the metric

$$d(x, y) = \left\{ \sum_{n=1}^{\infty} (x_n - y_n)^2 \right\}^{1/2},$$

where $x = (x_n), y = (y_n) \in \ell_2$.

Let $K = \{x \in \ell_2 : d(0, x) \leq 1\}$: K is the closed ball with centre 0 and radius 1. Although closed and bounded, K is not compact. For let $e^n = (\delta_j^n)_{j \in \mathbf{N}} \in \ell_2$, where δ_j^n is the Kronecker delta, equal to 1 when $j = n$ and zero otherwise. As the sequence (e^n) in ℓ_2 is such that $d(e^m, e^n) = \sqrt{2}$ if $m \neq n$, it has no convergent subsequence.

Theorem 2.3.18 *Let A be a non-empty subset of* ℓ_2. *Then A is compact if, and only if, it is closed, bounded and such that*

$$\sup_{x=(x_k) \in A} \sum_{k=n}^{\infty} x_k^2 \to 0 \text{ as } n \to \infty. \tag{2.3.1}$$

Proof Let A be compact. By Lemma 2.3.15 (i), A is closed and bounded. Let $\varepsilon > 0$. There exists a finite set $F = \{a^1, \ldots, a^p\} \subset A$ such that, for all $x \in A, d(x, F) < \varepsilon/2$. Choose $m \in \mathbf{N}$ such that

$$\max_{1 \leq q \leq p} \left\{ \sum_{k=m}^{\infty} |a_k^q|^2 \right\}^{1/2} < \varepsilon/2.$$

Let $x = (x_k) \in A$. There exists r, $1 \le r \le p$, such that $d(x, a^r) < \varepsilon/2$. Thus, by the Minkowski inequality,

$$\left\{ \sum_{k=m}^{\infty} |x_k|^2 \right\}^{1/2} \le \left\{ \sum_{k=m}^{\infty} |x_k - a_k^r|^2 \right\}^{1/2} + \left\{ \sum_{k=m}^{\infty} |a_k^r|^2 \right\}^{1/2}$$

$$\le d(x, a^r) + \left\{ \sum_{k=m}^{\infty} |a_k^r|^2 \right\}^{1/2} < \varepsilon,$$

and so

$$\sup_{x=(x_k) \in A} \left\{ \sum_{k=m}^{\infty} x_k^2 \right\}^{1/2} \le \varepsilon.$$

Conversely, suppose that A is closed and bounded and has the property (2.3.1). Then A is closed in the complete space ℓ_2 and hence is a complete subspace of ℓ_2. It remains to show that it is totally bounded. Given $\varepsilon > 0$, choose $n \in \mathbf{N}$ such that

$$\sup_{x=(x_k) \in A} \left\{ \sum_{k=n+1}^{\infty} x_k^2 \right\}^{1/2} < \varepsilon/2.$$

Since A is bounded, there exists a real number Λ such that, for all $x \in A$, $d(0, x) \le \Lambda$. Let $\lambda = \left\{ \sum_{k=1}^{\infty} k^{-2} \right\}^{1/2}$ and choose $m \in \mathbf{N}$ so that $m\varepsilon > 4\lambda\Lambda$. For each j, $1 \le j \le n$, let

$$F_j = \left\{ \left(-1 + \frac{2r}{jm} \right) \Lambda : r = 0, 1, \ldots, jm \right\}.$$

Put

$$F = \{x = (x_k) \in \ell_2 : x_k \in F_k \text{ if } 1 \le k \le n, \text{ and } x_k = 0 \text{ if } k > n\} :$$

F is a finite set in ℓ_2. Let $x = (x_k) \in A$. There exists $f = (f_k) \in F$ such that, if $1 \le k \le n$, then $|x_k - f_k| \le k^{-1}(2\Lambda/m)$. Hence

$$d^2(x, f) = \sum_{k=1}^{n} |x_k - f_k|^2 + \sum_{k=n+1}^{n} |x_k|^2$$

$$< (2\lambda\Lambda/m)^2 + (\varepsilon/2)^2 < \varepsilon^2.$$

As it is covered by finitely many balls of radius ε, it follows that A is totally bounded. $\qquad\square$

Corollary 2.3.19 *The Hilbert cube*

$$\mathcal{H} = \left\{ x = (x_k) \in \ell_2 : \text{for each } k \in \mathbf{N}, |x_k| \leq k^{-1} \right\}$$

is compact in ℓ_2.

Proof It is routine to check that $\mathcal{H}$ is closed and bounded in ℓ_2 and that (2.3.1) holds.   □

Pursuing the characterisation of compact sets in special spaces a little further, we consider next the position in spaces kindred to $C[0, 1]$. The best known characterisation in such spaces is the Arzelà-Ascoli theorem, which involves the concept of equicontinuity explained below.

Definition 2.3.20 Let (X, d) be a metric space. A set $\mathcal{F} \subset C(X)$ is said to be **equicontinuous at a point** $x \in X$ if, given any $\varepsilon > 0$, there exists $\delta > 0$ such that for all $y \in X$ with $d(x, y) < \delta$, we have $\sup_{f \in \mathcal{F}} |f(y) - f(x)| < \varepsilon$; if $\mathcal{F}$ is equicontinuous at every point of X we say that $\mathcal{F}$ is **equicontinuous on X**.

Lemma 2.3.21 *Let (X, d) be a compact metric space and let $\mathcal{F} \subset C(X)$. Then $\mathcal{F}$ is equicontinuous on X if, and only if, it is **uniformly equicontinuous on X** in the sense that given any $\varepsilon > 0$, there exists $\delta > 0$ such that for all $x, y \in X$ with $d(x, y) < \delta$, we have $\sup_{f \in \mathcal{F}} |f(y) - f(x)| < \varepsilon$.*

Proof Suppose that $\mathcal{F}$ is equicontinuous on X and let $\varepsilon > 0$. Then given any $x \in X$, there exists $\delta_x > 0$ such that $\sup_{f \in \mathcal{F}} |f(y) - f(x)| < \varepsilon/3$ if $d(x, y) < \delta_x$. By the compactness of X, there exist $x_1, \ldots, x_n \in X$ such that $X = \cup_{j=1}^n B(x_j, \delta_{x_j}/2)$; let $\delta = \frac{1}{2} \min \{\delta_{x_1}, \ldots, \delta_{x_n}\}$. Let $x, y \in X$, $d(x, y) < \delta$ and $f \in \mathcal{F}$. For some j, $x \in B(x_j, \delta_{x_j}/2)$. Thus x and y belong to $B(x_j, \delta_{x_j})$ and

$$|f(x) - f(y)| \leq |f(x) - f(x_j)| + |f(x_j) - f(y)| < 2\varepsilon/3.$$

It follows that $\sup_{f \in \mathcal{F}} |f(y) - f(x)| < \varepsilon$ whenever $d(x, y) < \delta$: $\mathcal{F}$ is uniformly equicontinuous on X.

The converse is obvious.   □

Theorem 2.3.22 (Arzelà-Ascoli) *Let (X, d) be a compact metric space and let $\mathcal{K} \subset C(X)$. Then $\mathcal{K}$ is compact if, and only if, it is closed, bounded and equicontinuous on X.*

Proof Suppose $\mathcal{K}$ is compact. Then it is certainly closed and bounded. To establish equicontinuity, let $\varepsilon > 0$ and let $f_1, \ldots, f_n \in \mathcal{K}$ be such that $\mathcal{K} \subset \cup_{k=1}^n B_{C(X)}(f_k, \varepsilon/3)$. Since each f_k is uniformly continuous on X, there exists $\delta > 0$ such that $|f_k(x) - f_k(y)| < \varepsilon/3$ if $d(x, y) < \delta$ and $k \in \{1, 2, \ldots, n\}$. Now let $f \in \mathcal{K}$ and $d(x, y) < \delta$: then $f \in B_{C(X)}(f_k, \varepsilon/3)$ for some k and

$$|f(x) - f(y)| \leq |f(x) - f_k(x)| + |f_k(x) - f_k(y)| + |f_k(y) - f(y)| < \varepsilon;$$

that is, $\mathcal{H}$ is equicontinuous on X.

Conversely, suppose that $\mathcal{H}$ is equicontinuous on X, closed and bounded. It is enough to prove that $\mathcal{H}$ is totally bounded. To do so, let $\varepsilon > 0$. Since X is compact, $\mathcal{H}$ is uniformly equicontinuous on X and thus there is a $\delta > 0$ such that, whenever $x, y \in X$ and $d(x, y) < \delta$,

$$\sup_{f \in \mathcal{H}} |f(x) - f(y)| < \varepsilon/4;$$

also, there exist $x_1, \ldots, x_n \in X$ such that

$$X = \cup_{i=1}^n U_i,$$

where $U_i = B_X(x_i, \delta)$ $(1 \le i \le n)$. Moreover, each $x \in X$ is such that, for some $i \in \{1, 2, \ldots, n\}$, $x \in U_i$ and

$$\sup_{f \in \mathcal{H}} |f(x) - f(x_i)| < \varepsilon/4.$$

Let $\Lambda = \sup_{f \in \mathcal{H}} d_\infty(f, 0)$, $I = \{\lambda \in \mathbf{R} : |\lambda| \le \Lambda\}$ and define $\theta : \mathcal{H} \to I^n \subset \mathbf{R}^n$ by

$$\theta(f) = (f(x_1), \ldots, f(x_n)).$$

Since I^n is totally bounded, so also is $\theta(\mathcal{H})$. Thus there exist $f_1, \ldots, f_m \in \mathcal{H}$ such that

$$\theta(\mathcal{H}) \subset \cup_{j=1}^m B_{\mathbf{R}^n}(\theta(f_j), \varepsilon/4)$$

and, to conclude the proof, we show that

$$\mathcal{H} \subset \cup_{j=1}^m B_{C(X)}(f_j, \varepsilon).$$

Let $f \in \mathcal{H}$. For some $j \in \{1, 2, \ldots, m\}$,

$$d_{\mathbf{R}^n}(\theta(f), \theta(f_j)) = \left\{ \sum_{i=1}^n |f(x_i) - f_j(x_i)|^2 \right\}^{1/2} < \varepsilon/4,$$

and therefore

$$\max_{1 \le i \le n} |f(x_i) - f_j(x_i)| < \varepsilon/4.$$

Let $x \in X$. Then, for some $i \in \{1, 2, \ldots, n\}$, $x \in U_i$ and

$$\max \left\{ |f(x) - f(x_i)|, |f_j(x) - f_j(x_i)| \right\} < \varepsilon/4.$$

Hence

$$|f(x) - f_j(x)| \le |f(x) - f(x_i)| + |f(x_i) - f_j(x_i)| + |f_j(x_i) - f_j(x)| < 3\varepsilon/4.$$

It follows that $d_\infty(f, f_j) < \varepsilon$ and $f \in B_{C(X)}(f_j, \varepsilon)$. The proof is complete. $\qquad\square$

Corollary 2.3.23 *Let X be a compact metric space and let $\mathscr{K} \subset C(X)$. Then $\mathscr{K}$ is relatively compact (that is, $\overline{\mathscr{K}}$ is compact) if, and only if, it is bounded and equicontinuous.*

Proof It is easy to prove that if $\mathscr{K}$ is equicontinuous on X, then so is $\overline{\mathscr{K}}$. The rest is obvious. $\qquad\square$

We next turn to continuous maps on compact spaces.

Theorem 2.3.24 *Let X_1 and X_2 be metric spaces and let $f : X_1 \to X_2$ be continuous.*

(i) *If E is a compact set in X_1, then $f(E)$ is a compact set in X_2.*
(ii) *If X_1 is compact and f is bijective, then f is a homeomorphism.*

Proof

(i) Suppose $E \ne \emptyset$; otherwise the result holds trivially. Let $\mathscr{V}$ be an open covering of $f(E)$. Then $\mathscr{U} = \{f^{-1}(V) : V \in \mathscr{V}\}$ is an open covering of E. Since E is compact in X_1, by Lemma 2.3.14, $\mathscr{U}$ contains a finite covering, $\{f^{-1}(V_1), \ldots, f^{-1}(V_n)\}$ say, of E. But this implies that $\{V_1, \ldots, V_n\} \subset \mathscr{V}$ is a finite covering of $f(E)$. Using Lemma 2.3.14 again, we see that $f(E)$ is a compact set in X_2.
(ii) Let U be an open set in X_1. Since $X_1 \backslash U$ is closed and therefore compact in X_1 (Lemma 2.3.15 (ii)), by the first part of the theorem, $f(X_1 \backslash U)$ is compact and therefore closed in X_2 (Lemma 2.3.15 (i)). Further, since $f(X_1 \backslash U) = X_2 \backslash f(U)$, it follows that $f(U)$ is open in X_2. Now, appeal to Remark 2.1.37 (ii) completes the proof. $\qquad\square$

Corollary 2.3.25 *Let X be a metric space, let $f : X \to \mathbf{R}$ be continuous, and let E be a non-empty compact set in X. Then $f(E)$ is bounded and both $\inf f(E)$ and $\sup f(E)$ belong to $f(E)$. In particular, there exist points u and v in E such that*

$$f(u) = \inf f(E) \text{ and } f(v) = \sup f(E).$$

Proof By Theorem 2.3.24, $f(E)$ is compact in $\mathbf{R}$; by Theorem 2.3.16, it is closed and bounded. The result now follows easily. $\qquad\square$

The novelty of Corollary 2.3.25 is in the attainment of bounds. It is utilised in proving the next result, about the distance of a point from a set and the distance between two sets.

Theorem 2.3.26 *Let A and B be non-empty subsets of a metric space (X, d), and let $d(x, A)$, $d(A, B)$ be defined as in Lemma 2.1.40. Then*

(i) *if A is compact, B is closed and $A \cap B = \emptyset$, then there exists $a \in A$ such that $d(a, B) = d(A, B) > 0$;*
(ii) *if $X = \mathbf{R}^n$, d is the Euclidean metric on $\mathbf{R}^n$, A is compact, B is closed and $A \cap B = \emptyset$, then there exist $a \in A$ and $b \in B$ such that $d(a, b) = d(A, B)$.*

Proof

(i) By Lemma 2.1.40 (iii), the map $x \longmapsto d(x, B) : X \to \mathbf{R}$ is continuous. Since A is compact, by Corollary 2.3.25 there exists $a \in A$ such that

$$d(a, B) = \inf \{d(x, B) : x \in A\}.$$

Using Lemma 2.1.40 (i), we therefore see that $d(a, B) = d(A, B)$. Now, if $d(a, B) = 0$ then $a \in \overline{B}$ (by Lemma 2.1.40 (ii)). Given that B is closed, it would follow that $a \in A \cap B$, which contradicts $A \cap B = \emptyset$.

(ii) By (i), there exists $a \in A$ such that $d(a, B) = d(A, B)$. Choose any $\widetilde{b} \in B$. Let $\widetilde{B} = \{y \in B : d(a, y) \leq d(a, \widetilde{b})\}$; note that $d(a, \widetilde{B}) = d(a, B)$. Since $\widetilde{B}$ is closed and bounded, by Theorem 2.3.16 it is compact. Thus, using (i), there exists $b \in \widetilde{B} \subset B$ such that $d(a, b) = d(b, \{a\}) = d(\widetilde{B}, \{a\}) = d(a, \widetilde{B}) = d(a, B)$. $\qquad\qquad\square$

Note that the conclusion of (i) is false if the set A is merely required to be closed, rather than compact. To illustrate this take $X = \mathbf{R}$, with the usual metric, let $A = \mathbf{N}$, $B = \{n - 1/n : n \in \mathbf{N}\}$. Plainly A and B are closed, and $A \cap B = \emptyset$; but, for all $n \in \mathbf{N}$, $d(A, B) \leq d(n - 1/n, n) = 1/n \to 0$ as $n \to \infty$, and hence $d(A, B) = 0$.

Corollary 2.3.25 has many uses, and it is worthwhile to note that key aspects of it apply to functions with properties similar to continuity but of weaker regularity. Looking back at Definition 2.1.27 it is clear that for a real-valued function f on a metric space X to be continuous at $x \in X$, it is necessary and sufficient that, given any $\varepsilon > 0$, there exists a neighbourhood U of x such that

(i) $f(x) - \varepsilon < f(u)$ whenever $u \in U$,

and

(ii) $f(u) < f(x) + \varepsilon$ whenever $u \in U$.

Taken separately, conditions (i) and (ii) define classes of functions of importance in their own right.

Definition 2.3.27 Let (X, d) be a metric space, $x \in X$ and f be a real-valued function on X. Then f is said to be **lower semi-continuous at** x if, given any $\varepsilon > 0$, there exists $\delta > 0$ such that

$$f(x) - \varepsilon < f(y) \text{ if } d(x, y) < \delta.$$

The function f is said to be **lower semi-continuous on** X if it is lower semi-continuous at each point of X. Similarly, f is called **upper semi-continuous at** x if, given any $\varepsilon > 0$, there exists $\delta > 0$ such that

$$f(y) < f(x) + \varepsilon \text{ if } d(x, y) < \delta;$$

upper semi-continuity on X is defined in the obvious way.

Plainly, given $x \in X$, a necessary and sufficient condition for f to be continuous at x is that it should be both lower semi-continuous at x and upper semi-continuous at x. Note that if f is lower semi-continuous at x, then $-f$ is upper semi-continuous at x.

Example 2.3.28

(1) Let X be a metric space and $x \in X$. A function $f : X \to \mathbf{R}$ is said to have a **relative minimum** (respectively, **relative maximum**) at x if there exists a neighbourhood U of x such that $f(x) \leq f(u)$ (respectively, $f(x) \geq f(u)$) whenever $u \in U$. It is clear that if f has a relative minimum (maximum) at x then it is lower (upper) semi-continuous at x.
(2) Let X be a metric space and $A \subset X$. Then A is open in X if, and only if, the characteristic function χ_A of A is lower semi-continuous on X. To see this, suppose that A is open. Then χ_A has a relative minimum at each point of X and so is lower semi-continuous on X. Conversely, let χ_A be lower semi-continuous on X. Omitting the trivial case of $A = \emptyset$, let $a \in A$. Then there is a neighbourhood V of a such that

$$\frac{1}{2} = \chi_A(a) - \frac{1}{2} < \chi_A(x) \text{ if } x \in V.$$

But this shows that $V \subset A$, that $A \subset \overset{o}{A}$ and that A is open.
(3) Let $f : [0, 1] \to \mathbf{R}$ be defined by $f(t) = 0$ if t is irrational, $f(t) = 1/q$ if $t = p/q$, where p and q are integers with no common factor greater than 1, and $q > 0$. Then f is upper semi-continuous on $[0, 1]$. Note that f is continuous and therefore upper semi-continuous at all irrational points of $[0, 1]$ (refer to Exercise 1.3.10/6). Further, it has a relative maximum at every rational point of $[0, 1]$.

Lemma 2.3.29 *Let X be a compact metric space and $f : X \to \mathbf{R}$ be lower semi-continuous on X. Then f has an **absolute minimum** on X : there exists $u \in X$ such that, for all $x \in X, f(u) \leq f(x)$.*

Proof We begin by showing that f is bounded below. Suppose otherwise. Then for each $n \in \mathbf{N}$, there exists $x_n \in X$ such that $f(x_n) \leq -n$. Hence

$$\lim_{n \to \infty} f(x_n) = -\infty. \tag{2.3.2}$$

Since X is compact, the sequence (x_n) has a convergent subsequence, $(x_{m(n)})$ say. Suppose $\lim_{n \to \infty} x_{m(n)} = x$. As f is lower semi-continuous at x, a neighbourhood V of x exists such that

$$f(x) - 1 < f(v) \ (v \in V).$$

Hence, for sufficiently large n,

$$f(x) - 1 < f(x_{m(n)}),$$

a conclusion incompatible with (2.3.2).

Now let

$$K = \inf\{f(x) : x \in X\};$$

it remains to show that $K \in f(X)$. For each $n \in \mathbf{N}$, there exists $u_n \in X$ such that $f(u_n) < K + n^{-1}$. Since X is compact, the sequence (u_n) has a convergent subsequence, $(u_{k(n)})$ say. Suppose $u_{k(n)} \to u$. Clearly

$$K \leq f(u_{k(n)}) < K + k(n)^{-1} \ (n \in \mathbf{N})$$

and so

$$K = \lim_{n \to \infty} f(u_{k(n)}).$$

Further, as f is lower semi-continuous at u, for each $\varepsilon > 0$ there exists a neighbourhood U of u such that

$$f(u) - \varepsilon < f(x) \text{ whenever } x \in U.$$

Thus for all $\varepsilon > 0$,

$$f(u) - \varepsilon \leq \lim_{n \to \infty} f(u_{k(n)}) = K,$$

and so $f(u) \leq K$. But by definition of K, $f(u) \geq K$. Thus $K = f(u) \in f(X)$. $\qquad \square$

Theorem 2.3.30 *Let (X_1, d_1), (X_2, d_2) be metric spaces, let (X_1, d_1) be compact and let $f : X_1 \to X_2$ be continuous. Then f is uniformly continuous.*

Proof When (X_2, d_2) is $\mathbf{R}$ equipped with the standard metric, this result follows from Lemma 2.3.21. To illustrate the ways in which compactness can be used we give here a proof by contradiction that is not so readily available for Lemma 2.3.21. Suppose f is not uniformly continuous. Then there exists $\varepsilon > 0$ such that, given any $\delta > 0$, there exist $x, y \in X_1$ with

$$d_1(x, y) < \delta \text{ and } d_2(f(x), f(y)) \geq \varepsilon.$$

Hence, for all $n \in \mathbf{N}$ there exist $x_n, y_n \in X_1$ such that

$$d_1(x_n, y_n) < 1/n \text{ and } d_2(f(x_n), f(y_n)) \geq \varepsilon.$$

Since X_1 is compact, there exist $x \in X_1$ and a subsequence $(x_{m(n)})$ of (x_n) such that $d_1(x_{m(n)}, x) \to 0$. As $d_1(x_{m(n)}, y_{m(n)}) < 1/m(n) \leq 1/n \to 0$, it follows that $d_1(y_{m(n)}, x) \to 0$. But given that f is continuous,

$$\varepsilon \leq d_2(f(x_{m(n)}), f(y_{m(n)})) \leq d_2(f(x_{m(n)}), f(x)) + d_2(f(x), f(y_{m(n)})) \to 0,$$

a contradiction. Hence f is uniformly continuous. $\qquad\square$

Corollary 2.3.31 *Let $f : \mathbf{R}^2 \to \mathbf{R}$ be continuous and have a continuous first partial derivative $\partial_2 f$ with respect to the second coordinate. For each $t \in \mathbf{R}$ put*

$$F(t) = \int_0^1 f(s, t)ds.$$

Then $F : \mathbf{R} \to \mathbf{R}$ is differentiable and, for all $t \in \mathbf{R}$,

$$F'(t) = \int_0^1 \partial_2 f(s, t)ds.$$

Proof Let $t_0, t \in \mathbf{R}$, $t \neq t_0$. Then

$$\left| \frac{F(t) - F(t_0)}{t - t_0} - \int_0^1 \partial_2 f(s, t_0)ds \right| \leq \int_0^1 \left| \frac{f(s, t) - f(s, t_0)}{t - t_0} - \partial_2 f(s, t_0) \right| ds.$$

Let $\varepsilon > 0$. By Theorem 2.3.16, $[0, 1] \times [t_0 - 1, t_0 + 1]$ is a compact subset of $\mathbf{R}^2$; thus the continuous map $\partial_2 f$ is uniformly continuous on this rectangle, by Theorem 2.3.30. Hence there exists $\delta \in (0, 1)$ such that, for all $s \in [0, 1]$ and all $v \in [t_0 - \delta, t_0 + \delta]$,

$$|\partial_2 f(s, v) - \partial_2 f(s, t_0)| < \varepsilon.$$

It follows that, if $s \in [0, 1]$ and $0 < |t - t_0| < \delta$, then (by the mean-value theorem) for some v strictly between t and t_0,

$$\left| \frac{f(s, t) - f(s, t_0)}{t - t_0} - \partial_2 f(s, t_0) \right| = |\partial_2 f(s, v) - \partial_2 f(s, t_0)| < \varepsilon.$$

Thus

$$\left| \frac{F(t) - F(t_0)}{t - t_0} - \int_0^1 \partial_2 f(s, t_0)ds \right| < \varepsilon$$

if $0 < |t - t_0| < \delta$, and the result follows. $\qquad\square$

This Corollary is particularly useful. To illustrate, we give the following example.

Example 2.3.32 For each $t \in \mathbf{R}$, put $g(t) = \left(\int_0^t e^{-s^2} ds \right)^2$. By the fundamental theorem of integral calculus,

$$g'(t) = 2e^{-t^2} \int_0^t e^{-s^2} ds.$$

Put $h(t) = \int_0^1 (1+s^2)^{-1} e^{-t^2(1+s^2)} ds$ $(t \in \mathbf{R})$. By Corollary 2.3.31,

$$h'(t) = -2t \int_0^1 e^{-t^2(1+s^2)} ds = -2te^{-t^2} \int_0^1 e^{-t^2 s^2} ds.$$

If $t \neq 0$, the substitution $u = st$ reduces this to

$$h'(t) = -2e^{-t^2} \int_0^t e^{-u^2} du \ (t \neq 0);$$

plainly $h'(t) = 0$ if $t = 0$. Thus, for all $t \in \mathbf{R}$, $g'(t) + h'(t) = 0$ and so

$$g(t) + h(t) = \text{ constant } = g(0) + h(0) = \int_0^1 (1+s^2)^{-1} ds = \pi/4.$$

Hence

$$\left| \frac{\pi}{4} - g(t) \right| = h(t) = e^{-t^2} \int_0^1 (1+s^2)^{-1} e^{-t^2 s^2} ds \le e^{-t^2} \to 0$$

as $t \to \infty$. It follows that

$$\lim_{t \to \infty} \left(\int_0^t e^{-s^2} ds \right)^2 = \frac{\pi}{4},$$

which gives the famous result that

$$\int_0^\infty e^{-s^2} ds = \frac{1}{2} \sqrt{\pi}.$$

We conclude this section by giving two applications of the ideas of completeness and compactness.

2.3.1 Application 1

This is to the theory of ordinary differential equations. Let I be a non-degenerate interval in $\mathbf{R}$; given any $n \in \mathbf{N}$, let $f^{(n)}$ be the nth derivative of f and put

$$C^n(I) = \left\{ f \in C(I) : f^{(n)} \text{ exists and is continuous on } I \right\}.$$

By a **linear ordinary differential equation of order** n **on** I we shall mean a problem, denoted by

$$x^{(n)} + a_1(t)x^{(n-1)} + \ldots + a_n(t)x = h(t), \tag{2.3.3}$$

of the following type:

Given $a_1, \ldots, a_n, h \in C(I)$, does there exist a function $x \in C^n(I)$ such that for all $t \in I$,

$$x^{(n)}(t) + a_1(t)x^{(n-1)}(t) + \ldots + a_n(t)x(t) = h(t) ?$$

If there is such a function, it is called a **solution** of (2.3.3). Equation (2.3.3) is called **homogeneous** or **non-homogeneous** according to whether $h = 0$ or $h \neq 0$. Given $t_0 \in I$, an **initial-value problem set at** t_0, **associated with** (2.3.3), is the problem of whether given $(\eta_1, \ldots, \eta_n) \in \mathbf{R}^n$, there is a solution ϕ of (2.3.3) such that

$$\left(\phi(t_0), \phi^{(1)}(t_0), \ldots, \phi^{(n-1)}(t_0) \right) = (\eta_1, \ldots, \eta_n).$$

This problem is symbolised by

$$\left. \begin{array}{l} x^{(n)} + a_1(t)x^{(n-1)} + \ldots + a_n(t)x \qquad = h(t), \\[2mm] x(t_0) = \eta_1, \ x^{(1)}(t_0) = \eta_2, \ \ldots, \ x^{(n-1)}(t_0) = \eta_n. \end{array} \right\} \tag{2.3.4}$$

Example 2.3.33

(i) Let $I = \mathbf{R}$. The problem

$$\ddot{x} - x = 0 \tag{2.3.5}$$

is a homogeneous ordinary differential equation of order 2. The function $t \longmapsto e^t$ is a solution. The problem

$$\ddot{x} - x = 0, \ x(0) = 1, \ \dot{x}(0) = 0$$

is an initial-value problem set at 0, associated with (2.3.5). Its unique solution is $t \longmapsto \cosh t$.

(ii) Let $I = (0, \infty)$. The problem

$$\ddot{x} + t^{-1}\dot{x} - t^{-2}x = \log t, \tag{2.3.6}$$

is an inhomogeneous equation of order 2, with $t \longmapsto (3\log t - 4)t^2/9$ as a solution. The problem

$$\ddot{x} + t^{-1}\dot{x} - t^{-2}x = \log t, \ x(1) = 1, \ \dot{x}(1) = 0$$

is an initial-value problem set at 1 associated with (2.3.6), and has the unique
solution

$$t \longmapsto t + \frac{4}{9}t^{-1} + \frac{1}{9}(3\log t - 4)t^2.$$

We shall now prove the existence and uniqueness of solutions of initial-value
problems, when $n = 2$ and I is closed and bounded.

Consider the initial-value problem

$$\left. \begin{array}{l} \ddot{x} + a_1(t)\dot{x} + a_2(t)x = h(t), \\[2mm] x(t_0) = \eta_0, \ \dot{x}(t_0) \ = \eta_1, \end{array} \right\} \tag{2.3.7}$$

where $a_1, a_2, h \in C(I)$ and $t_0 \in I$. Suppose $\phi \in C^2(I)$ is a solution of (2.3.7) and
let $u = \ddot{\phi}$. By Taylor's theorem with the integral form of the remainder,

$$\phi(t) = \eta_0 + (t - t_0)\eta_1 + \int_{t_0}^t (t - s)u(s)ds, \ \dot{\phi}(t) = \eta_1 + \int_{t_0}^t u(s)ds.$$

Substitution in (2.3.7) now shows that

$$u(t) = h(t) - \eta_1 a_1(t) - \{\eta_0 + (t - t_0)\eta_1\} a_2(t) - \int_{t_0}^t \{a_1(t) + (t - s)a_2(t)\} u(s)ds,$$

and so u satisfies the **integral equation** (of **Volterra** type)

$$u(t) = g(t) + \int_{t_0}^t k(t, s)u(s)ds, \tag{2.3.8}$$

where

$$g(t) = h(t) - \eta_1 a_1(t) - \{\eta_0 + (t - t_0)\eta_1\} a_2(t),$$

$$k(t, s) = -\{a_1(t) + (t - s)a_2(t)\}.$$

Thus the second derivative of any solution of (2.3.7) is a solution of (2.3.8).
On the other hand, if the integral equation (2.3.8) has a unique solution $w \in C(I)$,
let

$$\psi(t) = \eta_0 + (t - t_0)\eta_1 + \int_{t_0}^t (t - s)w(s)ds \quad (t \in I)$$

so that

$$\dot{\psi}(t) = \eta_1 + \int_{t_0}^t w(s)ds, \ \ddot{\psi}(t) = w(t)(t \in I).$$

Plainly ψ satisfies (2.3.7). Moreover, it is the unique such solution, for if ψ_1 were another, then $\ddot{\psi}_1 = w = \ddot{\psi}$ and so, by Taylor's theorem, $\psi_1 = \psi$.

It follows that the problem of the existence of a unique solution of (2.3.7) can be reduced to that of the existence of a unique solution of the integral equation (2.3.8). We now prove, with the aid of the contraction mapping theorem, that (2.3.8) does indeed have a unique solution.

Theorem 2.3.34 *Let I be closed and bounded, let $g \in C(I)$, $t_0 \in I$, put $D = I \times I$ and let $k : D \to \mathbf{R}$ be continuous. Then there is a unique $\phi \in C(I)$ satisfying the Volterra equation*

$$\phi(t) = g(t) + \int_{t_0}^{t} k(t, s)\phi(s)ds \quad (t \in I).$$

Proof Let $I = [a, b]$, $u \in C(I)$ and define ψ by

$$\psi(t) = g(t) + \int_{t_0}^{t} k(t, s)u(s)ds \quad (t \in I).$$

We claim that $\psi \in C(I)$. To prove this, first note that for fixed $t \in I$, the map $s \longmapsto k(t, s)u(s)$ belongs to $C(I) \subset \mathscr{R}(I)$. Now let $t_1 \in I$ and $\varepsilon > 0$. For each $t \in I$,

$$|\psi(t) - \psi(t_1)| \leq |g(t) - g(t_1)| + \left| \int_{t_0}^{t} k(t, s)u(s)ds - \int_{t_0}^{t_1} k(t_1, s)u(s)ds \right|$$

$$\leq |g(t) - g(t_1)| + \left| \int_{t_1}^{t} k(t, s)u(s)ds \right|$$

$$+ \left| \int_{t_0}^{t_1} \{k(t, s) - k(t_1, s)\} u(s)ds \right|.$$

Let

$$m = \sup\{|u(s)| : s \in I\}, \quad M = \sup\{|k(t, s)| : (t, s) \in D\}.$$

In view of Theorem 2.3.16 and Corollary 2.3.25, both m and M are finite. The continuity of g at t_1 and the uniform continuity of k on the compact set D (see Theorem 2.3.30) imply that there exists $\delta > 0$ such that

$$|g(t) - g(t_1)| < \varepsilon/3, \ Mm |t - t_1| < \varepsilon/3 \text{ and } m |k(t, s) - k(t_1, s)| (b - a) < \varepsilon/3$$

if $s, t \in I = [a, b]$ and $|t - t_1| < \delta$. It follows that

$$|\psi(t) - \psi(t_1)| < \varepsilon \text{ if } t \in I \text{ and } |t - t_1| < \delta,$$

and so ψ is continuous on I.

Next, define $T : C(I) \to C(I)$ by

$$Tu(t) = g(t) + \int_{t_0}^{t} k(t, s)u(s)ds \ (t \in I, \ u \in C(I)).$$

Since $C(I)$ is a complete metric space when equipped with the uniform metric d_∞, we claim that for some $k \in \mathbf{N}$, T^k is a contraction mapping and so propose to use Corollary 2.2.14 to show that T has a unique fixed point. For each $n \in \mathbf{N}_0$, let $P(n)$ be the proposition

$$\left| T^n u(t) - T^n v(t) \right| \le (M \left| t - t_0 \right|)^n \, d_\infty(u, v)/n! \text{ for all } u, v \in C(I) \text{ and all } t \in I.$$

Evidently $P(0)$ is true; and if $P(n)$ is true for some $n \in \mathbf{N}_0$, then

$$\left| T^{n+1} u(t) - T^{n+1} v(t) \right| = \left| \int_{t_0}^{t} k(t, s) \left\{ T^n u(s) - T^n v(s) \right\} ds \right|$$

$$\le M \left| \int_{t_0}^{t} (M \left| s - t_0 \right|)^n / n! ds \right| d_\infty(u, v)$$

$$\le (M \left| t - t_0 \right|)^{n+1} d_\infty(u, v)/(n + 1)!$$

for all $u, v \in C(I)$ and all $t \in I$, so that $P(n + 1)$ is true. Hence $P(n)$ is true for all $n \in \mathbf{N}$. Thus

$$d_\infty(T^n u, T^n v) \le \frac{(M(b - a))^n}{n!} d_\infty(u, v)$$

for all $u, v \in C(I)$ and all $n \in \mathbf{N}$. Choose $k \in \mathbf{N}$ so large that $(M(b - a))^k / k! < 1$; T^k is a contraction. Hence by Corollary 2.2.14, T has a unique fixed point, ϕ say, and

$$\phi(t) = T\phi(t) = g(t) + \int_{t_0}^{t} k(t, s)\phi(s)ds (t \in I).$$

The proof is complete.   □

As an immediate consequence of this theorem we have

Corollary 2.3.35 *Let I be closed and bounded. Then the initial-value problem* (2.3.7) *has a unique solution.*

Next we show how the Arzelà-Ascoli theorem may be used to prove a famous theorem, due to Peano , which establishes the existence of a solution of the initial-value problem for a non-linear differential equation.

Theorem 2.3.36 *Let* $t_0, x_0 \in \mathbf{R}$ *and* $a, b > 0$, *put* $I = [t_0, t_0+a], J = [x_0-b, x_0+b]$ *and suppose that* $f : I \times J \to \mathbf{R}$ *is continuous, with*

$$M = \max_{(t,x) \in I \times J} |f(t, x)| > 0;$$

put $c = \min(a, b/M)$. *Then there is a function* $x \in C^1([t_0, t_0 + c])$ *such that*

$$\dot{x}(t) = f(t, x(t)) \text{ for } t \in [t_0, t_0 + c], x(t_0) = x_0. \tag{2.3.9}$$

Proof Plainly x is a solution of (2.3.9) if, and only if,

$$x(t) = x_0 + \int_{t_0}^{t} f(s, x(s))ds, \ t \in [t_0, t_0 + c]. \tag{2.3.10}$$

For simplicity of exposition, suppose that $t_0 = 0$; the general case is handled similarly. Put $I_1 = [0, c]$ and for each $n \in \mathbf{N}$ define $x_n : I_1 \to \mathbf{R}$ by

$$x_n(t) = \begin{cases} x_0, & 0 \leq t \leq c/n, \\ x_0 + \int_0^{t-c/n} f(s, x_n(s))ds, & c/n < t \leq c. \end{cases}$$

The function x_n is well-defined: it is given by

$$x_n(t) = x_{j,n}(t) \text{ for } jc/n \leq t \leq (j+1)c/n \text{ and } j = 0, 1, \ldots, n-1,$$

where

$$x_{0,n}(t) = x_0 \ (0 \leq t \leq c/n),$$

$$x_{1,n}(t) = x_0 + \int_0^{t-c/n} f(s, x_0)ds \ (c/n < t \leq 2c/n)$$

and, for $j = 2, \ldots, n-1$ and $jc/n < t \leq (j+1)c/n$,

$$x_{j,n}(t) = x_0 + \sum_{k=1}^{j-1} \int_{(k-1)c/n}^{kc/n} f(s, x_{k-1,n}(s))ds$$

$$+ \int_{(j-1)c/n}^{t-c/n} f(s, x_{j-1,n}(s))ds.$$

It is clear that $x_n \in C(I_1)$. Moreover, for all $t \in I_1$ and all $n \in \mathbf{N}$,

$$|x_n(t) - x_0| \leq cM \leq b \text{ and } |x_n(t)| \leq |x_0| + b.$$

Hence the sequence (x_n) is uniformly bounded. In fact, it is equicontinuous, for given any $n \in \mathbf{N}$ and any $t_1, t_2 \in I_1$,

$$|x_n(t_1) - x_n(t_2)| \leq \left| \int_{t_1 - c/n}^{t_2 - c/n} f(s, x_n(s))ds \right| \leq M |t_2 - t_1|.$$

Hence by the Arzelà-Ascoli theorem (Theorem 2.3.22), there is a subsequence $(x_{k(n)})$ of (x_n) which is uniformly convergent on I_1, to x say. For all $t \in I_1$, as $k(n) \to \infty$,

$$\left| x_{k(n)}(t) - x_0 - \int_0^t f(s, x_{k(n)}(s))ds \right| = \left| \int_{t-c/k(n)}^t f(s, x_{k(n)}(s))ds \right|$$

$$\leq Mc/k(n) \to 0.$$

Since f is uniformly continuous on the compact set $I_1 \times J$, $f(s, x_{k(n)}(s))$ converges uniformly on $I_1 \times J$ to $f(s, x(s))$, and

$$\int_0^t f(s, x_{k(n)}(s))ds \to \int_0^t f(s, x(s))ds$$

as $k(n) \to \infty$. Thus

$$x(t) = x_0 + \int_0^t f(s, x(s))ds, \ t \in I_1,$$

and the proof is complete. $\square$

Note that there may well be more than one solution of the initial-value problem (2.3.9). For example, the initial-value problem

$$\dot{x}(t) = |x(t)|^{1/2} \ \text{for } t \in [0, 1], \ x(0) = 0,$$

has, apart from the zero function, a whole family of solutions given by

$$x(t) = \begin{cases} 0, & 0 \leq t \leq c, \\ (t-c)^2/4, & c < t \leq 1, \end{cases}$$

for any $c \in (0, 1)$. Sufficient conditions on the function f for uniqueness to be restored are given in Exercise 2.3.38/14 below.

2.3.2 Application 2

Here we revisit the Riemann integral and give a celebrated criterion for functions to be Riemann-integrable. To do this, we need the concept of a null set. A subset E of $\mathbf{R}$ is said to be a **null set** if, given any $\varepsilon > 0$, there is a sequence (I_n) of intervals I_n of length $l(I_n)$ such that $E \subset \cup_n I_n$ and $\sum_{n=1}^\infty l(I_n) < \varepsilon$. It is clear that every finite set is a null set, as is every subset of a null set. Somewhat less obviously, if (E_n) is a sequence of null sets, then $\cup_{n=1}^\infty E_n$ is a null set. To establish this, let $\varepsilon > 0$ and note that given any $n \in \mathbf{N}$, there is a sequence $(I_m^{(n)})$ of intervals such that $E_n \subset \cup_{m=1}^\infty I_m^{(n)}$, $\sum_{m=1}^\infty l(I_m^{(n)}) < \varepsilon/2^n$. The sequence $(I_m^{(n)})_{m,n \in \mathbf{N}}$ is countable and so may be arranged as a sequence $(J_k)_{k \in \mathbf{N}}$, with $\cup_{n=1}^\infty E_n \subset \cup_{k=1}^\infty J_k$, and

$$\sum_{k=1}^{\infty} l(J_k) \le \sum_{n=1}^{\infty} \varepsilon/2^n = \varepsilon.$$

This justifies our claim.

The criterion mentioned above is as follows.

Theorem 2.3.37 *Let* $a, b \in \mathbf{R}$*, with* $a < b$*, let* $f \in \mathscr{B}[a, b]$ *and set*

$$D_f = \{x \in [a, b] : f \text{ is not continuous at } x\}.$$

Then $f \in \mathscr{R}[a, b]$ *if, and only if,* D_f *is a null set.*

Proof We may clearly suppose that f is not the zero function. Let $M = \sup \{|f(x)| : x \in [a, b]\}$ and for each $n \in \mathbf{N}$ put

$E_n = \{x \in [a, b] :$ for all $\delta > 0$ there exist $s, t \in (x - \delta, x + \delta) \cap [a, b]$ such that $|f(s) - f(t)| > 1/n\}.$

Plainly f is not continuous at x if $x \in E_n$ for some n. On the other hand, if $x \in D_f$, then there is a sequence (x_k) in $[a, b]$, with $x_k \to x$, such that $f(x_k) \nrightarrow f(x)$. This implies that there exist $n \in \mathbf{N}$ and a subsequence of (x_k), still denoted by (x_k) for convenience, such that $|f(x_k) - f(x)| > 1/n$ for all $k \in \mathbf{N}$. Thus $x \in E_n$. It follows that

$$D_f = \bigcup_{n \in \mathbf{N}} E_n.$$

We claim that each E_n is compact. Since E_n is obviously bounded, it is sufficient to prove that it is closed. To do this, let $x \in \overline{E_n}$. Given $\delta > 0$, there exists $y \in E_n$ with $|x - y| < \delta/2$; and since $(y - \delta/2, y + \delta/2) \subset (x - \delta, x + \delta)$ and there are $s, t \in (y - \delta/2, y + \delta/2)$ with $|f(s) - f(t)| > 1/n$, it follows that $x \in E_n$, which establishes our claim.

Now suppose that $f \in \mathscr{R}[a, b]$. By Exercise 1.1.10/7, given $n \in \mathbf{N}$ and $\varepsilon > 0$, there is a partition $P = \{a = x_0, x_1, \ldots, x_m = b\} \in \mathscr{P}[a, b]$ such that

$$\left| \sum_{r=1}^{m} \{f(\xi_r) - f(\eta_r)\}(x_r - x_{r-1}) \right| < \varepsilon/n$$

whenever $\xi_r, \eta_r \in [x_{r-1}, x_r]$, for $r \in \{1, \ldots, m\}$. For each $r \in \{1, \ldots, m\}$ we may plainly choose ξ_r, η_r so that $f(\xi_r) \ge f(\eta_r)$; moreover, if $(x_{r-1}, x_r) \cap E_n \ne \emptyset$, we may ensure that $f(\xi_r) > f(\eta_r) + 1/n$. It now follows that the sum of the lengths of those intervals (x_{r-1}, x_r) with non-empty intersection with E_n is less than ε. Hence, since the length of degenerate intervals is zero, E_n is a null set; and as D_f is the countable union of the E_n, it also is a null set.

For the converse, suppose that D_f is a null set. Let $\varepsilon > 0$ and choose $n \in \mathbf{N}$ so that $n > 1/\varepsilon$. Since E_n is obviously null, there is a sequence (I_r) of open subintervals of the metric space $[a, b]$ which covers E_n, with $\sum_{r=1}^{\infty} l(I_r) < \varepsilon$. As E_n is compact, it is covered by a finite number of these intervals, say $J_1, \ldots, J_p$; and of course $\sum_{r=1}^{p} l(J_r) < \varepsilon$. An inductive argument shows that the set

$$[a, b] \setminus \bigcup_{r=1}^{p} J_r$$

consists of a finite collection of closed intervals, say $K_1, \ldots, K_q$; for each $j \in \{1, \ldots, q\}$, there exists $P_j \in \mathscr{P}(K_j)$ such that $|f(x) - f(y)| \le 1/n$ for all x, y in the same subinterval of P_j. Finally, let $P \in \mathscr{P}[a, b]$ consist of the points of $\bigcup_{j=1}^{q} P_j$ together with the endpoints of the intervals $J_1, \ldots, J_p$. Then, using Exercise 1.1.10/2, we see that the contribution to $U(P, f) - L(P, f)$ from the points of $P_1, \ldots, P_q$ can be estimated from above by

$$\frac{1}{n}(b - a) < \varepsilon(b - a).$$

The rest of $U(P, f) - L(P, f)$ arises from the endpoints of the J_r and may be estimated from above by

$$2M \sum_{r=1}^{p} l(J_r) < 2M\varepsilon.$$

Hence

$$U(P, f) - L(P, f) < \varepsilon(2M + b - a),$$

and so $f \in \mathscr{R}[a, b]$.   □

Note that this theorem gives an immediate proof of the fact, established earlier, that Riemann-integrability is preserved by taking sums and products.

Exercise 2.3.38

1. Let $I = [0, 1]$. Exhibit a subset of the metric space $C(I)$, endowed with the uniform metric, that is unbounded. Show that the mapping $f \longmapsto \int_0^1 f$ of $C(I)$ to $\mathbf{R}$ is uniformly continuous on $C(I)$.
2. Let (X, d) be a compact metric space and let $(F_i)_{i \in I}$ be a family of non-empty closed subsets of X with empty intersection. Prove that there is a positive number c such that for each $x \in X$, $d(x, F_i) \ge c$ for some $i \in I$.
3. Let (X, d) be a compact metric space such that for all $x, y, z \in X$, $d(x, y) \le \max\{d(x, z), d(y, z)\}$, and let $x_0 \in X$; let $x \in X$ be such that $d(x_0, x) = r > 0$. By assuming the contrary show that

   $$\sup\{d(x_0, y) : y \in B(x_0, r)\} < r \text{ and } \inf\{d(x_0, y) : y \in X, d(x_0, y) > r\} > r.$$

   Hence prove that $\{d(x_0, z) : z \in X\}$ is finite or countably infinite.
4. Let (X, d) be a compact metric space, let $T : X \to X$ be such that for all $x, y \in X$, $d(x, y) \le d(T(x), T(y))$, and let a, b be any points of X. By considering appropriate subsequences of $(T^n(a))$ and $(T^n(b))$, show that given any $\varepsilon > 0$, there is an integer k such that $d(a, T^k(a)) < \varepsilon$ and $d(b, T^k(b)) < \varepsilon$. Deduce that

$d(T(a), T(b)) = d(a, b)$ and that $T(X)$ is dense in X. Hence show that T maps X isometrically onto itself.

5. Let (X, d) be a compact metric space and suppose that $T : X \to X$ is such that $d(T(x), T(y)) < d(x, y)$ for all $x, y \in X$ with $x \neq y$. Prove that T has a unique fixed point.

6. (Dini's theorem: see also Exercise 1.7.17/18) Let (X, d) be a compact metric space and let (f_n) be a monotone sequence in $C(X)$ which is pointwise convergent to $f \in C(X)$. Prove that $f_n \to f$ in the uniform metric on $C(X)$.

7. Let $\alpha \in (0, 1]$. A real-valued function f on $[0, 1]$ is said to be Hölder-continuous with exponent α if there is a constant C such that for all $x, y \in [0, 1]$, $|f(x) - f(y)| \leq C |x - y|^\alpha$. Define

$$\|f\|_\alpha = \max_{x \in [0,1]} |f(x)| + \sup \frac{|f(x) - f(y)|}{|x - y|^\alpha},$$

where the supremum is taken over all $x, y \in [0, 1]$ with $x \neq y$. Prove that the set of all functions f with $\|f\|_\alpha \leq 1$ is a compact subset of $C[0, 1]$.

8. Let $\mathcal{K} = \{f \in C[0, 1] : d_\infty(f, 0) \leq 1\}$. Show that $\mathcal{K}$ is not compact in $C[0, 1]$.

9. Let (X, d) be a compact metric space and let (f_n) be a sequence in $C(X)$. Prove that if the set $\{f_n : n \in \mathbf{N}\}$ is equicontinuous, and for each $x \in X$ the sequence $(f_n(x))$ converges, then (f_n) is convergent in $C(X)$.

10. Let $f_n(t) = \sin \sqrt{t + 4n^2\pi^2}$ for $t \geq 0$, $n \in \mathbf{N}$. Prove that $\{f_n : n \in \mathbf{N}\}$ is a bounded and uniformly equicontinuous subset of $\mathscr{C}[0, \infty)$, but that it is not relatively compact. Prove also that the sequence (f_n) converges pointwise to 0 on $[0, \infty)$. [This shows that the Arzelà-Ascoli theorem and Exercise 9 may fail when X is not compact.]

11. Let $\mathcal{K} \subset C[0, 1]$. Suppose that each $f \in \mathcal{K}$ is differentiable on $(0, 1)$ and that there exists $M > 0$ such that $|f'(t)| \leq M$ for all $t \in (0, 1)$ and all $f \in \mathcal{K}$. Prove that $\mathcal{K}$ is equicontinuous.

12. Let X be a metric space, $x \in X$, and f be a real-valued function on X. Prove that f is lower semi-continuous at $x \in X$ if, and only if,

$$f(x) \leq \liminf_{n \to \infty} f(x_n) \text{ whenever } x_n \to x.$$

13. Let (X, d) be a metric space and let $f : X \to \mathbf{R}$ be bounded and lower semi-continuous. For each $n \in \mathbf{N}$, let $g_n : X \to \mathbf{R}$ be defined by

$$g_n(x) = \inf_{y \in X} \{f(y) + nd(x, y)\} \ (x \in X).$$

(i) Prove that (g_n) is an increasing sequence of continuous functions that converges pointwise to f.

(ii) Show that the set of points of continuity of f is residual in X and deduce that, if X is complete, then this set is dense in X.

14. Let $a, b, c \in \mathbf{R}$ with $a < b$ and $c \geq 0$, let u, v be non-negative continuous functions on $[a, b]$ and suppose that

$$v(t) \leq c + \int_a^t v(s)u(s)ds \text{ for } a \leq t \leq b.$$

Establish Gronwall's inequality:

$$v(t) \leq c \exp \left(\int_a^t u(s)ds \right) \text{ for } a \leq t \leq b,$$

so that if $c = 0$, then v is the zero function. Deduce that the initial-value problem (2.3.9) has a unique solution if the function f is Lipschitz-continuous in the sense that there is a constant K such that

$$|f(t, w_1) - f(t, w_2)| \leq K |w_1 - w_2| \text{ for all } t \in [t_0, t_0 + c] \text{ and all } w_1, w_2 \in J.$$

15. Let $\mathscr{U}$ be an open covering of a compact metric space X. Show that there is a positive number ε (called a Lebesgue number of $\mathscr{U}$) such that if $A \subset X$ and diam $A < \varepsilon$, then there exists $U \in \mathscr{U}$ that contains A.

16. Let (X, d) be a complete metric space and let $\mathscr{K}$ be the family of all non-empty compact subsets of X. The Hausdorff metric δ on $\mathscr{K}$ is defined by

$$\delta(A, B) = \max\{\sup_{a \in A} d(a, B), \sup_{b \in B} d(b, A)\} \, (A, B \in \mathscr{K}),$$

in the notation of Lemma 2.1.40. Show that

$$\delta(A, B) = \inf\{r > 0 : A \subset V_r(B), B \subset V_r(A)\},$$

where $V_r(A) = \{x \in X : d(x, A) < r\}$. Prove that δ is a metric on $\mathscr{K}$ and that $(\mathscr{K}, \delta)$ is complete. Show further that if X is compact, then so is $(\mathscr{K}, \delta)$. Prove that if for each $i \in \{1, \ldots, n\}$, A_i and B_i belong to $\mathscr{K}$, then

$$\delta \left(\cup_{i=1}^n A_i, \cup_{i=1}^n B_i \right) \leq \max_{1 \leq i \leq n} \delta(A_i, B_i).$$

Let $F : X \to X$ be a contraction; that is, there exists $r \in (0, 1)$ such that for all $x, y \in X$, $d(F(x), F(y)) \leq rd(x, y)$. Prove that for all $A, B \in \mathscr{K}$,

$$\delta(F(A), F(B)) \leq r\delta(A, B).$$

Now suppose that for each $i \in \{1, \ldots, n\}$, $F_i : X \to X$ is a contraction. Define $\mathscr{F} : \mathscr{K} \to \mathscr{K}$ by $\mathscr{F}(A) = \cup_{i=1}^n F_i(A)$ $(A \in \mathscr{K})$, show that $\mathscr{F}$ is a contraction on $(\mathscr{K}, \delta)$ and hence prove that there is a unique $K \in \mathscr{K}$ such that

$$K = \cup_{i=1}^{n} F_i(K).$$

By taking $X = [0, 1]$ (with the metric inherited from $\mathbf{R}$), $n = 2$, $F_1(x) = x/3$ and $F_2(x) = (2 + x)/3$ ($x \in [0, 1]$), deduce that $\lim_{n \to \infty} \mathscr{F}^n([0, 1])$ exists in $(\mathscr{K}, \delta)$ and so defines a compact non-empty subset of $[0, 1]$. This is the Cantor set.

2.4 Connectedness

In this section we isolate those metric spaces with the following property: if a map $f : X \to \mathbf{R}$ is continuous, then its range, $f(X)$, is an interval. The motivation for this stems from the well-known intermediate-value theorem.

We begin with a characterisation of those subsets of $\mathbf{R}$ which are intervals.

Lemma 2.4.1 *A subset S of $\mathbf{R}$ is an interval if, and only if, it has the following* **intermediate-value property** *(abbreviated as ivp):*

$$\text{if } x, y \in S \text{ and } x < z < y, \text{ then } z \in S.$$

Proof If S has at most one element, it is a degenerate interval and the result holds by default.

Suppose that S has at least two elements. If it is an interval then it clearly has the ivp. To establish the converse we distinguish four cases:

(i) $\inf S = a > -\infty$, $\sup S = b < \infty$. Evidently $S \subset [a, b]$; we claim that $(a, b) \subset S$. For suppose that $x \in (a, b)$. Then there exist $c, d \in S$ such that $a \le c < x < d \le b$ and hence, by the ivp, $x \in S$. Thus $(a, b) \subset S \subset [a, b]$ and S is an interval.

(ii) $\inf S = a > -\infty$, $\sup S = \infty$. Here $S \subset [a, \infty)$. If $x \in (a, \infty)$, then there are $c, d \in S$ such that $a \le c < x < d < \infty$ and, as before, $x \in S$. Thus $(a, \infty) \subset S \subset [a, \infty)$ and S is an interval.

(iii) $\inf S = -\infty$, $\sup S = b < \infty$.

(iv) $\inf S = -\infty$, $\sup S = \infty$.

We omit the proofs in cases (iii) and (iv) as they are similar to that of case (ii).□

We can now give equivalent forms of the property with which we began this section.

Theorem 2.4.2 *Let X be a metric space. The following three statements are equivalent:*

(i) *The only subsets of X which are both open and closed are $\emptyset$ and X.*

(ii) *There do not exist two non-empty disjoint open subsets of X whose union is X.*

(iii) *The range of each continuous map $f : X \to \mathbf{R}$ is an interval.*

Proof Suppose that (i) holds and that (ii) does not. Then there are non-empty open subsets U, V of X such that $U \cap V = \emptyset$ and $U \cup V = X$. This implies that $U = {}^c V$ and so U is closed. Thus $\emptyset \neq U \neq X$ and U is both open and closed, contradicting (i).

Now suppose that (ii) holds and (iii) does not. Then there is a continuous map $f : X \to \mathbf{R}$ such that $f(X)$ is not an interval. Hence, in view of Lemma 2.4.1, there exist $x, y \in X$ and $\lambda \in \mathbf{R}$ such that $f(x) < \lambda < f(y)$ and, for all $z \in X, f(z) \neq \lambda$. Let $U = f^{-1}((-\infty, \lambda))$ and $V = f^{-1}((\lambda, \infty))$. These sets are non-empty, disjoint, open and their union is X, contradicting (ii).

Finally, suppose (iii) holds and (i) does not. Then there is a set U which is both open and closed in X and $\emptyset \neq U \neq X$. Define $f : X \to \mathbf{R}$ by $f(x) = 1$ if $x \in U$, $f(x) = 0$ otherwise. Since $f^{-1}(W) \in \{\emptyset, U, {}^c U, X\}$ if $W \subset \mathbf{R}$, it follows that $f^{-1}(W)$ is open in X whenever W is open in $\mathbf{R}$. Hence f is continuous, but its range is not an interval and (iii) is contradicted.   □

This leads us to formulate the following definition.

Definition 2.4.3 A metric space X is said to be **connected** if it is not expressible as a union of two non-empty, disjoint open subsets of itself; it is said to be **disconnected** if it is not connected.

Of course, any of the equivalences of Theorem 2.4.2 could have been used to define a connected space. There is some loss of motivation in not choosing (iii), but the compensation is that we have an intrinsic and functional definition.

We now turn to subsets of a metric space.

Definition 2.4.4 A subset of a metric space X is said to be a **connected set** in X if it is either empty or it is connected as a subspace of X; it is said to be a **disconnected set** in X if it is not a connected set in X.

Let E be a subspace of a metric space X. By definition, E is a disconnected space if, and only if, there are non-empty sets O_1 and O_2, each open in E, such that $O_1 \cap O_2 = \emptyset$ and $O_1 \cup O_2 = E$. If $\mathscr{U}$ denotes the family of all the sets open in X, then $\{U \cap E : U \in \mathscr{U}\}$ is the family of all the sets open in the metric space E. It follows that E is a disconnected space if, and only if, there are sets U and V, each open in X, such that

$$U \cap E \neq \emptyset, \ V \cap E \neq \emptyset$$

and

$$(U \cap E) \cap (V \cap E) = \emptyset, \ (U \cap E) \cup (V \cap E) = E.$$

With the observation that
(a) $(U \cap E) \cap (V \cap E) = \emptyset$ if, and only if, $U \cap V \cap E = \emptyset$,
and
(b) $(U \cap E) \cup (V \cap E) = E$ if, and only if, $E \subset U \cup V$,
this means that we have established the following theorem.

Theorem 2.4.5 *Let E be a subset of a metric space X. Then E is a disconnected set in X if, and only if, there are sets U and V, each open in X, such that*

$$U \cap E \neq \emptyset, \ V \cap E \neq \emptyset,$$

$$U \cap V \cap E = \emptyset, \ E \subset U \cup V.$$

In practice, given a set E in a metric space X, Theorem 2.4.5 provides a basic test for its disconnectedness. In the event that the set E is known to be disconnected, a condition stronger in form than the test-condition of Theorem 2.4.5 holds. This appears next.

Theorem 2.4.6 *Let E be a subset of a metric space X. Then E is a disconnected set in X if, and only if, there are **disjoint** open sets U and V in X such that $U \cap E \neq \emptyset$, $V \cap E \neq \emptyset$ and $E \subset U \cup V$.*

Proof Let E be disconnected in X. Then there are sets U_1 and V_1, each open in X, such that $E \cap U_1 \neq \emptyset, E \cap V_1 \neq \emptyset, E \cap U_1 \cap V_1 = \emptyset$ and $E \subset U_1 \cup V_1$. Moreover, given any $u \in E \cap U_1$, there exists $r(u) > 0$ such that $B(u, r(u)) \subset U_1$; also, given any $v \in E \cap V_1$, there exists $r(v) > 0$ such that $B(v, r(v)) \subset V_1$. Put

$$U = \bigcup_{u \in E \cap U_1} B(u, r(u)/2), \ V = \bigcup_{v \in E \cap V_1} B(v, r(v)/2).$$

It is clear that U and V are open, that $E \cap U = E \cap U_1 \neq \emptyset$ and $E \cap V = E \cap V_1 \neq \emptyset$, and that $E = (E \cap U_1) \cup (E \cap V_1) \subset U \cup V$. It remains to prove that $U \cap V = \emptyset$. To obtain a contradiction, suppose that $U \cap V \neq \emptyset$. Let $w \in U \cap V$. Then there are points $u \in E \cap U_1, v \in E \cap V_1$ such that $d(u, w) < \frac{1}{2}r(u), d(v, w) < \frac{1}{2}r(v)$, where d is the metric on X. Thus

$$d(u, v) \leq d(u, w) + d(w, v) \leq \frac{1}{2}\{r(u) + r(v)\} \leq \max \ \{r(u), r(v)\}.$$

It follows that either $v \in U_1$ or $u \in V_1$. Whichever is the case, $U_1 \cap V_1 \cap E \neq \emptyset$, and we have a contradiction.

The converse is obvious. $\square$

Example 2.4.7

(i) In every metric space (X, d) any set containing only one point is obviously connected; any finite set with at least two points is disconnected. Thus if $S = \{a, b\} \subset X$ and $a \neq b$, for example, we may take $U = B(a, r), V = B(b, r)$, where $r = \frac{1}{2}d(a, b)$, and note that U and V are open, $U \cap V = \emptyset, U \cap S \neq \emptyset$, $V \cap S \neq \emptyset$ and $S \subset U \cup V$.

(ii) In any discrete metric space every subset with more than one point is disconnected, as every subset of the space is both open and closed.

(iii) Let X be a metric space, let A, B be non-empty, disjoint, closed sets in X and let
$E = A \cup B$. Then E is disconnected. To see this, put $U = {}^cA$, $V = {}^cB$ so that
U and V are open in X. Then $U \cap E = B \neq \emptyset$, $V \cap E = A \neq \emptyset$, $U \cap V \cap E =$
${}^cA \cap {}^cB \cap (A \cup B) = (A \cup B)^c \cap (A \cup B) = \emptyset$ and $E = A \cup B \subset {}^cB \cup {}^cA = U \cup V$.
To illustrate this, let $X = \mathbf{R}^2$, $A = \{(x, y) \in \mathbf{R}^2 : x \geq 0, xy = 1\}$, $B = \{(x, y) \in$
$\mathbf{R}^2 : y = 0\}$. Then $A \cup B$ is disconnected in $\mathbf{R}^2$.

(iv) Let X be a metric space and let A, B be non-empty, open, disjoint sets in X with
union X. Then if C is a connected subset of X, either $C \subset A$ or $C \subset B$. For
otherwise $C \cap A \neq \emptyset$, $C \cap B \neq \emptyset$, $C \cap A \cap B = \emptyset$ and $C \subset A \cup B$, and the
connectedness of C is contradicted.

(v) A metric space X is connected if, and only if, given any $x, y \in X$, there is a
connected subset A of X such that $x, y \in A$. To prove this, suppose first that
given any $x, y \in X$, there is a connected subset A of X such that $x, y \in A$. If X
were not connected, there would be disjoint, open, non-empty sets U, V with
union X. By (iv), either $A \subset U$ or $A \subset V$, and we have a contradiction. The
converse is obvious.

The connected subsets of $\mathbf{R}$, equipped with the usual metric, can be classified
completely.

Theorem 2.4.8 *Let $S \subset \mathbf{R}$. The following three statements are equivalent.*

(i) *S is connected.*
(ii) *S has the intermediate-value property.*
(iii) *S is an interval.*

Proof Suppose that S is connected yet fails to have the intermediate-value property.
Then there are real numbers x, y and z with $x, y \in S$, $x < z < y$ and $z \notin S$. Put
$U = \{t \in \mathbf{R} : t < z\}$ and $V = \{t \in \mathbf{R} : t > z\}$. Then U and V are open, $S \cap U \neq \emptyset$,
$S \cap V \neq \emptyset$, $U \cap V = \emptyset$ and $S \subset U \cup V$. Thus S is disconnected and we have a
contradiction. Hence (i) implies (ii).
   Conversely, suppose that S has the intermediate-value property and is disconnected. Then there are disjoint open sets U, V in $\mathbf{R}$ and points $x, y \in S$ with $x < y$ such
that $x \in S \cap U$, $y \in S \cap V$ and $S \subset U \cup V$. Let $z := \sup\{U \cap [x, y]\}$. Plainly $z \in \overline{U}$
and, as $\overline{U}$ is contained in the closed set $\mathbf{R} \setminus V$, $z \notin V$. Since $z \in [x, y] \subset S \subset U \cup V$,
it follows that $z \in U$. Since U is open and $z \neq y$, there exists $z_1 > z$ such that
$[z, z_1] \subset U \cap [x, y]$. But this contradicts the definition of z. Hence (ii) implies (i).
   The rest of the proof follows from Lemma 2.4.1.   □

Corollary 2.4.9 *Let $S \subset \mathbf{R}$, $S \neq \emptyset$. Then S is an interval if, and only if, $f(S)$ has
the ivp whenever $f : S \to \mathbf{R}$ is continuous. [The 'only if' part of this result is called
the **intermediate-value** theorem.]*

Proof By Theorem 2.4.8, S is an interval if, and only if, S is connected; by Theorem 2.4.2, this is so if, and only if, $f(S)$ is an interval whenever $f : S \to \mathbf{R}$ is
continuous; and now the result follows from Lemma 2.4.1.   □

Corollary 2.4.10 *Let $a, b \in \mathbf{R}$, with $a < b$, and let $f : [a, b] \to [a, b]$ be continuous. Then f has a fixed point; that is, there exists $c \in [a, b]$ such that $f(c) = c$.*

Proof If $f(a) = a$ or $f(b) = b$ there is nothing to prove. We shall therefore assume that $f(a) > a$ and $f(b) < b$. Define $g : [a, b] \to \mathbf{R}$ by $g(x) = x - f(x)$, $x \in [a, b]$. Then g is continuous, $g(a) < 0$ and $g(b) > 0$. By Corollary 2.4.9, there exists $c \in [a, b]$ such that $g(c) = 0$; that is, $f(c) = c$. $\qquad\Box$

This elementary fixed-point result may be extended to higher dimensions with considerably greater effort: see Chap. 3 for the two-dimensional version.

Under a continuous map connectedness is preserved. Amongst other uses this fact allows new connected sets to be generated from old.

Theorem 2.4.11 *Let X and Y be metric spaces and let $f : X \to Y$ be continuous. Then $f(E)$ is a connected subset of Y whenever E is a connected subset of X.*

Proof Suppose that E is connected and yet $f(E)$ is not. Then there are disjoint open sets U, V in Y such that $U \cap f(E) \neq \emptyset$, $V \cap f(E) \neq \emptyset$ and $f(E) \subset U \cup V$. It follows that $f^{-1}(U) \cap E \neq \emptyset$, $f^{-1}(V) \cap E \neq \emptyset$, $E \subset f^{-1}(U) \cup f^{-1}(V)$ and, since $U \cap V = \emptyset$, $f^{-1}(U) \cap f^{-1}(V) = \emptyset$. As f is continuous, $f^{-1}(U)$ and $f^{-1}(V)$ are also open in X. Thus E is disconnected and we have a contradiction. $\qquad\Box$

Corollary 2.4.12 *Let $S = \{(x, y) \in \mathbf{R}^2 : x^2 + y^2 = r^2\}$, where $r > 0$. Let $f : S \to \mathbf{R}$ be continuous (S inherits the Euclidean metric from $\mathbf{R}^2$). Then there exists $\mathbf{u} = (u, v) \in S$ such that $f(\mathbf{u}) = f(-\mathbf{u})$.*

Proof Note that S is connected: it is the image of the interval $[0, 2\pi]$ under the continuous map $t \longmapsto (r \cos t, r \sin t)$.

Let $g : S \to \mathbf{R}$ be defined by

$$g(\mathbf{p}) = f(\mathbf{p}) - f(-\mathbf{p}).$$

Then g is continuous: if $\mathbf{p}_n \in S$ ($n \in \mathbf{N}$) and $\mathbf{p}_n \to \mathbf{p} \in S$, then $f(\mathbf{p}_n) \to f(\mathbf{p})$ and $f(-\mathbf{p}_n) \to f(-\mathbf{p})$, so that $g(\mathbf{p}_n) \to g(\mathbf{p})$. Since S is connected, it follows from Theorem 2.4.2 that $g(S)$ is an interval. This interval is symmetric about the origin: if $\theta \in g(S)$, then $g(\mathbf{s}) = \theta$ for some $\mathbf{s} \in S$ and so $-\theta = -g(\mathbf{s}) = g(-\mathbf{s}) \in g(S)$. Hence $0 \in g(S)$ and there exists $\mathbf{u} \in S$ with $g(\mathbf{u}) = 0$; that is, $f(\mathbf{u}) = f(-\mathbf{u})$. $\qquad\Box$

The use of the term connected in the context of metric spaces may seem remote from the everyday sense in which the term is employed. That sense, in which elements are linked or joined, does have a specialised counterpart for which the technical expression is path-connected. Three definitions introduce this.

Definition 2.4.13 Let X be a metric space and let $a, b \in \mathbf{R}$, with $a < b$. A continuous map $\gamma : [a, b] \to X$ is called a **path in X** with **parameter interval** $[a, b]$. The points $\gamma(a), \gamma(b)$ are called the **initial** and **terminal** points, respectively, of γ; γ is said to **join** its initial and terminal points; γ is a **closed** path if $\gamma(a) = \gamma(b)$; γ is a **simple**

path if $\gamma(s) \neq \gamma(t)$ whenever $s, t \in [a, b]$, $s \neq t$ and $\{s, t\} \neq \{a, b\}$. The range $\gamma^* = \gamma([a, b])$ of γ is called the **track** of γ. If $\gamma^* \subset E \subset X$ we refer to γ as a **path in E**.

Without loss of generality, any path may be chosen to have parameter interval $[0, 1]$: make the obvious change of variable $t \mapsto (1 - t)a + tb : [0, 1] \to [a, b]$.

Example 2.4.14 The function $\gamma : [0, 1] \to \mathbf{R}^2$ defined by $\gamma(t) = (\cos \pi t, \sin \pi t)$ is a path in $\mathbf{R}^2$ which joins its initial point $(1, 0)$ to its terminal point $(-1, 0)$ and has track

$$\gamma^* = \{(x, y) \in \mathbf{R}^2 : x^2 + y^2 = 1, y \geq 0\}.$$

Observe that different paths may have the same track: the path $\nu : [0, 1] \to \mathbf{R}^2$ given by $\nu(t) = (\cos \pi t^2, \sin \pi t^2)$ has the same track as γ, though $\nu \neq \gamma$.

Paths in $\mathbf{R}^n$ of a particular character are singled out.

Definition 2.4.15 Given $a, b \in \mathbf{R}$ with $a < b$, a map $\gamma : [a, b] \to \mathbf{R}^n$ is said to be a **polygonal path** if points $x^{(0)}, x^{(1)}, \ldots, x^{(k)} \in \mathbf{R}^n$ and a partition $\{a = t_0, t_1, \ldots, t_k = b\}$ of $[a, b]$ exist such that

$$\gamma(t) = (t_j - t_{j-1})^{-1} \left\{ (t_j - t)x^{(j-1)} + (t - t_{j-1})x^{(j)} \right\}$$

whenever $t_{j-1} \leq t \leq t_j$ and $j \in \{1, 2, \ldots, k\}$; if, in addition, γ is such that for each $j \in \{1, 2, \ldots, k\}$ there is a line passing through $x^{(j-1)}$ and $x^{(j)}$ parallel to a coordinate axis, then it is said to be a **p-path**. In the elementary case of $k = 1$, when

$$\gamma(t) = (b - a)^{-1}\{(b - t)x^{(0)} + (t - a)x^{(1)}\} \ (a \leq t \leq b),$$

the path γ is referred to as a **line segment** and may be denoted by $[x^{(0)}, x^{(1)}]$. This terminology and symbolism is used also for γ^*, the track of γ, and the intended meaning has to be understood by context.

Elementary reasoning shows that a polygonal path is continuous and therefore a path in the sense of Definition 2.4.13. Also, the track of a polygonal path (or p-path) γ is a union of line segments: $\gamma^* = \cup_{j=1}^k [x^{(j-1)}, x^{(j)}]$.

Definition 2.4.16 A subset E of a metric space X is called **path-connected** if, given any $x, y \in E$, there is a path in E with initial point x and terminal point y. If $X = \mathbf{R}^n$, E is said to be **polygonally connected** if, given any $x, y \in E$, there is a polygonal path in E which joins x to y.

Example 2.4.17

(i) Let $a \in \mathbf{R}$. Then $\mathbf{R}\backslash\{a\}$, with the metric inherited from $\mathbf{R}$, is not path-connected. For let $x, y \in \mathbf{R}$, with $x < a < y$, and suppose there is a path $\gamma : [0, 1] \to \mathbf{R}\backslash\{a\}$ joining x to y. By the intermediate-value theorem, $\gamma(t) = a$ for some $t \in [0, 1]$, and we have a contradiction.

(ii) Let $\mathbf{a} \in \mathbf{R}^2$. Then $\mathbf{R}^2\backslash\{\mathbf{a}\}$, with the metric inherited from $\mathbf{R}^2$, is path-connected. For let $\mathbf{x}, \mathbf{y} \in \mathbf{R}^2\backslash\{\mathbf{a}\}$, $\mathbf{x} \neq \mathbf{y}$. Then, if $\mathbf{x}, \mathbf{y}$ and $\mathbf{a}$ are not collinear, the line segment joining $\mathbf{x}$ to $\mathbf{y}$ is a path in $\mathbf{R}^2\backslash\{\mathbf{a}\}$; and if these three points are collinear, $\mathbf{x}$ may be joined to $\mathbf{y}$ by a p−path in $\mathbf{R}^2\backslash\{\mathbf{a}\}$ whose track is a union of at most three line segments, each parallel to one of the coordinate axes. The same argument shows that when $n > 2$, removal of one point from $\mathbf{R}^n$ leaves the set path-connected.

Proposition 2.4.18 *Let X and Y be homeomorphic metric spaces. Then X is path-connected if, and only if, Y is path-connected.*

Proof Let $\phi : X \to Y$ be a homeomorphism, suppose that X is path-connected and let $y_1, y_2 \in Y$. Then $y_1 = \phi(x_1)$, $y_2 = \phi(x_2)$ for some $x_1, x_2 \in X$; let $\gamma : [0, 1] \to X$ be a path joining x_1 to x_2. Then $\phi \circ \gamma$ is a path in Y joining y_1 to y_2, and so Y is path-connected. The result is now clear. □

Corollary 2.4.19 *If $n > 1$, $\mathbf{R}$ and $\mathbf{R}^n$ are not homeomorphic.*

Proof Suppose the result is false. Then for some $n > 1$, there is a homeomorphism $\phi : \mathbf{R} \to \mathbf{R}^n$. Let $a \in \mathbf{R}$: then the restriction of ϕ to $\mathbf{R}\backslash\{a\}$ is a homeomorphism of $\mathbf{R}\backslash\{a\}$ onto $\mathbf{R}^n\backslash\{\phi(a)\}$. But by Example 2.4.17 (i) and (ii), $\mathbf{R}\backslash\{a\}$ is not path-connected while $\mathbf{R}^n\backslash\{\phi(a)\}$ is path-connected. This contradicts Proposition 2.4.18 and completes the proof. □

We remark that it is also true that if $m, n \in \mathbf{N}$ and $m \neq n$, then $\mathbf{R}^m$ and $\mathbf{R}^n$ are not homeomorphic. However, this is much harder to prove.

Next we relate the notions of connectedness and path-connectedness.

Theorem 2.4.20 *Let E be a path-connected subset of a metric space X. Then E is a connected set in X.*

Proof Suppose E is not connected. Then there are disjoint open sets U, V in X such that

$$U \cap E \neq \emptyset, \ V \cap E \neq \emptyset \text{ and } E \subset U \cup V.$$

Let $x \in U \cap E$, $y \in V \cap E$; as E is path-connected, there is a path $\gamma : [0, 1] \to E$ with initial point x and terminal point y. Since γ is continuous, $\gamma^{-1}(U)$ and $\gamma^{-1}(V)$ are open sets in $[0, 1]$; also $\gamma^{-1}(U) \cup \gamma^{-1}(V) = [0, 1]$ and $\gamma^{-1}(U) \cap \gamma^{-1}(V) = \emptyset$. Thus $[0, 1]$ is not connected, contradicting Theorem 2.4.8. □

Example 2.4.21

(i) Every open ball in $\mathbf{R}^n$ ($n \geq 1$) is connected, as is $\mathbf{R}^n$ itself.
    To see this, let $\mathbf{a} \in \mathbf{R}^n$ and $r > 0$. We show that $B(\mathbf{a}, r)$ is path-connected and therefore connected. Let d denote the Euclidean metric on $\mathbf{R}^n$. Let $\mathbf{x}, \mathbf{y} \in B(\mathbf{a}, r)$ and let $\gamma : [0, 1] \to \mathbf{R}^n$ be given by $\gamma(t) = (1 - t)\mathbf{x} + t\mathbf{y}$. We claim that γ is a path in $B(\mathbf{a}, r)$ joining $\mathbf{x}$ to $\mathbf{y}$. Evidently γ is continuous: if $t_n \in [0, 1]$ ($n \in \mathbf{N}$)

and $t_n \to t \in [0, 1]$, then $d(\gamma(t_n), \gamma(t)) = |t_n - t| d(x, y) \to 0$. Moreover, $\gamma^* \subset B(\mathbf{a}, r)$: for all $t \in [0, 1]$,

$$\gamma(t) - \mathbf{a} = (1 - t)(\mathbf{x} - \mathbf{a}) + t(\mathbf{y} - \mathbf{a})$$

and

$$d(\gamma(t), \mathbf{a}) = \left(\sum_{j=1}^{n} \{(1 - t)(x_j - a_j) + t(y_j - a_j)\}^2 \right)^{1/2}$$

$$\leq (1 - t)d(\mathbf{x}, \mathbf{a}) + td(\mathbf{y}, \mathbf{a}) < r.$$

The rest is clear.

(ii) The converse of Theorem 2.4.20 is false: not every connected set is path-connected. To illustrate this, take $X = \mathbf{R}^2$ and

$$E = \{(0, y) : -1 \leq y \leq 1\} \cup \left\{ \left(x, \sin \frac{\pi}{x} \right) : 0 < x \leq 1 \right\} = A \cup B, \text{ say.}$$

The set B is the image of $(0, 1]$ under the continuous map $t \longmapsto \left(t, \sin \frac{\pi}{t} \right)$ and so is connected. We claim that $B \subset A \cup B \subset \bar{B}$: granted this, it follows from Exercise 2.4.33/1 that E is connected. To establish our claim, let $(0, y) \in A$ and let $\varepsilon > 0$; let $n \in \mathbf{N}$ be so large that $1/n < \varepsilon$. Since $\sin \left(2n \pm \frac{1}{2} \right) \pi = \pm 1$, there exists $t \in \left[\frac{2}{4n+1}, \frac{2}{4n-1} \right]$ such that $\sin \frac{\pi}{t} = y$. The point $\left(t, \sin \frac{\pi}{t} \right)$ belongs to B and its distance from $(0, y)$ is less than ε; thus $A \cup B \subset \bar{B}$ and the claim is justified.

However, E is not path-connected. For suppose $\gamma : [0, 1] \to E$ is a path in E with initial and terminal points $(0, 0)$ and $(1, 0)$, respectively; write $\gamma(t) = (\gamma_1(t), \gamma_2(t))$, $t \in [0, 1]$. Then $\gamma^{-1}(A)$ is a closed set contained in $[0, 1]$ and containing 0; thus $b := \sup \gamma^{-1}(A) \in \gamma^{-1}(A)$ and $0 \leq b < 1$. Suppose that $\gamma_2(b) \leq 0$. Then given any $\delta > 0$ with $b + \delta \leq 1$, we have $\gamma_1(b + \delta) > 0$, and there exists $n \in \mathbf{N}$ such that

$$0 = \gamma_1(b) < 2/(4n + 1) < \gamma_1(b + \delta);$$

also, by the intermediate-value theorem, there exists t such that $b < t < b + \delta$ and $\gamma_1(t) = 2/(4n + 1)$. Hence $\gamma_2(t) = 1$ and $\gamma_2(t) - \gamma_2(b) \geq 1$. The same kind of argument may be used if $\gamma_2(b) \geq 0$, and we conclude that γ_2 is not continuous at b. This contradiction shows that E is not path-connected.

(iii) The closure of a path-connected set need not be path-connected. For with the notation of (ii), B is plainly path-connected but E, and hence $\bar{B}$, are not.

(iv) For any $n \in \mathbf{N}$, the *unit sphere* S^n in $\mathbf{R}^{n+1}$ is a connected subset of $\mathbf{R}^{n+1}$. To see this, note that by Example 2.4.17 (ii), $\mathbf{R}^{n+1}\setminus\{0\}$ is path-connected; by Theorem 2.4.20 it is connected. Define $f : \mathbf{R}^{n+1}\setminus\{0\} \to S^n$ by

$$f(x_1, \ldots, x_n) = (x_1, \ldots, x_n)/(x_1^2 + \ldots + x_n^2)^{1/2}.$$

Since f is clearly continuous and surjective, it follows from Theorem 2.4.11 that S^n is connected.

In view of Example 2.4.21 (ii) above, it is a relief to know that provided that we restrict ourselves to open subsets of $\mathbf{R}^n$, the notions of connectedness and path-connectedness coincide. The next lemma prepares for this result.

Lemma 2.4.22 *Let* $x = (x_1, \ldots, x_n)$, $y = (y_1, \ldots, y_n) \in \mathbf{R}^n$. *Then there is a map* $\gamma : [0, 1] \to \mathbf{R}^n$ *which is a p-path in* $\mathbf{R}^n$ *joining* x *to* y *such that*

$$d(\gamma(s), \gamma(t)) \le d(x, y) \ (s, t \in [0, 1])$$

where d *is the Euclidean metric on* $\mathbf{R}^n$.

Proof Let $e^{(1)}, \ldots, e^{(n)}$ be the vectors of the natural basis for $\mathbf{R}^n$. Let $p^{(0)} = x$ and

$$p^{(j)} = x + \sum_{k=1}^{j}(y_k - x_k)e^{(k)} \ (j = 1, \ldots, n).$$

Define $\gamma : [0, 1] \to \mathbf{R}^n$ by

$$\gamma(s) = (j - ns)p^{(j-1)} + (ns - j + 1)p^{(j)}$$

if $j - 1 \le ns \le j$ and j is a positive integer not exceeding n. It is routine to verify that, for all $s \in [0, 1]$,

$$\gamma(s) = x + \sum_{k=1}^{n} \Psi_k(s)(y_k - x_k)e^{(k)},$$

where

$$\Psi_k(s) = \min\{\max\{ns - k + 1, 0\}, 1\}.$$

Hence γ is a p-path in $\mathbf{R}^n$ joining x to y; moreover, since $0 \le \Psi_k(s), \Psi_k(t) \le 1$ and therefore $|\Psi_k(s) - \Psi_k(t)| \le 1$, we have for all $s, t \in [0, 1]$,

$$d(\gamma(s), \gamma(t)) = \left\{\sum_{k=1}^{n}|\Psi_k(s) - \Psi_k(t)|^2 |y_k - x_k|^2\right\}^{1/2} \le d(x, y).$$

$\square$

Theorem 2.4.22 *Let G be an open set in $\mathbf{R}^n$. Then the following statements are equivalent.*

(i) *G is connected.*

(ii) *G is polygonally connected; further, given any $x, y \in G$ there is a p-path in G joining them.*

(iii) *G is path-connected.*

Proof Suppose that $G \neq \emptyset$; otherwise, the result holds trivially. It is obvious that (ii) implies (iii); also, Theorem 2.4.20 shows that (iii) implies (i). It remains to prove that (i) implies (ii).

Suppose that G is connected, let $a \in G$ and let

$$A := \{x \in G : \text{there is a p-path in } G \text{ joining } a \text{ to } x\}.$$

To show that G is polygonally connected it is enough to prove that $A = G$. First we prove that A is open. Let $x \in A$ and let $\mu : [0, 1] \to \mathbf{R}^n$ be a p-path in G joining a to x. Since $x \in G$, there exists $r > 0$ such that $B(x, r) \subset G$. Let $y \in B(x, r)$. By Lemma 2.4.22, there is a map $\nu : [0, 1] \to \mathbf{R}^n$ which is a p-path in $B(x, r)$, and hence in G, joining x to y. Let $\gamma : [0, 1] \to \mathbf{R}^n$ be defined by

$$\gamma(t) = \begin{cases} \mu(2t) & \text{if } 0 \leq t \leq \frac{1}{2}, \\ \nu(2t - 1) & \text{if } \frac{1}{2} \leq t \leq 1. \end{cases}$$

Evidently γ is a p-path in G joining a to y. Hence $y \in A$. It follows that $B(x, r) \subset A$ and that A is open.

Next we show that $G \backslash A$ is open. Let $z \in G \backslash A$ and let $r' > 0$ be such that $B(z, r') \subset G$. It is enough to prove that $B(z, r') \subset G \backslash A$. To obtain a contradiction, suppose that this is not the case. Then there exist $w \in B(z, r') \cap A$ and a p-path in G joining a to w. Further, this path may be extended, by means of a construction similar to that of the previous paragraph, to a p-path in G joining a to z. It follows that $z \in A \cap (G \backslash A)$, an impossibility.

Finally, note that $a \in A$, $G = A \cup (G \backslash A)$ and $A \cap (G \backslash A) = \emptyset$. Thus, since G is connected, $G \backslash A = \emptyset$ and $A = G$.   □

Next we turn to components: the idea is that even if a set is not connected, it is made up of connected subsets; components are the largest such subsets.

Definition 2.4.24 Let E be a non-empty subset of a metric space X. A subset D of E is called a **component** of E if it is a maximal connected subset of E, that is, if (i) D is connected, and (ii) whenever D_1 is connected and $D \subset D_1 \subset E$, it follows that $D = D_1$.

To prove the basic theorem about components, the following lemma will be very useful.

Lemma 2.4.25 *Let E be a non-empty subset of a metric space X and let $\mathscr{F}$ be a non-empty family of connected subsets of E with one point in common; that is, there exists $a \in \cap \mathscr{F}$. Then $A := \cup \mathscr{F}$ is a connected subset of E.*

Proof In view of Lemma 2.1.5 (iii) it is enough to show that A is a connected subset of X. Suppose that this is not so. Then there are disjoint open sets U, V in X such that $A \cap U \neq \emptyset$, $A \cap V \neq \emptyset$ and $A \subset U \cup V$. Since each $F \in \mathscr{F}$ is connected and $F \subset U \cup V$, either $F \cap U = \emptyset$ or $F \cap V = \emptyset$. As $A \cap U \neq \emptyset$, there exists $F_1 \in \mathscr{F}$ such that $F_1 \cap U \neq \emptyset$ and so $F_1 \cap V = \emptyset$. Since $A \cap V \neq \emptyset$, there exists $F_2 \in \mathscr{F}$ such that $F_2 \cap V \neq \emptyset$ and so $F_2 \cap U = \emptyset$. Hence $a \in F_1 \cap F_2 = F_1 \cap F_2 \cap (U \cup V) \subset (F_2 \cap U) \cup (F_1 \cap V) = \emptyset$, and we have a contradiction. $\square$

Theorem 2.4.26 *Let E be a non-empty subset of a metric space X. Then*

(i) *each $a \in E$ lies in a component of E (so that E is the union of its components);*
(ii) *distinct components of E are disjoint.*

Proof Let $a \in E$ and let $\mathscr{F}$ be the family of all connected subsets of E which contain a. Plainly $\mathscr{F} \neq \emptyset$, since $\{a\} \in \mathscr{F}$. By Lemma 2.4.25, $A := \cup \mathscr{F}$ is connected and contains a. Now A is a component of E : for, if $A \subset A_1 \subset E$ and A_1 is connected, then $A_1 \in \mathscr{F}$ and so $A_1 = A$. This proves (i).

Regarding (ii), let A_1 and A_2 be components of E, suppose that $A_1 \neq A_2$ and that $a \in A_1 \cap A_2$. By Lemma 2.4.25 , $A_1 \cup A_2$ is connected. But in that event, since A_1 and A_2 are components, it follows that $A_1 = A_1 \cup A_2 = A_2$, a contradiction. $\square$

Theorem 2.4.27 *Let G be a non-empty open subset of $\mathbf{R}^n$. Then G has countably many components, each of which is open.*

Proof Let A be a component of G and let $a \in A$. Since G is open, there exists $\varepsilon > 0$ such that $B(a, \varepsilon) \subset G$. Now $B(a, \varepsilon)$ is path-connected and thus connected: hence, by Lemma 2.4.25, $A \cup B(a, \varepsilon)$ is connected. As A is a component this implies that $A \cup B(a, \varepsilon) = A$. Hence $B(a, \varepsilon) \subset A$ and A is open.

The set $\mathbf{Q}^n$ is a countable subset of $\mathbf{R}^n$ and may be written as $\{p_k : k \in \mathbf{N}\}$. Given any component A of G, there exists a least $k \in \mathbf{N}$ such that $p_k \in A$. By Theorem 2.4.26, to distinct components there correspond distinct k, and so the components may be put in one-to-one correspondence with a subset of $\mathbf{N}$. The proof is complete. $\square$

Corollary 2.4.28 *Let $G \subset \mathbf{R}$ be open. Then $G = \bigcup_{n=1}^{\infty} I_n$, where the I_n are pairwise disjoint open intervals.*

Companion to the notion of a component of a set there is that of a path-component.

Definition 2.4.29 A **path-component** of a subset A of a metric space X is a maximal path-connected subset of A.

This idea has useful consequences, given below. Note that, plainly, distinct path-components are disjoint.

Theorem 2.4.30 *Each path-component of a metric space X is open (and therefore also closed) if, and only if, each point of X has a path-connected neighbourhood. The space X is path-connected if, and only if, it is connected and each $x \in X$ has a path-connected neighbourhood.*

Proof Suppose that each path-component of X is open, and let $x \in X$. Let C be the path-component containing $x : C$ is a neighbourhood of x and is path-connected. Conversely, suppose that each point of X has a path-connected neighbourhood, let C be a path-component and let $x \in C$. Then there is a path-connected neighbourhood $U(x)$ of x, and since C is a maximal path-connected set containing x, $U(x) \subset C$. Thus $C = \bigcup_{x \in C} U(x)$ is open. Since $X \backslash C$ is the union of the remaining open path-components, it is open: thus C is closed.

If X is path-connected it is connected, by Theorem 2.4.20, and, of course, X is a path-connected neighbourhood of its points. Conversely, suppose that X is connected and that each $x \in X$ has a path-connected neighbourhood. Then each path-component is both open and closed; and since X is connected, this path-component must be X.   □

To conclude this section we show that connectedness and path-connectedness are preserved on taking products.

Theorem 2.4.31 *Let X_1, X_2 be connected (respectively, path-connected) metric spaces. Then the metric space $X_1 \times X_2$ (see Example 2.1.2 (ix)) is connected (respectively, path-connected).*

Proof First suppose that X_1 and X_2 are connected and let $(a_1, a_2), (b_1, b_2) \in X_1 \times X_2$. Then $\{a_1\} \times X_2$ and $X_1 \times \{b_2\}$ are connected subsets of $X_1 \times X_2$ as they are homeomorphic (even isometric) to X_2 and X_1 respectively; moreover, they have a common point, (a_1, b_2). By Lemma 2.4.25 their union is connected: thus there is a connected set containing (a_1, a_2) and (b_1, b_2). The connectedness of $X_1 \times X_2$ now follows from Example 2.4.7 (v).

Now suppose that X_1 and X_2 are path-connected and again let $(a_1, a_2), (b_1, b_2) \in X_1 \times X_2$. There is a path $\gamma_1 : [0, 1] \rightarrow X_1$ joining a_1 to b_1, and hence there is a path $\widetilde{\gamma}_1 : [0, 1] \rightarrow X_1 \times X_2$ joining (a_1, b_2) to (b_1, b_2), given by $\widetilde{\gamma}_1(t) = (\gamma_1(t), b_2)$. Similarly, there is a path $\widetilde{\gamma}_2 : [0, 1] \rightarrow X_1 \times X_2$ joining (a_1, a_2) to (a_1, b_2). The path $\widetilde{\gamma} : [0, 1] \rightarrow X_1 \times X_2$ defined by

$$\widetilde{\gamma}(t) = \begin{cases} \widetilde{\gamma}_2(2t) & \text{if } 0 \leq t \leq \frac{1}{2}, \\ \widetilde{\gamma}_1(2t - 1) & \text{if } \frac{1}{2} \leq t \leq 1 \end{cases}$$

joins (a_1, a_2) to (b_1, b_2), and so $X_1 \times X_2$ is path-connected.   □

Corollary 2.4.32 *The torus $T := S^1 \times S^1$ is connected.*

Proof From Example 2.4.21 (iv) we see that S^1 is connected. The corollary now follows from Theorem 2.4.31.   □

Of course, the torus as defined here is a subset of $\mathbf{R}^4$ and is endowed with the inherited metric. In fact, T is homeomorphic to the subset $\widetilde{T}$ of $\mathbf{R}^3$ obtained by revolution of the circle $\{(0, y, z) : (y - 1)^2 + z^2 = 1/4\}$ about the z-axis. For $\widetilde{T}$ is given parametrically by

$$x = \left(1 + \frac{\cos\theta}{2}\right)\cos\phi, \ y = \left(1 + \frac{\cos\theta}{2}\right)\sin\phi,$$

$$z = \frac{\sin\theta}{2}(0 \le \theta < 2\pi, \ 0 \le \phi < 2\pi),$$

and the map

$$((\cos\theta, \sin\theta), (\cos\phi, \sin\phi)) \longmapsto \left(\left(1 + \frac{\cos\theta}{2}\right)\cos\phi, \left(1 + \frac{\cos\theta}{2}\right)\sin\phi, \frac{\sin\theta}{2}\right)$$

is a homeomorphism of T onto $\widetilde{T}$. This map, f, is given by

$$f\left((a, b), (c, d)\right) = \left(\left(1 + \frac{1}{2}a\right)c, \left(1 + \frac{1}{2}a\right)d, \frac{1}{2}b\right)$$

and is evidently continuous. It is bijective, with

$$f^{-1}(p, q, r) = \left(2\left(-1 + \sqrt{p^2 + q^2}\right), 2r, \frac{p}{\sqrt{p^2 + q^2}}, \frac{q}{\sqrt{p^2 + q^2}}\right)$$

since $p = \left(1 + \frac{1}{2}a\right)c, q = \left(1 + \frac{1}{2}a\right)d, r = \frac{1}{2}b$, and so

$$p^2 + q^2 = \left(1 + \frac{1}{2}a\right)^2, \ \frac{1}{2}a = -1 + \sqrt{p^2 + q^2}$$

since $1 + \frac{1}{2}a \ge \frac{1}{2}$. Plainly f^{-1} is continuous, and so f is a homeomorphism.

Exercise 2.4.33

1. Let A be a connected subset of a metric space and suppose that $A \subset B \subset \bar{A}$. Prove that B is connected. Deduce that the components of a closed set are closed.
2. Let $\mathbf{R}^2$ be endowed with the Euclidean metric and let S be a subset of $\mathbf{R}^2$ which is both open and closed. Prove that either $S = \emptyset$ or $S = \mathbf{R}^2$.
3. Let E and F be subsets of $\mathbf{R}^2$ (endowed with the Euclidean metric) defined by

$$E = \{(x, y) : x^2 + y^2 \le 1\} \cup \{(x, y) : (x - 2)^2 + y^2 < 1\},$$
$$F = \{(x, y) : x^2 + y^2 < 1\} \cup \{(1 + 1/n, 0) : n \in \mathbf{N}\}.$$

Determine whether E or F is connected. What are the components of these sets?

4. Let $n \in \mathbf{N}$ and let $GL(n, \mathbf{R})$ be the set of all non-singular $n \times n$ matrices; identify $GL(n, \mathbf{R})$ with a subset of $\mathbf{R}^{n^2}$ in an obvious way and give it the inherited metric. Prove that $GL(n, \mathbf{R})$ is not connected.

5. Let A and B be path-connected subsets of a metric space such that $A \cap B \neq \emptyset$. Prove that $A \cup B$ is path-connected.

6. Let E and F be metric spaces, with E path-connected, and let $f : E \to F$ be continuous. Prove that $f(E)$ is path-connected.

7. Let K be the subset of $[0, 1]$ consisting of all numbers of the form $\sum_{n=0}^{\infty} 3^{-n} c_n$, with $c_n \in \{0, 2\}$ for all $n \in \mathbf{N}_0$. This set is called the **Cantor** set (see Exercise 2.3.38/16). Show that K is compact, that $[0, 1] \backslash K$ is a countable union of disjoint intervals, and that the sum of the lengths of these intervals is 1. Show that given any $x \in K$, the connected component of K which contains x is $\{x\}$.

8. Let $S = [0, 1] \times [0, 1]$, let K be as in the question above and let $f : K \to S$ be the map which to each $x \in K$, with $x = \sum_{n=0}^{\infty} 3^{-n} c_n$, assigns the element $\left(\sum_{n=0}^{\infty} 2^{-n} b_{2n+1}, \sum_{n=0}^{\infty} 2^{-n} b_{2n} \right)$, where $b_m = c_m/2$ $(m \in \mathbf{N}_0)$. Show that f is well-defined, and that it is surjective and continuous. Deduce that there is a continuous surjective map $g : [0, 1] \to S$. (This is Peano's **space-filling** curve.)

2.5 Simple-Connectedness

Our interest here is in those path-connected metric spaces which, loosely speaking, may be viewed as without holes. To bring precision to this, the notion of homotopy is introduced. Throughout this section the closed interval $[0, 1]$ will be denoted by I; and if X is a metric space the product $X \times I$ is assumed to be equipped with the metric of Example 2.1.2 (ix).

Definition 2.5.1 Let X and Y be metric spaces and let $f_0, f_1 : X \to Y$ be continuous. We say that the maps f_0 and f_1 are **homotopic**, and write $f_0 \simeq f_1$, if there is a continuous map $F : X \times I \to Y$ such that, for all $x \in X$,

$$F(x, 0) = f_0(x) \text{ and } F(x, 1) = f_1(x).$$

Such a map F is called a **homotopy** between f_0 and f_1.

Example 2.5.2 Let X be a metric space and let $f_0, f_1 : X \to \mathbf{R}^n$ be continuous. Define $F : X \times I \to \mathbf{R}^n$ by

$$F(x, t) = (1 - t)f_0(x) + tf_1(x), \ (x, t) \in X \times I.$$

Then it is easy to verify that F is a homotopy between f_0 and f_1.

With regard to the homotopy F in Definition 2.5.1, if we set $f_t(x) = F(x, t)$, then $\{f_t : t \in I\}$ is a one-parameter family of continuous maps from X to Y, and we may think of the homotopy as a continuous deformation of f_0 into f_1.

Definition 2.5.3 Let X and Y be metric spaces and let A be a subset of X. Let $f_0, f_1 : X \to Y$ be continuous maps such that $f_0(a) = f_1(a)$ for all $a \in A$, that is, $f_0 |_A = f_1 |_A$. If there is a homotopy F between f_0 and f_1 such that, for all $a \in A$ and all $t \in I$,

$$F(a, t) = f_0(a) = f_1(a),$$

or equivalently $f_t |_A = f_1 |_A$ for all $t \in I$, then we say that f_0 and f_1 are **homotopic relative to** A and write $f_0 \simeq f_1$ rel A. [Note: if A is empty, then $\simeq$ rel A and $\simeq$ coincide.]

Example 2.5.4 Let $A = \{0, 1\}$. Let $f_0, f_1 : I \to \mathbf{R}^n$ be paths in $\mathbf{R}^n$ such that $f_0(0) = f_1(0)$ and $f_0(1) = f_1(1)$: the paths have a common initial point and a common terminal point so that $f_0 |_A = f_1 |_A$. Consideration of $F : I \times I \to \mathbf{R}^n$ given by

$$F(s, t) = (1 - t)f_0(s) + tf_1(s)$$

shows that $f_0 \simeq f_1$ rel $\{0, 1\}$.

Theorem 2.5.5 *Let X and Y be metric spaces and A be a subset of X. Then $\simeq$ rel A is an equivalence relation in $C(X, Y)$, the family of continuous maps from X to Y.*

Proof The steps which follow show that $\simeq$ rel A is reflexive, symmetric and transitive.

(1) If $f \in C(X, Y)$, then $f \simeq f$ rel A.
The continuous map $F : X \times I \to Y$ given by $F(x, t) = f(x)$ verifies this claim.
(2) If $f, g \in C(X, Y)$ and $f \simeq g$ rel A, then $g \simeq f$ rel A.

By hypothesis, there exists a homotopy F relative to A between f and g. Let $G : X \times I \to Y$ be defined by

$$G(x, t) = F(x, 1 - t).$$

As it is a composition of continuous maps, G is continuous. Moreover, for all $x \in X$,

$$G(x, 0) = g(x), \quad G(x, 1) = f(x);$$

also, for all $a \in A$ and $t \in I$,

$$G(a, t) = g(a) = f(a).$$

Hence $g \simeq f$ rel A.
(3) If $f, g, h \in C(X, Y), f \simeq g$ rel A and $g \simeq h$ rel A, then $f \simeq h$ rel A.

Given that there are homotopies F and G relative to A between f and g, and g and h, respectively, let $H : X \times I \to Y$ be defined by

$$H(x, t) = \begin{cases} F(x, 2t), & 0 \leq t \leq 1/2, \\ G(x, 2t - 1), & 1/2 \leq t \leq 1. \end{cases}$$

Since $F(x, 1) = g(x) = G(x, 0)$ for all $x \in X$, there is consistency of definition on $X \times \{1/2\}$ and, by appeal to the glueing lemma (Lemma 2.1.35), it follows that H is continuous. Further, for all $x \in X$,

$$H(x, 0) = f(x), H(x, 1) = h(x);$$

also, for all $a \in A$ and $t \in I$,

$$H(a, t) = f(a) = h(a).$$

Hence $f \simeq h$ rel A.   □

Corollary 2.5.6 *Let $f \in C(X, Y)$ and denote by $\langle f \rangle$ the equivalence class associated with f :*

$$\langle f \rangle = \{g \in C(X, Y) : g \simeq f \text{ rel } A\}.$$

The family of equivalence classes $\{\langle f \rangle : f \in C(X, Y)\}$ constitutes a partition of $C(X, Y)$, by which we mean that no equivalence class is empty, their union exhausts $C(X, Y)$ and, for all $f, g \in C(X, Y)$, the classes $\langle f \rangle$ and $\langle g \rangle$ are either disjoint or identical.

Proof We leave this to the reader, noting that it is a special case of a general result concerning equivalence classes: see, for example, [19], p. 50.   □

We now show that relative to the composition of functions, homotopy is well-behaved.

Theorem 2.5.7 *Let X, Y and Z be metric spaces and A be a subset of X. Let f_0, $f_1 : X \to Y$ and $g_0, g_1 : Y \to Z$ be continuous maps such that $f_0 \simeq f_1$ rel A and $g_0 \simeq g_1$ rel $f_0(A)$. Then*

$$g_0 \circ f_0 \simeq g_1 \circ f_1 \text{ rel } A.$$

Proof Let $F : X \times I \to Y$ and $G : Y \times I \to Z$ be homotopies establishing that $f_0 \simeq f_1$ rel A and $g_0 \simeq g_1$ rel $f_0(A)$, respectively. The map $g_0 \circ F : X \times I \to Z$ is continuous; also, for all $x \in X$,

$$(g_0 \circ F)(x, 0) = (g_0 \circ f_0)(x), \quad (g_0 \circ F)(x, 1) = (g_0 \circ f_1)(x),$$

and, for all $a \in A$ and $t \in I$,

$$(g_0 \circ F)(a, t) = (g_0 \circ f_0)(a) = (g_0 \circ f_1)(a).$$

Hence $g_0 \circ f_0 \simeq g_0 \circ f_1$ rel A. Next, consider the map $H : X \times I \to Z$ defined by $H(x, t) = G(f_1(x), t)$. It is continuous; moreover, for all $x \in X$,

$$H(x, 0) = (g_0 \circ f_1)(x), \quad H(x, 1) = (g_1 \circ f_1)(x),$$

and for all $a \in A$ and $t \in I$,

$$H(a, t) = (g_0 \circ f_1)(a) = (g_1 \circ f_1)(a).$$

Thus $g_0 \circ f_1 \simeq g_1 \circ f_1$ rel A. Finally, by Theorem 2.5.5, $g_0 \circ f_0 \simeq g_1 \circ f_1$ rel A. $\quad\square$

In the next section we appeal to the simplest aspect of this theorem, when $g_0 = g_1$.

2.5.1 Homotopies Between Paths

Let X be a metric space. For present purposes we shall think of $C(I, X)$ as the set of all paths in X, each path being assumed to have $I = [0, 1]$ as its parameter interval. For brevity, the symbol $\sim$ will be used for the relation $\simeq$ rel $\{0, 1\}$ on $C(I, X)$. Hence $f_0 \sim f_1$, to be read f_0 is equivalent to f_1, is understood to mean that $f_0(0) = f_1(0)$, $f_0(1) = f_1(1)$ and that a continuous map $F : I \times I \to X$ exists such that

$$F(s, 0) = f_0(s), \quad F(s, 1) = f_1(s) \ (s \in I)$$

and

$$F(0, t) = f_0(0), \quad F(1, t) = f_0(1) \ (t \in I).$$

The homotopy F may be viewed as continuously deforming f_0 into f_1 through a family of paths with prescribed endpoints.

Definition 2.5.8 Let f and g be paths in a metric space X such that $f(1) = g(0)$. The product path $f * g : I \to X$ is defined by

$$(f * g)(s) = \begin{cases} f(2s), & 0 \le s \le 1/2, \\ g(2s - 1), & 1/2 \le s \le 1. \end{cases}$$

Similarly, if $f_1, f_2, \ldots, f_n : I \to X$ are paths in X such that, for $1 \le j \le n - 1$, $f_j(1) = f_{j+1}(0)$, then the product path $f_1 * f_2 * \ldots * f_n : I \to X$ is defined by

$$(f_1 * f_2 * \ldots * f_n)(s) = \begin{cases} f_1(ns), & 0 \le s \le 1/n, \\ f_2(ns - 1), & 1/n \le s \le 2/n, \\ \ldots & \ldots \\ f_j(ns - j + 1), & (j - 1)/n \le s \le j/n, \\ \ldots & \ldots \\ f_n(ns - n + 1), & (n - 1)/n \le s \le 1. \end{cases}$$

Evidently $f * g$ is a path joining $f(0)$ to $g(1)$; likewise, $f_1 * f_2 * \ldots * f_n$ is a path joining $f_1(0)$ to $f_n(1)$.

Theorem 2.5.9 *Let f, f', g and g' be paths in a metric space X; suppose that $f \sim f', g \sim g'$ and $f * g$ is defined. Then $f * g \sim f' * g'$.*

Proof Let f, f' join x to y and g, g' join y to z : since $f * g$ is defined, $f(1) = y = g(0)$. As $f \sim f'$, there exists a continuous map $F : I \times I \to X$ such that

$$F(s, 0) = f(s), \ F(s, 1) = f'(s) \ (s \in I)$$

and

$$F(0, t) = x, \ F(1, t) = y \ (t \in I).$$

Similarly, since $g \sim g'$, there is a continuous map $G : I \times I \to X$ such that

$$G(s, 0) = g(s), \ G(s, 1) = g'(s) \ (s \in I)$$

and

$$G(0, t) = y, \ G(1, t) = z \ (t \in I).$$

Let $H : I \times I \to X$ be defined by

$$H(s, t) = \begin{cases} F(2s, t), & 0 \le s \le 1/2, \ 0 \le t \le 1, \\ G(2s - 1, t), & 1/2 \le s \le 1, \ 0 \le t \le 1. \end{cases}$$

Since $F(1, t) = y = G(0, t)$ for all $t \in I$, there is consistency of definition on the line segment $\{1/2\} \times I$. The glueing lemma ensures that H is continuous; further, for all $s \in I$,

$$H(s, 0) = (f * g)(s), \ H(s, 1) = (f' * g')(s)$$

and, for all $t \in I$,

$$H(0, t) = x, \ H(1, t) = z.$$

Hence $f * g \sim f' * g'$. $\square$

By dividing the unit square $I \times I$ into n vertical strips rather than 2, the following generalisation of the last theorem may be established: details are left to the reader.

Theorem 2.5.10 *Let $f_1, f_2, \ldots, f_n$ and $f'_1, f'_2, \ldots, f'_n$ be paths in a metric space X; suppose that, for $1 \le j \le n$, $f_j \sim f'_j$ and that the product path $f_1 * f_2 * \ldots * f_n$ is defined. Then*

$$f_1 * f_2 * \ldots * f_n \sim f'_1 * f'_2 * \ldots * f'_n.$$

Theorem 2.5.11 *Let $f_1, f_2, \ldots, f_n$ ($n \ge 3$) be paths in a metric space X such that the product $f_1 * f_2 * \ldots * f_n$ is defined. Suppose $1 \le k \le n - 1$ and let the map $\phi : I \to I$ be given by*

$$\phi(s) = \begin{cases} 2sk/n, & 0 \le s \le 1/2, \\ (n - k)(2s - 1)/n + k/n, & 1/2 \le s \le 1. \end{cases}$$

Then

(i) $(f_1 * f_2 * \ldots * f_n) \circ \phi = (f_1 * f_2 * \ldots * f_k) * (f_{k+1} * \ldots * f_n);$
(ii) $(f_1 * f_2 * \ldots * f_k) * (f_{k+1} * \ldots * f_n) \sim f_1 * f_2 * \ldots * f_n;$

and

(iii) *setting* $n = 3$, $(f_1 * f_2) * f_3 \sim f_1 * (f_2 * f_3)$.

Proof

(i) Illustrated below, the map ϕ is continuous and strictly increasing; $\phi(s) \leq k/n$ if $0 \leq s \leq 1/2$, $\phi(s) > k/n$ if $1/2 < s \leq 1$.

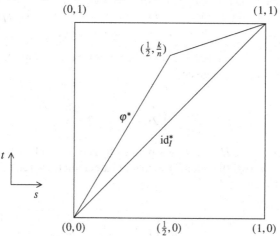

Hence, for all $s \in I$,

$$((f_1 * f_2 * \ldots * f_n) \circ \phi)(s)$$
$$= f_j(n\phi(s) - j + 1) \text{ if } (j-1)/n \leq \phi(s) \leq j/n \text{ and } 1 \leq j \leq n$$
$$= \begin{cases} f_j(2ks - j + 1), & j - 1 \leq 2sk \leq j, \ 1 \leq j \leq k, \\ f_j((n-k)(2s-1) + k - j + 1), & j - k - 1 \leq (2s-1)(n-k) \leq j - k, \\ & k + 1 \leq j \leq n, \end{cases}$$

$$= \begin{cases} f_j(2ks - j + 1), & j - 1 \leq 2sk \leq j, \ 1 \leq j \leq k, \\ f_{k+j'}((n-k)(2s-1) - j' + 1), & j' - 1 \leq (2s-1)(n-k) \leq j', \\ & 1 \leq j' \leq n - k, \end{cases}$$

$$= ((f_1 * f_2 * \ldots * f_k) * (f_{k+1} * \ldots * f_n))(s).$$

(ii) Consideration of the map $H : I \times I \to I$ given by

$$H(s, t) = (1 - t)\phi(s) + ts$$

shows that ϕ and id_I (the identity map on I) are homotopic relative to $\{0, 1\}$. Hence, using Theorem 2.5.7,

$$(f_1 * f_2 * \ldots * f_k) * (f_{k+1} * \ldots * f_n) = (f_1 * f_2 * \ldots * f_n) \circ \phi$$
$$\sim (f_1 * f_2 * \ldots * f_n) \circ id_I$$
$$= f_1 * f_2 * \ldots * f_n.$$

(ii) By (ii), both the product paths $f_1 * (f_2 * f_3)$ and $(f_1 * f_2) * f_3$ are equivalent to $f_1 * f_2 * f_3$. Since the relation $\sim$ is transitive, it follows that $f_1 * (f_2 * f_3) \sim (f_1 * f_2) * f_3$. $\square$

Theorem 2.5.12 *Let X be a metric space, let $x, y \in X$ and let e_x, e_y be the constant paths in X defined by $e_x(s) = x$ and $e_y(s) = y$ ($s \in I$). Let f be a path in X such that $f(0) = x$ and $f(1) = y$. Then $e_x * f \sim f$ and $f * e_y \sim f$.*

Proof As each equivalence has a similar proof we give only that which involves $e_x * f$. Let $\psi : I \to I$ be given by

$$\psi(s) = \begin{cases} 0, & 0 \le s \le 1/2, \\ 2s - 1, & 1/2 \le s \le 1. \end{cases}$$

The continuous map $H : I \times I \to I$ defined by $H(s, t) = (1 - t)\psi(s) + ts$ enables us to see that ψ and id_I (the identity map on I) are homotopic relative to $\{0, 1\}$. Hence, noting that for all $s \in I$,

$$(f \circ \psi)(s) = \begin{cases} x, & 0 \le s \le 1/2, \\ f(2s - 1), & 1/2 \le s \le 1, \end{cases}$$
$$= (e_x * f)(s),$$

application of Theorem 2.5.7 shows that

$$e_x * f = f \circ \psi \sim f \circ id_I = f,$$

as required. $\square$

Theorem 2.5.13 *Let X be a metric space, f be a path in X and $\widehat{f}$ be the path defined by $\widehat{f}(s) = f(1 - s)$ ($s \in I$); $\widehat{f}$ is termed the **reverse** of f. Let $f(0) = x$ and $f(1) = y$. Then*

$$f * \widehat{f} \sim e_x, \quad \widehat{f} * f \sim e_y,$$

where e_x, e_y are the constant paths given by $e_x(s) = x$, $e_y(s) = y$ ($s \in I$), respectively.

Proof Since the rôles of f and $\widehat{f}$ can be interchanged, it is sufficient to prove that $f * \widehat{f} \sim e_x$. Let $\tau, \theta : I \to I$ be given by

$$\tau(s) = \begin{cases} 2s, & 0 \le s \le 1/2, \\ 2(1 - s), & 1/2 \le s \le 1, \end{cases}$$

and $\theta(s) = 0$ $(s \in I)$. The map

$$(s, t) \mapsto (1 - t)\tau(s) : I \times I \to I$$

shows that $\tau \simeq \theta$ rel $\{0, 1\}$. Since

$$(f * \widehat{f})(s) = \begin{cases} f(2s), & 0 \le s \le 1/2, \\ \widehat{f}(2s - 1), & 1/2 \le s \le 1 \end{cases}$$

$$= \begin{cases} f(2s), & 0 \le s \le 1/2, \\ f(2(1 - s)), & 1/2 \le s \le 1, \end{cases}$$

$$= (f \circ \tau)(s),$$

so that $f * \widehat{f} = f \circ \tau$, application of Theorem 2.5.7 shows that $f * \widehat{f} = f \circ \tau \sim f \circ \theta$
$= e_x$.   □

Definition 2.5.14 A **closed path** (or **loop**) in a metric space X is a path whose initial
and terminal points coincide: this common point is called its **base point**. Thus, if
$x \in X$ and f is a path in X such that $f(0) = f(1) = x$, then f is a closed path in X
with base point x.

Remark 2.5.15

(i) Each $x \in X$ is a base point for at least one closed path in X, namely e_x, given
   by $e_x(s) = x$ $(s \in I)$, the path constant at x.
(ii) If $x, y \in X$ and there is a path f joining x to y then, with $\widehat{f}$ denoting the path
   given by $\widehat{f}(s) = f(1 - s)$, $f * \widehat{f}$ is a closed path with base point x and $\widehat{f} * f$ is a
   closed path with base point y.

The definition to follow introduces a new type of homotopy, specific to closed
paths, called **free homotopy**. For closed paths f, g in a metric space X recall that the
statement $f \simeq g$ rel $\{0, 1\}$, more simply denoted $f \sim g$, means that f and g have a
common base point and that a continuous map $H : I \times I \to X$ exists such that
   (1) $H(s, 0) = f(s), H(s, 1) = g(s)$ $(s \in I)$
   and
   (2) $H(0, t) = f(0) = H(1, t)$ $(t \in I)$.
Note that the one-parameter family $\{h_t\}$ of paths determined by H is made up
of closed paths with a common base point. The notion of free homotopy relaxes
condition (2).

Definition 2.5.16 Let f and g be closed paths in a metric space X. Then f is said to
be **freely homotopic** to g if there is a continuous map $H : I \times I \to X$ such that

(i) $H(s, 0) = f(s), H(s, 1) = g(s)$ $(s \in I)$

and

(ii) $H(0, t) = H(1, t)$ $(t \in I)$.

Note that the paths h_t, where $h_t(s) = H(s, t)$, are closed but are not required to have the same base point; the path $t \longmapsto h_t(0) : I \to X$ is not required to be a constant map. A simple example of a free homotopy occurs when the base point of a closed path is 'shifted' to another point on its track: see Exercise 2.5.30/4.

Theorem 2.5.17 *Let f and g be closed paths in a metric space X such that f is freely homotopic to g under a homotopy $F : I \times I \to X$. Let v be the path in X from $f(0)$ to $g(0)$ defined by $v(s) = F(0, s)$ ($s \in I$). Then, with $\widehat{v}$ given by $\widehat{v}(s) = v(1 - s)$ ($s \in I$),*

$$f \sim v * g * \widehat{v}.$$

Proof Let $x = f(0)$, $y = g(0)$ so that v is a path joining x to y and $\widehat{v}$ is its reverse: the figure below is a guide.

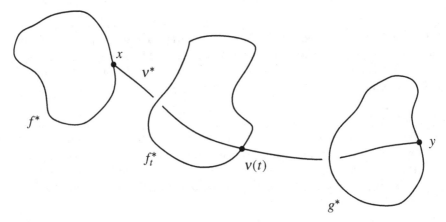

Recall that

$$(v * g * \widehat{v})(s) = \begin{cases} v(3s), & 0 \le s \le 1/3, \\ g(3s - 1), & 1/3 \le s \le 2/3, \\ v(3(1 - s)), & 2/3 \le s \le 1. \end{cases}$$

For fixed $t \in I$, consider the path γ_t given by

$$\gamma_t(s) = \begin{cases} v(3s), & 0 \le s \le t/3, \\ f_t((3 - 2t)^{-1}(3s - t)), & t/3 \le s \le 1 - t/3, \\ v(3(1 - s)), & 1 - t/3 \le s \le 1, \end{cases}$$

where f_t is that path such that $f_t(s) = F(s, t)$. Loosely speaking, γ_t proceeds from x to $v(t)$, circuits the track of f_t and then retraces its steps back from $v(t)$ to x. We show that $\{\gamma_t : t \in I\}$ determines a suitable homotopy. Define $G : I \times I \to X$ by

$$G(s, t) = \begin{cases} F(0, 3s), & 0 \le s \le t/3, \\ F\left((3 - 2t)^{-1}(3s - t), t\right), & t/3 \le s \le 1 - t/3, \\ F(0, 3(1 - s)), & 1 - t/3 \le s \le 1. \end{cases}$$

Let K_1, K_2, K_3 be the subsets of $I \times I$ defined by the inequalities $0 \le s \le t/3$, $t/3 \le s \le 1 - t/3$ and $1 - t/3 \le s \le 1$ respectively and indicated below.

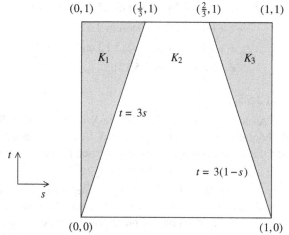

It is plain that each of the following maps is continuous:
$$(s, t) \longmapsto (0, 3s) \longmapsto F(0, 3s) : K_1 \to I \times I \to X,$$
$$(s, t) \longmapsto ((3 - 2t)^{-1}(3s - t), t) \longmapsto F((3 - 2t)^{-1}(3s - t), t) : K_2 \to I \times I \to X$$
and
$$(s, t) \longmapsto (0, 3(1 - s)) \longmapsto F(0, 3(1 - s)) : K_3 \to I \times I \to X.$$
Thus each $G|_{K_i}$ is continuous. Since G is consistently defined on the line segments $K_1 \cap K_2$ and $K_2 \cap K_3$, and each K_i is closed, it follows from the glueing lemma that G is continuous. Now

$$G(s, 0) = F(s, 0) = f(s) \ (s \in I),$$

$$G(s, 1) = \begin{cases} F(0, 3s) = v(3s), & 0 \le s \le 1/3, \\ F(3s - 1, 1) = g(3s - 1), & 1/3 \le s \le 2/3, \\ F(0, 3(1 - s)) = v(3(1 - s)), & 2/3 \le s \le 1, \end{cases}$$
$$= (v * g * \widehat{v})(s) \ (s \in I)$$

and

$$G(0, t) = x = G(1, t) \ (t \in I).$$

Thus $f \sim v * (g * \widehat{v})$, as required. $\qquad \square$

Theorem 2.5.18 *Let X be a metric space and let f be a closed path in X with base point x. Then $f \sim e_x$ if, and only if, f is freely homotopic to a constant path in X.*

Proof If $f \sim e_x$ the result is obvious. Conversely, suppose that for some $y \in X, f$ is freely homotopic, under a homotopy $F : I \times I \to X$, to the constant path e_y. Let v

be the path in X from x to y given by $v(s) = F(0, s)$ $(s \in I)$. Then

$$f \sim v * e_y * \widehat{v} \text{ (by Theorem 2.5.17)}$$
$$\sim v * (e_y * \widehat{v}) \text{ (by Theorem 2.5.11)}$$
$$\sim v * \widehat{v} \text{ (by Theorems 2.5.12 and 2.5.9)}$$
$$\sim e_x \text{ (by Theorem 2.5.13).}$$

Hence by Theorem 2.5.5, $f \sim e_x$.   □

Definition 2.5.19 A closed path in a metric space X is said to be **null-homotopic in X** if it is freely homotopic to a constant path in X. A metric space X is said to be **simply-connected** if it is path-connected and each closed path in X is null-homotopic in X.

Remark 2.5.20 By Theorem 2.5.18, X is simply-connected if it is path-connected and $f \sim e_{f(0)}$ for each closed path f in X. Intuitively, a simply-connected space may be viewed as one within which each pair of points can be joined by a path and each closed path is continuously shrinkable to a point. No closed path can 'encompass a hole' in the space.

Example 2.5.21

(i) Let K be a subset of a metric space X. Suppose that X is also a linear space and that K is convex, so that $tx + (1 - t)y \in K$ whenever $x, y \in K$ and $t \in I$. Then K is simply-connected: its convexity implies that it is path-connected; moreover, if γ_0 is any closed path in K and $z \in K$, then γ_0 is freely homotopic to the constant path γ_1, where $\gamma_1(s) = z$ for all $s \in I$, under the homotopy $H : I \times I \to K$ defined by $H(s, t) = (1 - t)\gamma_0(s) + tz$. Hence each ball in $\mathbf{R}^n$ is simply-connected.

(ii) We shall see in Chap. 3, once the notion of winding number has been developed, that neither a circle nor an annulus in $\mathbf{R}^2$ is simply-connected.

The next two theorems reinforce the definition above and have application in the chapter to follow.

Theorem 2.5.22 *Let x and y be points in a simply-connected metric space X and let f, g be paths in X which join x to y. Then $f \sim g$.*

Proof Let paths e_x, e_y and $\widehat{g}$ be given by $e_x(s) = x$, $e_y(s) = y$ and $\widehat{g}(s) = g(1 - s)$. Note that $\widehat{g} * f$ is a closed path with base point y and, since X is simply connected, $\widehat{g} * f \sim e_y$. Since the relation $\sim$ is transitive, the steps below yield the result:

$$f \sim e_x * f \text{ (by Theorem 2.5.12)}$$
$$\sim (g * \widehat{g}) * f \text{ (by Theorems 2.5.9 and 2.5.13)}$$
$$\sim g * (\widehat{g} * f) \text{ (by Theorem 2.5.11)}$$
$$\sim g * e_y \text{ (by Theorem 2.5.9)}$$
$$\sim g \text{ (by Theorem 2.5.12).}$$

□

Theorem 2.5.23 *Let X, Y be metric spaces and let $X \times Y$ be endowed with the usual metric (see Example 2.1.2 (ix)). Then $X \times Y$ is simply-connected if, and only if, both X and Y are simply-connected.*

Proof Suppose that X and Y are simply-connected and let $\gamma = (\gamma_1, \gamma_2)$ be a closed path in $X \times Y$. Then γ_1 and γ_2 are closed paths which are null-homotopic in X and Y respectively. Let maps $F_1 : I \times I \to X$ and $F_2 : I \times I \to Y$ establish these homotopies. Then the map $F : I \times I \to X \times Y$ given by $F(s, t) = (F_1(s, t), F_2(s, t))$ shows that γ is null-homotopic in $X \times Y$. Since, by Theorem 2.4.31, $X \times Y$ is path-connected it follows that $X \times Y$ is simply-connected.

Conversely, suppose that $X \times Y$ is simply-connected. Elementary considerations show that X and Y are path-connected. Let γ_1 and γ_2 be closed paths in X and Y respectively and define $\gamma : I \to X \times Y$ by $\gamma(t) = (\gamma_1(t), \gamma_2(t))$. Then γ is a closed path which is null-homotopic in $X \times Y$ under a homotopy $H = (H_1, H_2)$, say. Since the maps H_1 and H_2 are themselves homotopies which, respectively, establish that γ_1 is null-homotopic in X and γ_2 is null-homotopic in Y, the spaces X and Y are simply-connected. $\qquad\square$

2.5.2 The Fundamental Group

Definition 2.5.24 Let X be a metric space and $x \in X$. Let $\mathscr{L}(x)$ denote the family of all closed paths in X with base point x :

$$\mathscr{L}(x) = \{f \in C(I, X) : f(0) = f(1) = x\}.$$

By Theorem 2.5.5, the relation $\sim$ is an equivalence relation in $C(I, X)$ and therefore in $\mathscr{L}(x)$. For $f \in \mathscr{L}(x)$, let $\langle f \rangle$ denote the equivalence class associated with f :

$$\langle f \rangle = \{g \in \mathscr{L}(x) : g \sim f\}.$$

The set

$$\pi(X, x) = \{\langle f \rangle : f \in \mathscr{L}(x)\}$$

equipped with the product defined by

$$\langle f \rangle \langle g \rangle = \langle f * g \rangle$$

is called the **fundamental group of X at the base point** x.

We must justify this terminology by showing that $\pi(X, x)$ is indeed a group.

(i) Theorem 2.5.9 shows that the product of equivalence classes is well-defined:

$$\langle f \rangle = \langle f' \rangle, \ \langle g \rangle = \langle g' \rangle \ \Rightarrow f \sim f', \ g \sim g' \Rightarrow f * g \sim f' * g' \Rightarrow \langle f * g \rangle$$
$$= \langle f' * g' \rangle.$$

(ii) By Theorem 2.5.11, the product of equivalence classes is associative:

$$(\langle f \rangle \langle g \rangle) \langle h \rangle = \langle f * g \rangle \langle h \rangle = \langle (f * g) * h \rangle = \langle f * (g * h) \rangle = \langle f \rangle \langle g * h \rangle$$
$$= \langle f \rangle (\langle g \rangle \langle h \rangle).$$

(iii) Theorem 2.5.12 confirms that $\langle e_x \rangle$ is the identity:

$$\langle e_x \rangle \langle f \rangle = \langle e_x * f \rangle = \langle f \rangle = \langle f * e_x \rangle = \langle f \rangle \langle e_x \rangle.$$

(iv) Theorem 2.5.13 shows that given $\langle f \rangle$ in $\pi(X, x)$, its inverse $\langle f \rangle^{-1} = \langle \widehat{f} \rangle$:

$$\langle f \rangle \langle \widehat{f} \rangle = \langle f * \widehat{f} \rangle = \langle e_x \rangle = \langle \widehat{f} * f \rangle = \langle \widehat{f} \rangle \langle f \rangle.$$

Given distinct points x and y in X, it is natural to ask whether there is any relationship between $\pi(X, x)$ and $\pi(X, y)$. It turns out that one exists if x and y can be joined by a path in X.

Theorem 2.5.25 *Let x and y be points in a metric space X and let α be a path in X such that $\alpha(0) = x$, $\alpha(1) = y$. Then $\pi(X, x)$ and $\pi(X, y)$ are isomorphic.*

Proof As usual, let $\widehat{\alpha}(s) = \alpha(1 - s)$ $(s \in I)$. Using the notation of Definition 2.5.24, note that if $f \in \mathcal{L}(x)$, then $\widehat{\alpha} * f * \alpha \in \mathcal{L}(y)$. Consider the map

$$\phi_\alpha : \pi(X, x) \to \pi(X, y)$$

defined (see (i), below) by

$$\phi_\alpha(\langle f \rangle) = \langle \widehat{\alpha} * f * \alpha \rangle.$$

Routine use of Theorems 2.5.9 to 2.5.13 shows that
(i) for all $f, g \in \mathcal{L}(x)$,

$$\langle f \rangle = \langle g \rangle \ \Leftrightarrow \ \phi_\alpha(\langle f \rangle) = \phi_\alpha(\langle g \rangle);$$

(ii) for all $u \in \mathcal{L}(y)$,

$$\phi_\alpha(\langle \alpha * u * \widehat{\alpha} \rangle) = \langle u \rangle;$$

(iii) for all $f, g \in \mathcal{L}(x)$,

$$\phi_\alpha(\langle f \rangle) \phi_\alpha(\langle g \rangle) = \phi_\alpha(\langle f \rangle \langle g \rangle).$$

Detailed proof of (i) to (iii) is left to the reader, but by way of illustration of the procedures to be adopted we indicate how to deal with (iii). For all $f, g \in \mathcal{L}(x)$,

$$
\begin{aligned}
\phi_\alpha(\langle f \rangle)\phi_\alpha(\langle g \rangle) &= \langle(\widehat{\alpha} * f * \alpha) * (\widehat{\alpha} * g * \alpha)\rangle = \langle \widehat{\alpha} * f * \alpha * \widehat{\alpha} * g * \alpha\rangle \\
&= \langle(\widehat{\alpha} * f) * (\alpha * \widehat{\alpha} * g * \alpha)\rangle \\
&= \langle(\widehat{\alpha} * f) * ((\alpha * \widehat{\alpha}) * (g * \alpha))\rangle \\
&= \langle(\widehat{\alpha} * f) * (e_x * (g * \alpha))\rangle = \langle(\widehat{\alpha} * f) * (g * \alpha)\rangle \\
&= \langle \widehat{\alpha} * ((f * g) * \alpha)\rangle = \langle \widehat{\alpha} * (f * g) * \alpha\rangle = \phi_\alpha(\langle f \rangle \langle g \rangle).
\end{aligned}
$$

Statements (i) and (ii) show that ϕ_α is well-defined and bijective; (iii) shows that it is a homomorphism. Hence $\pi(X, x)$ and $\pi(X, y)$ are isomorphic groups. $\qquad\square$

This theorem has immediate corollaries.

Corollary 2.5.26 *Let x and y belong to a path-connected metric space X. Then $\pi(X, x)$ and $\pi(X, y)$ are isomorphic.*

Note that different paths between x and y may generate different isomorphisms.

Corollary 2.5.27 *A metric space X is simply-connected if, and only if, it is path-connected and $\pi(X, x) = \{\langle e_x \rangle\}$ for some (and thus each) $x \in X$.*

To conclude this section, we show that fundamental groups at two points, one from each of two homeomorphic, path-connected metric spaces, are isomorphic. The next result is key in this: it does not require the hypothesis of path-connectedness.

Theorem 2.5.28 *Let X and Y be homeomorphic metric spaces. Let $x \in X$ and suppose that $\psi : X \rightarrow Y$ is a homeomorphism. Then $\pi(X, x)$ and $\pi(Y, \psi(x))$ are isomorphic groups.*

Proof With the notation of Definition 2.5.24, if $f, g \in \mathcal{L}(x)$, then evidently $\psi \circ f$, $\psi \circ g \in \mathcal{L}(\psi(x))$. Further, use of Theorem 2.5.7 shows that

$$
f \sim g \text{ in } \mathcal{L}(x) \Leftrightarrow \psi \circ f \sim \psi \circ g \text{ in } \mathcal{L}(\psi(x)). \tag{2.5.1}
$$

Consider the map $\Psi : \pi(X, x) \rightarrow \pi(Y, \psi(x))$ given by

$$
\Psi(\langle f \rangle) = \langle \psi \circ f \rangle \; (f \in \mathcal{L}(x)).
$$

Because of (2.5.1), the map Ψ is well-defined and injective; it is surjective since $\Psi\left(\langle \psi^{-1} \circ u \rangle\right) = \langle u \rangle$ for each $u \in \mathcal{L}(\psi(x))$; moreover, it is a homomorphism as

$$
\Psi(\langle f \rangle)\Psi(\langle g \rangle) = \langle(\psi \circ f) * (\psi \circ g)\rangle = \langle \psi \circ (f * g)\rangle = \Psi(\langle f * g \rangle) = \Psi(\langle f \rangle \langle g \rangle)
$$

whenever $f, g \in \mathcal{L}(x)$. Thus Ψ is a group isomorphism and $\pi(X, x)$ and $\pi(Y, \psi(x))$ are isomorphic groups. $\qquad\square$

Corollary 2.5.29 *Let X and Y be homeomorphic metric spaces, each of which is path-connected. Then, for arbitrary choice of $x \in X$ and $y \in Y$, the groups $\pi(X, x)$ and $\pi(Y, y)$ are isomorphic.*

Proof The result follows from Theorem 2.5.28 and Corollary 2.5.26.   □

The message of the corollary is that homeomorphic, path-connected spaces give rise to isomorphic fundamental groups.

Exercise 2.5.30

1. Let S^n be the unit sphere in $\mathbf{R}^{n+1}$ (see Example 2.4.21 (iv)), let $f : S^n \to S^n$ be continuous, and suppose that, for all $x \in S^n$, $f(x) \neq -x$. Show that $f \simeq \mathrm{id}_{S^n}$, where id_{S^n} is the identity map on S^n. [Consider the map $H : S^n \times I \to S^n$ defined by

$$H(x, t) = \frac{(1 - t)f(x) + tx}{\|(1 - t)f(x) + tx\|},$$

where $\|u\| = \left(\sum_{j=1}^{n+1} u_j^2\right)^{1/2}$ for $u = (u_1, \ldots, u_{n+1}) \in \mathbf{R}^{n+1}$.]

2. Let x and y be points in a metric space X, and let $\mu, \nu : I \to X$ be paths in X from x to y. Show that $\mu \sim \nu$ if, and only if, $\mu * \widehat{\nu} \sim e_x$.

3. Give examples of closed paths f, g in $\mathbf{R}^2$ such that $(f * f) * f \neq f * (f * f)$ and $(g * g) * g = g * (g * g)$.

4. Let f be a closed path in a metric space X; let $a \in I$ and define $g : I \to X$ by

$$g(s) = \begin{cases} f(s + a) & \text{if } 0 \leq s \leq 1 - a, \\ f(a + s - 1) & \text{if } 1 - a \leq s \leq 1. \end{cases}$$

Show that g is a closed path in X, that $g^* = f^*$ and that $H : I \times I \to X$ defined by

$$H(s, t) = \begin{cases} f(s + ta) & \text{if } 0 \leq s \leq 1 - ta, \\ f(ta + s - 1) & \text{if } 1 - ta \leq s \leq 1 \end{cases}$$

establishes a free homotopy between f and g.

5. Generalise Example 2.5.2: let X be a metric space, Y be a subspace of $\mathbf{R}^n$ (a non-empty subset of $\mathbf{R}^n$ endowed with the metric inherited from $\mathbf{R}^n$, not to be confused with a linear subspace) and $f_0, f_1 : X \to Y$ be continuous maps such that, for all $(x, t) \in X \times I$, $(1 - t)f_0(x) + tf_1(x) \in Y$. Show that $f_0 \simeq f_1$.

6. Two metric spaces X and Y are said to be **homotopy-equivalent** (written $X \simeq Y$) if there exist continuous maps $f : X \to Y$ and $g : Y \to X$ such that $g \circ f \simeq \mathrm{id}_X$ and $f \circ g \simeq \mathrm{id}_Y$, where $\mathrm{id}_X : X \to X$ and $\mathrm{id}_Y : Y \to Y$ are the identity maps. Prove that homotopy-equivalence is an equivalence relation on the family of all metric spaces. Note that homeomorphic spaces are homotopy-equivalent; also, as illustrated below, the converse need not hold.

7. (i) Let X and Y be the subspaces of $\mathbf{R}^2$ given by $X = S^1$ and $Y = S^1 \cup \{(x, 0) : 1 \leq x \leq 2\}$. Prove that X and Y are homotopy-equivalent but not homeomorphic.

[Hint: consider maps $f : X \to Y$, $g : Y \to X$ defined respectively by $f(x) = x$ if $x \in X$, $g(y) = y$ if $y \in S^1$, $g(y) = (1, 0)$ if $y \in Y \backslash S^1$.]

(ii) Let X and Y be subspaces of $\mathbf{R}^2$ given by $X = S^1$ and $Y = \mathbf{R}^2 \backslash \{0\}$. Show that X and Y are homotopy-equivalent but not homeomorphic. [Hint: consider the map $f : X \to Y$ given by $f(x) = x$, and the map $g : Y \to X$ defined by $g(y) = |y|^{-1} y$, where $|y| = (y_1^2 + y_2^2)^{1/2}$ for $y = (y_1; y_2) \in \mathbf{R}^2$.]

8. A metric space X is called **contractible** if the identity map $id : X \to X$ is homotopic to a constant map. Prove that X is contractible if, and only if, X is homotopy-equivalent to a space consisting of a single point. Show that every convex, non-empty subset of $\mathbf{R}^n$ is contractible.

Chapter 3
Complex Analysis

The theory of complex analysis, which is based on the fundamental work of Cauchy, forms a most attractive, beautiful and useful part of elementary analysis. Quite apart from its structural beauty, it also has the quality of unexpectedness which distinguishes outstanding pieces of mathematics from the rest and which is responsible for a good deal of the charm of the subject. Reverting to Chap. 1, it is surely delightful that the theory of *complex* analysis gives rise to diverse *real-variable* results such as

$$\int_0^\infty \frac{x}{1+x^5}dx = \frac{\pi}{5\sin\frac{2\pi}{5}} \text{ and } \sum_{n=0}^\infty \frac{(-1)^n}{(2n+1)^3} = \frac{\pi^3}{32}.$$

We develop the cornerstones of this theory quite rapidly in this chapter.

To begin with, we introduce basic concepts, including power series, branches of the argument and the logarithm, the winding number for arbitrary paths in the plane and integrals over contours. Some of these topics are discussed in more detail than is common in books at this level: indeed the notion of the winding number is often not introduced at all! This groundwork enables a global (homology) form of Cauchy's theorem to be established, a result that is central to the determination of necessary and sufficient conditions under which an analytic function has a primitive. Moreover, it leads to the residue theorem with applications not only to the evaluation of definite integrals but also to Rouché's theorem, the open mapping theorem and the inverse function theorem. Further reward for the early preparation comes in Sects. 3.8 and 3.9 in which the Riemann mapping theorem and the Jordan curve theorem are shown to arise in an aesthetically pleasing way from these foundations. In particular, the work on the winding number is necessary for the straightforward and natural proof of the Jordan curve theorem that is given: the related concept of the index is inadequate for this purpose as it is not defined for general paths with no smoothness.

R. H. Dyer and D. E. Edmunds, *From Real to Complex Analysis*, Springer Undergraduate Mathematics Series, DOI: 10.1007/978-3-319-06209-9_3, © Springer International Publishing Switzerland 2014

3.1 Complex Numbers

Definition 3.1.1 A **complex number** is an ordered pair (a, b) of real numbers. If $z_1 = (a_1, b_1)$ and $z_2 = (a_2, b_2)$ are complex numbers, we write $z_1 = z_2$ if, and only if, $a_1 = a_2$ and $b_1 = b_2$. The **set of all complex numbers** is denoted by $\mathbf{C}$. **Addition** and **multiplication** of elements of $\mathbf{C}$ are defined by the rules

$$z_1 + z_2 = (a_1 + a_2, b_1 + b_2), z_1 z_2 = (a_1 a_2 - b_1 b_2, a_1 b_2 + a_2 b_1);$$

if $a^2 + b^2 \neq 0$ the **inverse** of $z = (a, b)$ is defined by $z^{-1} = \left(\frac{a}{a^2+b^2}, -\frac{b}{a^2+b^2} \right)$.

Plainly, addition and multiplication of complex numbers are both commutative and associative; the definition of z^{-1} is so chosen that $z z^{-1} = (1, 0)$ if $z \neq (0, 0)$. Note that the adjective 'ordered' in the definition simply means that (a, b) and (b, a) are regarded as distinct objects if $a \neq b$. Moreover, the definition evidently implies that for all $a_1, a_2 \in \mathbf{R}$,

$$(a_1, 0) + (a_2, 0) = (a_1 + a_2, 0) \text{ and } (a_1, 0)(a_2, 0) = (a_1 a_2, 0).$$

This indicates that complex numbers of the form $(a, 0)$ have the same arithmetical properties as the corresponding real numbers a; we therefore shall identify $(a, 0)$ with a, so that $\mathbf{R}$ may be identified with a subset of $\mathbf{C}$. In fact, it is plain that $\mathbf{C}$ is a field and that $\mathbf{R}$ may be identified with a subfield of $\mathbf{C}$.

Definition 3.1.2 The complex number $(0, 1)$ will be denoted by i.

Proposition 3.1.3 *Let ii be denoted by i^2. Then $i^2 = (-1, 0)$; with the identification made above, $i^2 = -1$.*

Proof By Definition 3.1.1, $i^2 = (0, 1)(0, 1) = (-1, 0)$. □

Proposition 3.1.4 *Given any $a, b \in \mathbf{R}$, $(a, b) = a + ib$.*

Proof We have

$$(a, b) = (a, 0) + (0, b) = (a, 0) + (0, 1)(b, 0) = a + ib. \qquad □$$

Because of this result we shall usually write the complex number (a, b) in the form $a + ib$.

Definition 3.1.5 Given any complex number $z = x + iy$ ($x, y \in \mathbf{R}$), the complex number $x - iy$ is called the **conjugate** of z and is written $\bar{z}$; x is called the **real part** of z and is denoted by re z; y is the **imaginary part** of z and is denoted by im z.

Proposition 3.1.6 *Let $w, z \in \mathbf{C}$. Then*

(i) $\overline{w + z} = \overline{w} + \overline{z}$;
(ii) $\overline{wz} = \overline{w}\,\overline{z}$;
(iii) $z + \overline{z} = 2 \operatorname{re} z$, $z - \overline{z} = 2i \operatorname{im} z$;
(iv) $z\overline{z}$ *is real, and positive save when $z = 0$.*

Proof (i) Put $w = u + iv$, $z = x + iy$, where $u, v, x, y \in \mathbf{R}$. Then

$$\overline{w + z} = \overline{(u + x) + i(v + y)} = u + x - i(v + y) = (u - iv) + (x - iy) = \overline{w} + \overline{z}.$$

(ii) $\overline{wz} = \overline{(ux - vy) + i(uy + vx)} = (ux - vy) - i(uy + vx) = (u - iv)(x - iy) = \overline{w}\overline{z}$.

(iii) $z + \overline{z} = x + iy + x - iy = 2x = 2 \operatorname{re} z$ and $z - \overline{z} = x + iy - (x - iy) = 2iy = 2i \operatorname{im} z$.

(iv) $z\overline{z} = (x + iy)(x - iy) = x^2 + y^2 \geq 0$; $z\overline{z} = 0$ if, and only if, $x = y = 0$. □

Definition 3.1.7 Let $z \in \mathbf{C}$. The **absolute value**, $|z|$, of z is the non-negative square root of $z\overline{z}$.

Thus $|z| = (z\overline{z})^{1/2}$, so that if $z = x + iy$ ($x, y \in \mathbf{R}$), $|z| = (x^2 + y^2)^{1/2}$. When z is real, $z = x$, say,

$$|x| = (x^2)^{1/2} = \begin{cases} x, & x \geq 0, \\ -x, & x < 0. \end{cases}$$

Proposition 3.1.8 *Let $w, z \in \mathbf{C}$. Then*

(i) $|z| > 0$ *unless $z = 0$, $|0| = 0$;*
(ii) $|z| = |\overline{z}|$;
(iii) $|wz| = |w|\,|z|$;
(iv) $|\operatorname{re} z| \leq |z|$, $|\operatorname{im} z| \leq |z|$;
(v) $|w + z| \leq |w| + |z|$.

Proof The proofs of (i)–(iv) are elementary, and are left to the reader. As for (v), with $w = u + iv$, $z = x + iy$ as usual, we have

$$\begin{aligned} |w + z|^2 &= (u + x)^2 + (v + y)^2 = (u^2 + v^2) + (x^2 + y^2) + 2(ux + vy) \\ &\leq |w|^2 + |z|^2 + 2(u^2 + v^2)^{1/2}(x^2 + y^2)^{1/2} \\ &= (|w| + |z|)^2. \end{aligned}$$

□

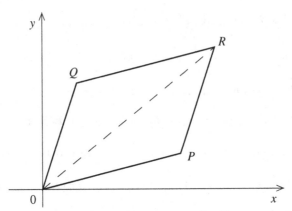

Property (v) is the **triangle inequality** and has a familiar geometrical interpretation. Represent complex numbers by points in the plane in the natural way, so that $z = x + iy$ is represented by the point P with coordinates (x, y), and $w = u + iv$ by the point Q with coordinates (u, v). Completion of the parallelogram $OPRQ$ gives the point R which corresponds to $w + z$, and the triangle inequality amounts to the familiar result that $OR \leq OP + OQ$, pictured above.

We now define powers of complex numbers.

Definition 3.1.9 Let $z \in \mathbf{C}$ and $n \in \mathbf{Z}$. Define

$$z^0 = 1, z^{n+1} = z^n z (n \geq 0),$$
$$z^{-1} = (x - iy)/(x^2 + y^2) \ (z = x + iy \neq 0), \ z^{-n} = (z^{-1})^n \ (z \neq 0, n > 0).$$

The usual laws of exponents hold.

Proposition 3.1.10 Let $m, n \in \mathbf{Z}$. Then $z^m z^n = z^{m+n}$ $(z \neq 0)$, $z_1^n z_2^n = (z_1 z_2)^n$ $(z_1, z_2 \neq 0)$.

The proof is by induction.

The set of all complex numbers may be furnished with a natural metric.

Proposition 3.1.11 Define $d\colon \mathbf{C} \times \mathbf{C} \to \mathbf{R}$ by $d(w, z) = |w - z|$. Then d is a metric on $\mathbf{C}$.

Proof The result is clear, in view of Proposition 3.1.8.   □

As sets, $\mathbf{R}^2$ and $\mathbf{C}$ are identical. They are distinguished by the algebraic structures they carry: $\mathbf{R}^2$ is a vector space over the reals and $\mathbf{C}$ is a field. Of course, the metric space $(\mathbf{C}, d)$ is identical with $(\mathbf{R}^2, \tilde{d})$, where $\tilde{d}$ is the Euclidean metric on $\mathbf{R}^2$. In view of this, we may take over all the properties of $(\mathbf{R}^2, \tilde{d})$ established in Chap. 2 to $(\mathbf{C}, d)$. Thus a sequence (z_n) of complex numbers is said to converge to $z \in \mathbf{C}$ if, and only if, $d(z_n, z) = |z_n - z| \to 0$ as $n \to \infty$; and (z_n) converges to z if, and only if, $\operatorname{re} z_n \to \operatorname{re} z$ and $\operatorname{im} z_n \to \operatorname{im} z$ as $n \to \infty$; a sequence (z_n) in $\mathbf{C}$ is a Cauchy sequence if, and only if, $d(z_n, z_m) = |z_n - z_m| \to 0$ as $m, n \to \infty$; $(\mathbf{C}, d)$ is a complete metric space. Henceforth we shall assume that $\mathbf{C}$ is endowed with the metric d, so that matters of convergence, continuity, openness etc. are to be understood in this sense.

Proposition 3.1.12 *Let (z_n) and (z_n') be sequences in $\mathbf{C}$ such that $z_n \to z \in \mathbf{C}$ and $z_n' \to z' \in \mathbf{C}$. Then $z_n + z_n' \to z + z'$ and $z_n z_n' \to zz'$; and if, in addition, $z \neq 0$ and for all $n \in \mathbf{N}$, $z_n \neq 0$, we have $z_n^{-1} \to z^{-1}$.*

Proof Use of the triangle inequality shows that

$$\left| z_n + z_n' - (z + z') \right| \leq |z_n - z| + |z_n' - z'| \to 0;$$

moreover, that

$$\begin{aligned}
\left| z_n z_n' - zz' \right| &= \left| z_n(z_n' - z') + z'(z_n - z) \right| \\
&= \left| (z_n - z)(z_n' - z') + z(z_n' - z') + z'(z_n - z) \right| \\
&\leq \left| (z_n - z)(z_n' - z') \right| + |z| \left| z_n' - z' \right| + |z'| \left| z_n - z \right| \to 0.
\end{aligned}$$

Hence $z_n + z_n' \to z + z'$ and $z_n z_n' \to zz'$. Regarding the inverses we have $|z_n| \geq \frac{1}{2} |z|$ for all $n \geq K$, say. Hence

$$\left| z_n^{-1} - z^{-1} \right| = |z - z_n| / |z_n z| \leq 2 |z - z_n| / |z|^2$$

for all $n \geq K$, and the proof is complete. $\qquad\qquad\qquad\qquad\qquad\square$

Convergence of series of complex numbers is dealt with just as in the real case.

Definition 3.1.13 Let (z_n) be a sequence of complex numbers and put $w_n = \sum_{r=1}^{n} z_r$ $(n \in \mathbf{N})$. If (w_n) converges to a limit $w \in \mathbf{C}$, we say that w is the **sum** of the series $\sum_{1}^{\infty} z_n$ and that the series $\sum_{1}^{\infty} z_n$ converges; w_n is called the nth **partial sum** of the series and we write $w = \sum_{1}^{\infty} z_n$. If $\sum_{1}^{\infty} |z_n|$ converges, $\sum_{1}^{\infty} z_n$ is said to converge **absolutely**.

Proposition 3.1.14 *If $\sum_{1}^{\infty} |z_n|$ converges, so does $\sum_{1}^{\infty} z_n$.*

Proof Put $W_n = \sum_{r=1}^{n} |z_r|$, $w_n = \sum_{r=1}^{n} z_r$ $(n \in \mathbf{N})$. If $m > n$, then

$$|w_m - w_n| = \left| \sum_{r=n+1}^{m} z_r \right| \leq \sum_{r=n+1}^{m} |z_r| = W_m - W_n \to 0$$

as $m, n \to \infty$. Thus (w_n) is a Cauchy sequence, which converges as $(\mathbf{C}, d)$ is complete. $\qquad\qquad\qquad\qquad\qquad\square$

The ratio and root tests for convergence hold for series of complex numbers just as for the real situation.

Theorem 3.1.15 (The ratio test) *Let (a_n) be a sequence of non-zero complex numbers and let*

$$r = \liminf_{n \to \infty} \left| \frac{a_{n+1}}{a_n} \right|, \; R = \limsup_{n \to \infty} \left| \frac{a_{n+1}}{a_n} \right|,$$

where $0 \le r \le R \le \infty$. Then $\sum a_n$ converges absolutely if $R < 1$, and diverges if $r > 1$. No information is given if $r \le 1 \le R$.

Proof First suppose that $R < 1$ and let λ be such that $R < \lambda < 1$. From the definition of R we know that there exists $N \in \mathbf{N}$ with $|a_{n+1}/a_n| < \lambda$ for all $n \ge N$. Hence $|a_{n+1}| < \lambda |a_n| \le \lambda^{n+1-N} |a_N|$ if $n \ge N$, and so comparison with the convergent series $\lambda^{-N} |a_N| \sum_{n=N}^{\infty} \lambda^n$ shows that $\sum a_n$ is absolutely convergent.

Next, suppose that $r > 1$. Then there exists $M \in \mathbf{N}$ such that $|a_{n+1}| > |a_n|$ for all $n \ge M$, and so $a_n \nrightarrow 0$ as $n \to \infty$: $\sum a_n$ must be divergent.

Finally, the series $\sum n^{-1}$ and $\sum n^{-2}$ both have $r = R = 1$, but the first is divergent while the second converges. $\qquad \square$

As an immediate consequence of this theorem we have

Corollary 3.1.16 *Let (a_n) be a sequence of non-zero complex numbers and suppose that $\lim_{n \to \infty} |a_{n+1}/a_n|$ exists and equals l, where $0 \le l \le \infty$. Then $\sum a_n$ is absolutely convergent if $l < 1$, and diverges if $l > 1$.*

Theorem 3.1.17 (The root test) *Let (a_n) be a sequence of complex numbers and let $r = \limsup_{n \to \infty} |a_n|^{1/n}$. Then $\sum a_n$ is absolutely convergent if $r < 1$, and diverges if $r > 1$. No information is given if $r = 1$.*

Proof Suppose that $r < 1$ and let $\lambda \in (r, 1)$. Then there exists $N \in \mathbf{N}$ such that $|a_n| < \lambda^n$ for all $n \ge N$. Comparison with $\sum \lambda^n$ now shows that $\sum a_n$ is absolutely convergent. On the other hand, if $r > 1$, then $|a_n| > 1$ for infinitely many n and so $a_n \nrightarrow 0$ as $n \to \infty$; thus $\sum a_n$ is divergent.

The last part follows from consideration of $\sum n^{-1}$ and $\sum n^{-2}$, both of which have $r = 1$. $\qquad \square$

Having dealt with sequences and series of complex numbers, we now turn to complex-valued functions. The definitions of pointwise and uniform convergence, and the principal results associated with these concepts, are exactly the same as for real-valued functions, which were dealt with in 1.7. For convenience we shall, however, give the basic definitions and results here.

Definition 3.1.18 Let S be a non-empty set and let (f_n) be a sequence of functions $f_n : S \to \mathbf{C}$. The sequence (f_n) is said to converge **pointwise on** S if there is a function $f : S \to \mathbf{C}$ such that for all $s \in S$, $f_n(s) \to f(s)$; (f_n) converges **uniformly on** S if there is a function $f : S \to \mathbf{C}$ such that given any $\varepsilon > 0$, there exists $N \in \mathbf{N}$ with $|f_n(s) - f(s)| < \varepsilon$ for all $s \in S$ and all $n \ge N$.

Theorem 3.1.19 (Cauchy's general principle of uniform convergence) *Let S be a non-empty set and let (f_n) be a sequence of functions $f_n : S \to \mathbf{C}$. Then (f_n) converges uniformly on S if, and only if, for all $\varepsilon > 0$, there exists $N \in \mathbf{N}$ with $|f_n(s) - f_m(s)| < \varepsilon$ for all $s \in S$, all $m \geq N$ and all $n \geq N$.*

The proof is exactly the same as that of Theorem 1.7.3.

Theorem 3.1.20 *Let X be a metric space and let (f_n) be a sequence of continuous functions $f_n : X \to \mathbf{C}$ which converges uniformly on X to $f : X \to \mathbf{C}$. Then f is continuous on X.*

This is a special case of Theorem 2.1.42.

Corollary 3.1.21 *If X is a compact metric space, then $C(X, \mathbf{C})$ is complete.*

Given that every element of $C(X, \mathbf{C})$ is bounded, the proof mimics that of Theorem 2.2.6 and Corollary 2.2.7.

Definition 3.1.22 Let S be a non-empty set, let (f_n) be a sequence of functions $f_n : S \to \mathbf{C}$ and put $u_n(s) = \sum_{r=1}^{n} f_r(s)$ $(s \in S, n \in \mathbf{N})$; u_n is the nth partial sum of the series $\sum f_r$. If the sequence (u_n) converges pointwise (uniformly) on S, the series $\sum f_r$ is said to converge pointwise (uniformly) on S.

Theorem 3.1.23 (The Weierstrass M-test) *Let S be a non-empty set, let (f_n) be a sequence of functions $f_n : S \to \mathbf{C}$ and suppose there are constants M_n such that for all $s \in S$ and all $n \in \mathbf{N}$, $|f_n(s)| \leq M_n$; suppose also that $\sum M_n$ converges. Then $\sum f_n$ converges uniformly on S.*

The proof is precisely the same as that of Theorem 1.7.5.

Exercise 3.1.24

1. State the axioms for a field and check that, when endowed with the usual operations of addition and multiplication, the set $\mathbf{C}$ of all complex numbers forms a field. Show that it is not an ordered field.
2. Prove that for all $z_1, z_2 \in \mathbf{C}$, $||z_1| - |z_2|| \leq |z_1 - z_2|$.
3. Show that for any $n \in \mathbf{N}$ and any $z_1, \ldots, z_n \in \mathbf{C}$,

$$|z_1 + \cdots + z_n| \leq |z_1| + \cdots + |z_n|,$$

and show that the conditions $z_j \bar{z}_k = |z_j| |z_k|$ $(j, k = 1, 2, \ldots, n)$ are necessary and sufficient for equality.
4. Let $a, b \in \mathbf{C}$. Show that $\operatorname{im}(a\bar{b}) = 0$ if, and only if, there exist real numbers λ and μ, not both zero, such that $\lambda a = \mu b$.
5. Let $a_1, \ldots, a_n$ and $b_1, \ldots, b_n$ be complex numbers. Prove that

$$\left| \sum_{1}^{n} a_k b_k \right|^2 \leq \left(\sum_{1}^{n} |a_k|^2 \right) \left(\sum_{1}^{n} |b_k|^2 \right).$$

6. Let p be a polynomial of degree n, with $p(z) = \sum_0^n a_k z^k$ $(z \in \mathbf{C})$, and suppose that $a_0 > a_1 > \ldots > a_n > 0$. Prove that if $p(z) = 0$, then $|z| > 1$. (Consider $(1 - z)p(z)$.)

3.2 Analytic Functions: The Cauchy-Riemann Equations

We remind the reader that words such as **open, closed, connected** as applied to subsets of $\mathbf{C}$ are to be interpreted in the sense of the metric space $(\mathbf{C}, d)$, where $d(w, z) = |w - z|$ $(w, z \in \mathbf{C})$. Sets which are open and connected are of sufficient importance to warrant special terminology.

Definition 3.2.1 A subset of $\mathbf{C}$ which is non-empty, open and connected is called a **region**.

Our main concern in this section is with differentiable functions, and we begin with the definition.

Definition 3.2.2 Let G be an open subset of $\mathbf{C}$ and let $z_0 \in G$. A function $f : G \to \mathbf{C}$ is said to be **differentiable at** z_0 if there is a complex number λ such that $\lim_{z \to z_0} \frac{f(z) - f(z_0)}{z - z_0} = \lambda$; that is, if given any $\varepsilon > 0$, there exists $\delta > 0$ such that

$$\left| \frac{f(z) - f(z_0)}{z - z_0} - \lambda \right| < \varepsilon \text{ if } 0 < |z - z_0| < \delta.$$

If this limit exists it is unique, is denoted by $f'(z_0)$ and is called the **derivative of** f **at** z_0. If $f'(z)$ exists at each point z in some neighbourhood of z_0, f is said to be **analytic at** z_0. If f is differentiable at each point of an open subset U of G, then f is said to be **analytic** (or **holomorphic**) **in** U. The family of all functions $f : G \to \mathbf{C}$ which are analytic in G is denoted by $H(G)$; a function that belongs to $H(\mathbf{C})$ is said to be **entire**.

Just as in the case of real-valued functions of a real variable, natural properties of differentiable functions can be established.

Lemma 3.2.3 *Let G be an open, non-empty subset of $\mathbf{C}$, let $c \in \mathbf{C}$ and define functions $f, g : G \to \mathbf{C}$ by $f(z) = z$, $g(z) = c$ for all $z \in G$. Then $f'(z) = 1$ and $g'(z) = 0$ for all $z \in G$.*

The simple proof is left to the reader.

Theorem 3.2.4 *Let G be an open subset of $\mathbf{C}$, let $z_0 \in G$ and let $f, g : G \to \mathbf{C}$ be differentiable at z_0. Then*

(i) *f is continuous at z_0;*
(ii) *$(f + g)'(z_0) = f'(z_0) + g'(z_0)$;*

(iii) $(fg)'(z_0) = f'(z_0)g(z_0) + f(z_0)g'(z_0)$;

(iv) *if* $g(z_0) \neq 0$, *then* $(f/g)'(z_0) = \{f'(z_0)g(z_0) - f(z_0)g'(z_0)\}/g^2(z_0)$;

(v) *if* $f, g \in H(G)$, *then* $f + g$ *and* fg *belong to* $H(G)$.

The proof is omitted as it is exactly the same as the proof of these assertions for real-valued functions of a real variable.

The chain rule for the differentiation of composite functions holds just as in the real case.

Theorem 3.2.5 *Let* G, Ω *be open subsets of* **C**, *let* $f \in H(G)$, $g \in H(\Omega)$ *and suppose that* $f(G) \subset \Omega$. *Then* $g \circ f \in H(G)$ *and*

$$(g \circ f)'(w) = g'(f(w))f'(w) \text{ for all } w \in G.$$

Proof Let $w \in G$. For $z \in G, z \neq w$,

$$\frac{g(f(z)) - g(f(w))}{z - w} = \Phi(z)\Psi(z),$$

where $\Phi, \Psi : G \to$ **C** are defined by

$$\Phi(z) = \begin{cases} \frac{g(f(z)) - g(f(w))}{f(z) - f(w)} & \text{if } f(z) \neq f(w), \\ g'(f(w)) & \text{if } f(z) = f(w), \end{cases}$$

and

$$\Psi(z) = \begin{cases} \frac{f(z) - f(w)}{z - w} & \text{if } z \neq w, \\ f'(w) & \text{if } z = w. \end{cases}$$

Evidently Ψ is continuous at w; so is Φ, but this needs proof. Given $\varepsilon > 0$, there exists $\delta > 0$ such that

$$\left| \frac{g(\sigma) - g(f(w))}{\sigma - f(w)} - g'(f(w)) \right| < \varepsilon$$

if $0 < |\sigma - f(w)| < \delta$. Since f is continuous at w, there exists $\mu > 0$ such that $|f(z) - f(w)| < \delta$ if $|z - w| < \mu$. Hence $|\Phi(z) - g'(f(w))| < \varepsilon$ if $|z - w| < \mu$, and so Φ is continuous at w, with $\lim_{z \to w} \Phi(z) = g'(f(w))$. Since Φ and Ψ are continuous at w,

$$(g \circ f)'(w) = \lim_{z \to w} \frac{g(f(z)) - g(f(w))}{z - w} = \Phi(w)\Psi(w) = g'(f(w))f'(w).$$

□

Theorem 3.2.6 *Let* G *and* Ω *be open subsets of* **C**. *Let* $f : G \to$ **C** *and* $g : \Omega \to$ **C** *be continuous functions such that* $f(G) \subset \Omega$ *and* $g(f(z)) = z$ *for all* $z \in G$.

Then if $g \in H(\Omega)$ and for all $w \in \Omega$, $g'(w) \neq 0$, the function f is in $H(G)$ and $f'(z) = 1/g'(f(z))$ for all $z \in G$.

Proof Let $a \in G$. Then if $z \in G$ and $z \neq a$, it follows that $f(z) \neq f(a)$, for otherwise $g(f(z)) = g(f(a))$ and so $z = a$. Thus for $z \in G$, with $z \neq a$,

$$1 = \frac{g(f(z)) - g(f(a))}{z - a} = \frac{g(f(z)) - g(f(a))}{f(z) - f(a)} \cdot \frac{f(z) - f(a)}{z - a},$$

so that

$$\frac{f(z) - f(a)}{z - a} = \frac{1}{\Psi(z)},$$

where

$$\Psi(z) = \frac{g(f(z)) - g(f(a))}{f(z) - f(a)} \quad \text{if } z \neq a,\ \Psi(a) = g'(f(a)).$$

We claim that $\Psi : G \to \mathbf{C}$ is continuous at a. Accepting this for the moment, and noting that $\Psi(a) \neq 0$, it follows that

$$\lim_{z \to a} \frac{f(z) - f(a)}{z - a} = \frac{1}{\Psi(a)},$$

so that f is differentiable at a and $f'(a) = 1/g'(f(a))$. That $f \in H(G)$ is now plain.

It remains to show that $\lim_{z \to a} \Psi(z) = g'(f(a))$. Let $\varepsilon > 0$. Then there exists $\eta > 0$ such that

$$\left| \frac{g(w) - g(f(a))}{w - f(a)} - g'(f(a)) \right| < \varepsilon \qquad \text{if } 0 < |w - f(a)| < \eta.$$

Moreover, there exists $\delta > 0$ such that $0 < |f(z) - f(a)| < \eta$ if $0 < |z - a| < \delta$. Hence

$$\left| \frac{g(f(z)) - g(f(a))}{f(z) - f(a)} - g'(f(a)) \right| < \varepsilon \qquad \text{if } 0 < |z - a| < \delta,$$

and our claim is justified.   □

This result will be used later to show that branches of the logarithm are analytic. Next, the notion of partial derivatives of real-valued functions is needed.

Definition 3.2.7 Let G be an open subset of $\mathbf{C}$, let $x_0 + iy_0 \in G$ and let $u : G \to \mathbf{R}$; denote the value of u at $x + iy \in G$ by $u(x, y)$ (identifying $x + iy$ with $(x, y) \in \mathbf{R}^2$ in the usual way). If the function $x \longmapsto u(x, y_0)$ is differentiable at x_0 we say that the **first partial derivative of** u **with respect to** x **at** (x_0, y_0) exists, and that its value is the derivative of $x \longmapsto u(x, y_0)$ at x_0, denoted by $u_1(x_0, y_0)$ (or by $u_x(x_0, y_0)$ or

$\frac{\partial u}{\partial x}(x_0, y_0)$). The first partial derivative of u with respect to y at (x_0, y_0) is defined analogously, and is denoted by $u_2(x_0, y_0)$, $u_y(x_0, y_0)$ or $\frac{\partial u}{\partial y}(x_0, y_0)$.

In other words,

$$u_1(x_0, y_0) = \lim_{x \to x_0} \frac{u(x, y_0) - u(x_0, y_0)}{x - x_0}$$

and

$$u_2(x_0, y_0) = \lim_{y \to y_0} \frac{u(x_0, y) - u(x_0, y_0)}{y - y_0}.$$

For example, if $G = \mathbf{C}$ and $u(x, y) = e^x \cos y$, then for all $(x_0, y_0) \in \mathbf{C}$, $u_1(x_0, y_0) = e^{x_0} \cos y_0$ and $u_2(x_0, y_0) = -e^{x_0} \sin y_0$.

Theorem 3.2.8 *Let G be an open subset of $\mathbf{C}$, let $z_0 = (x_0, y_0) \in G$ and let $f : G \to \mathbf{C}$ be differentiable at z_0; define maps $u, v : G \to \mathbf{R}$ by $u(x, y) = \operatorname{re} f(z)$, $v(x, y) = \operatorname{im} f(z)$ $(z = x + iy \in G)$. Then the partial derivatives u_1, u_2, v_1, v_2 all exist at (x_0, y_0) and*

$$f'(z_0) = u_1(x_0, y_0) + iv_1(x_0, y_0) = v_2(x_0, y_0) - iu_2(x_0, y_0).$$

*In particular, we have the so-called **Cauchy-Riemann** equations*

$$u_1(x_0, y_0) = v_2(x_0, y_0), u_2(x_0, y_0) = -v_1(x_0, y_0).$$

Proof For any $h = h_1 + ih_2 \neq 0$ such that $z_0 + h \in G$ we have

$$f(z_0 + h) - f(z_0) = \{u(x_0 + h_1, y_0 + h_2) - u(x_0, y_0)\}$$
$$+ i \{v(x_0 + h_1, y_0 + h_2) - v(x_0, y_0)\}.$$

Since f is differentiable at z_0, it follows that given any $\varepsilon > 0$, there exists $\delta > 0$ such that

$$\left| \frac{f(z_0 + h) - f(z_0)}{h} - f'(z_0) \right| < \varepsilon \qquad \text{if } 0 < |h| < \delta.$$

Thus if $h_2 = 0$ and $0 < |h_1| < \delta$, we have, writing $U = \frac{u(x_0 + h_1, y_0) - u(x_0, y_0)}{h_1}$ and $V = \frac{v(x_0 + h_1, y_0) - v(x_0, y_0)}{h_1}$,

$$\left| U + iV - f'(z_0) \right| < \varepsilon.$$

Hence

$$\lim_{h_1 \to 0} \frac{u(x_0 + h_1, y_0) - u(x_0, y_0)}{h_1} = \operatorname{re} f'(z_0)$$

and

$$\lim_{h_1 \to 0} \frac{v(x_0 + h_1, y_0) - v(x_0, y_0)}{h_1} = im f'(z_0).$$

Thus u_1 and v_1 exist at (x_0, y_0) and $f'(z_0) = u_1(x_0, y_0) + iv_1(x_0, y_0)$. In a similar way, putting $h_1 = 0$ this time, we see that v_2 and u_2 exist at (x_0, y_0) and that $f'(z_0) = v_2(x_0, y_0) - iu_2(x_0, y_0)$. Comparison of these two forms of $f'(z_0)$ gives the Cauchy-Riemann equations. □

We remark that it may happen that the Cauchy-Riemann equations hold at a point but that the function concerned is not differentiable at that point: consider the function $f : \mathbf{C} \to \mathbf{C}$ defined by $f(z) = \sqrt{|xy|}$ $(z = x + iy \in \mathbf{C})$ and the point 0.

Corollary 3.2.9 *If $f \in H(G)$, then the functions u_1, u_2, v_1 and v_2 are continuous in G and satisfy the Cauchy-Riemann equations in G.*

Proof That the Cauchy-Riemann equations hold throughout G is clear from Theorem 3.2.8; that the functions u_1, u_2, v_1 and v_2 are continuous in G follows from a remarkable result (to be proved later: Theorem 3.6.10) that f' is analytic and thus continuous in G. □

It is here that we first begin to see how much stronger the notion of differentiability is for complex-valued functions than for real-valued ones.

The converse of Corollary 3.2.9 also holds, as we now prove.

Theorem 3.2.10 *Let G be an open subset of $\mathbf{C}$ and let $u : G \to \mathbf{R}$ and $v : G \to \mathbf{R}$ have first-order partial derivatives u_1, u_2, v_1 and v_2 which are continuous in G and satisfy the Cauchy-Riemann equations at each point of G:*

$$u_1(x, y) = v_2(x, y), u_2(x, y) = -v_1(x, y) \text{ for all points } (x, y) \text{ in } G.$$

Then the function $f : G \to \mathbf{C}$ defined by $f(z) = u(x, y) + v(x, y)$ $(z = (x, y) \in G)$ is analytic in G.

Proof Let $z = (x, y) \in G$ and let $r > 0$ be so small that $z + h \in G$ if $|h| < r$. Then if $h = (h_1, h_2)$ and $0 < |h| < r$,

$$\begin{aligned}
f(z + h) - f(z) &= \{u(x + h_1, y + h_2) - u(x, y + h_2)\} + \{u(x, y + h_2) - u(x, y)\} \\
&\quad + i\{v(x + h_1, y + h_2) - v(x, y + h_2)\} + i\{v(x, y + h_2) - v(x, y)\} \\
&= h_1 u_1(x + h_1', y + h_2) + h_2 u_2(x, y + h_2') \\
&\quad + i\{h_1 v_1(x + h_1'', y + h_2) + h_2 v_2(x, y + h_2'')\}
\end{aligned}$$

for some h_1', h_1'', h_2', h_2'' with $|h_1'|, |h_1''| \le |h_1|$ and $|h_2'|, |h_2''| \le |h_2|$, the final step following from the mean-value theorem. Use of the Cauchy-Riemann equations gives

$$\left| f(z+h) - f(z) - h\{u_1(x, y) + iv_1(x, y)\} \right|$$
$$\leq \left| h_1\{u_1(x+h'_1, y+h_2) - u_1(x, y)\} \right| + \left| h_2\{u_2(x, y+h'_2) - u_2(x, y)\} \right|$$
$$+ \left| h_1\{v_1(x+h''_1, y+h_2) - v_1(x, y)\} \right|$$
$$+ \left| h_2\{v_2(x, y+h''_2) - v_2(x, y)\} \right|, \tag{3.2.1}$$

and in view of the continuity of u_1, u_2, v_1 and v_2 at (x, y), given any $\varepsilon > 0$, there exists $\delta > 0$ such that the right-hand side of (3.2.1) is less than or equal to

$$\frac{1}{2}\varepsilon(|h_1| + |h_2|) \leq \varepsilon(|h_1|^2 + |h_2|^2)^{1/2} = \varepsilon |h| \quad \text{if } |h| < \delta.$$

Hence

$$\left| \frac{f(z+h) - f(z)}{h} - u_1(x, y) - iv_1(x, y) \right| \leq \varepsilon \quad \text{if } 0 < |h| < \delta,$$

and so $f'(z) = u_1(x, y) + iv_1(x, y)$ and f is analytic in G. □

To conclude this section we prove that an analytic function with zero derivative everywhere in an open set is constant, provided that the open set is connected: the connectedness is essential, as it rules out the situation in which the function takes different constant values on disjoint open sets.

Theorem 3.2.11 *Let G be a region in $\mathbf{C}$, let $f \in H(G)$ and suppose that for all $z \in G$, $f'(z) = 0$. Then f is constant in G.*

Proof By Theorem 3.2.8, the functions u_1, u_2, v_1 and v_2 all vanish identically in G. Since G is open and connected, by Theorem 2.4.23 it follows that given any $a, b \in G$, there is a polygonal path γ in G joining a to b and with line segments parallel to the coordinate axes, so that the track of γ is $\gamma^* = \bigcup_{k=1}^{n}[z_{k-1}, z_k]$, say, where $z_0 = a$, $z_n = b$ and $[z_{k-1}, z_k]$ denotes the line segment joining z_{k-1} to z_k. We prove that $f(a) = f(b)$ (and hence that f is constant, since a and b are arbitrary points in G) by showing that $f(z_0) = f(z_1) = \cdots = f(z_n)$. Put $z_k = x_k + iy_k$ $(0 \leq k \leq n)$. Either $x_k = x_{k-1}$ or $y_k = y_{k-1}$ $(k = 1, \ldots, n)$ as the segments are parallel to the axes. Suppose that for a particular k, $y_k = y_{k-1}$ (the other case is handled similarly). Then by the mean-value theorem, for some x', x'' between x_k and x_{k-1} we have

$$f(z_k) - f(z_{k-1}) = u(x_k, y_k) - u(x_{k-1}, y_k) + i\{v(x_k, y_k) - v(x_{k-1}, y_k)\}$$
$$= \{u_1(x', y_k) + iv_1(x'', x_k)\}(x_k - x_{k-1})$$
$$= 0.$$

The result follows. □

Exercise 3.2.12

1. Let $f : \mathbf{C} \to \mathbf{C}$ be continuous and such that for all $w, z \in \mathbf{C}$,

$$f(w + z) = f(w) + f(z).$$

Show that there are complex numbers a and b such that for all $z \in \mathbf{C}$,

$$f(z) = az + b\bar{z}.$$

If in addition, $f \in H(\mathbf{C})$, prove that $b = 0$.

2. Investigate the continuity at $z = 0$ of the functions f and g defined by

$$f(z) = \text{re}(z^2)/|z|^2 \ (z \neq 0), \ f(0) = 0;$$
$$g(z) = (\text{re}(z^2))^2/|z|^2 \ (z \neq 0), \ g(0) = 0.$$

3. Let $f : \mathbf{C}\backslash\{0\} \rightarrow \mathbf{C}$ be defined by

$$f(z) = x^3 y(y - ix)/(x^6 + y^2) \ (z = x + iy \neq 0).$$

Prove that for all $\alpha \in \mathbf{R}$,

$$\lim_{r \to 0} f(r \cos \alpha + ir \sin \alpha)/r = 0,$$

but that $\lim_{z \to 0} f(z)/z$ does not exist.

4. Prove that, regarded as functions from $\mathbf{C}$ to itself,

   (i) $z \longmapsto |z|$ is continuous but nowhere differentiable;
   (ii) $z \longmapsto |z|^2$ is differentiable at 0, but at no other point of $\mathbf{C}$.

5. Let G be a region in $\mathbf{C}$ and let $f \in H(G)$. Prove that if re f (im f or $|f|$) is constant, then f is constant.

6. Find a function $f \in H(\mathbf{C})$ such that for all $z = x + iy \in \mathbf{C}$,

$$\text{re } f(z) = e^x \cos y + xy.$$

7. Let G be a region in $\mathbf{C}$ with $0 \notin G$; let $\mathscr{F}$ be the family of all functions $f \in H(G)$ such that for all $z = x + iy \in G$,

$$|f(z)| = e^y/|z|^2.$$

Find one function $f_0 \in \mathscr{F}$, and by consideration of $|f(z)|/|f_0(z)|$, where f is any member of $\mathscr{F}$, show that $\mathscr{F} = \{\lambda f_0 : \lambda \in \mathbf{C}, |\lambda| = 1\}$.

3.3 Power Series

A central goal, achieved in Sect. 3.6, is to show that analytic functions are precisely those which can be represented by power series. We begin with some results which will have a familiar appearance in view of their similarity to real-variable statements.

Let $z_0 \in \mathbf{C}$. A **power series centred at** z_0 is a series of functions $\sum f_n$ with which is associated a sequence $(a_n)_{n \in \mathbf{N}_0}$ of complex numbers such that $f_n(z) = a_n(z - z_0)^n$ ($z \in \mathbf{C}$, $n \in \mathbf{N}_0$). To be understood by context, although with some abuse of notation, the series is usually denoted by $\sum_{n=0}^{\infty} a_n(z - z_0)^n$. Initially we set $z_0 = 0$ and deal with power series centred at 0, termed simply power series. Frequently the symbol $\sum$ will be used in place of $\sum_{n=0}^{\infty}$.

Given a power series $\sum a_n z^n$, a matter of first interest is the nature of its **convergence set**, that is, the set

$$S := \{ z \in \mathbf{C} : \sum a_n z^n \text{ is convergent} \}.$$

Since $0 \in S$, $S \neq \emptyset$; if S is bounded, it is contained in a closed ball centred at the origin. Let

$$R := \begin{cases} \sup\{|z| : z \in S\} & \text{if } S \text{ is bounded,} \\ \infty & \text{if } S \text{ is unbounded;} \end{cases}$$

R is called the **radius of convergence of** $\sum a_n z^n$. It will be shown that if $R = 0$ then $S = \{0\}$; if $0 < R < \infty$ then

$$\{ z \in \mathbf{C} : |z| < R \} \subset S \subset \{ z \in \mathbf{C} : |z| \leq R \};$$

and if $R = \infty$ then $S = \mathbf{C}$. The set $\{ z \in \mathbf{C} : |z| < R \}$ is referred to as the **disc of convergence of** $\sum a_n z^n$. Note that nothing is said about the convergence or divergence of $\sum a_n z^n$ when $|z| = R$.

In preparation for the next lemma, observe that if $z_0 \in S$ then, since $\lim_{n \to \infty} |a_n z_0^n| = 0$, the set $\{ a_n z_0^n : n \in \mathbf{N}_0 \}$ is bounded.

Lemma 3.3.1 (Weierstrass) *Let $(a_n)_{n \in \mathbf{N}_0}$ be a sequence of complex numbers and suppose there is a complex number $z_0 \neq 0$ such that $\{ a_n z_0^n : n \in \mathbf{N}_0 \}$ is bounded. Then for all $z \in \mathbf{C}$ with $|z| < |z_0|$, the series $\sum_{n=0}^{\infty} a_n z^n$ is convergent, indeed is absolutely convergent. Further, if $r < |z_0|$ then the series is uniformly convergent on $\{ z \in \mathbf{C} : |z| \leq r \}$.*

Proof By hypothesis, there exists a positive real number M such that for all $n \in \mathbf{N}_0$, $|a_n z_0^n| \leq M$. Thus for all $n \in \mathbf{N}_0$ and all $z \in \mathbf{C}$ with $|z| < |z_0|$,

$$|a_n z^n| = |a_n z_0^n| \, |z/z_0|^n \leq M \, |z/z_0|^n .$$

Hence, by comparison with the geometric series $\sum |z/z_0|^n$, the series $\sum a_n z^n$ is absolutely convergent if $|z| < |z_0|$. If $|z| \leq r < |z_0|$, then $|a_n z^n| \leq M \, |r/z_0|^n$, and use of the Weierstrass M-test gives the desired uniform convergence. $\qquad \square$

Theorem 3.3.2 *Let $(a_n)_{n \in \mathbf{N}_0}$ be a sequence of complex numbers. Let R be the radius of convergence of the power series $\sum a_n z^n$ and let S be its convergence set. Then*

(i) *$R = 0$ implies that $S = \{0\}$;*

(ii) $0 < R < \infty$ *implies that*

$$\{z \in \mathbf{C} : |z| < R\} \subset S \subset \{z \in \mathbf{C} : |z| \leq R\};$$

(iii) $R = \infty$ *implies that* $S = \mathbf{C}$.

Moreover, if $|z| < R$ *then* $\sum a_n z^n$ *converges absolutely; also, if* $0 < r < R$, *then* $\sum a_n z^n$ *converges uniformly on* $\{z \in \mathbf{C} : |z| \leq r\}$.

Proof (i) Suppose that $R = 0$. Since $|z| \leq R$ whenever $z \in S$, it follows that $S = \{0\}$.
(ii) Suppose that $0 < R < \infty$. Plainly, $S \subset \{z \in \mathbf{C} : |z| \leq R\}$. To obtain the other inclusion, suppose $z \in \mathbf{C}$ and $|z| < R$. Then there exists $w \in S$ such that $|z| < |w| < R$ and so, in consequence of Lemma 3.3.1 $\sum a_n z^n$ is absolutely convergent. Hence $\{z \in \mathbf{C} : |z| < R\} \subset S$.
(iii) Suppose $R = \infty$. Reasoning similar to that used to establish (ii) shows that $S = \mathbf{C}$. This is left to the reader.

Lastly, the absolute convergence of $\sum a_n z^n$ if $|z| < R$ was noted in the treatment of case (ii) and a parallel argument establishes it in case (iii). Further, if $0 < r < R$, then there exists $z_0 \in S$ such that $r < |z_0| < R$ and appeal to Lemma 3.3.1 shows that $\sum a_n z^n$ converges uniformly on $\{z \in \mathbf{C} : |z| \leq r\}$. $\square$

Theorem 3.3.3 *Let* $(a_n)_{n \in \mathbf{N}_0}$ *be a sequence of complex numbers, let* R *denote the radius of convergence of* $\sum a_n z^n$ *and let* S *denote its convergence set.*

(i) *Suppose* $0 \leq R < \infty$. *Then the sequence* $(a_n z^n)_{n \in \mathbf{N}_0}$ *is unbounded if* $|z| > R$: *symbolically,*

$$\sup \left\{ |a_n z^n| : n \in \mathbf{N}_0 \right\} = \infty \, (|z| > R).$$

(ii) *Let*

$$T := \{z \in \mathbf{C} : \text{ the sequence } (a_n z^n)_{n \in \mathbf{N}_0} \text{ is bounded}\}.$$

Then $S \subset T \subset \{z \in \mathbf{C} : |z| \leq R\}$ *if* $0 \leq R < \infty$; *and* $S = T = \mathbf{C}$ *if* $R = \infty$. *Also,*

$$R = \begin{cases} \sup\{|z| : z \in T\} & \text{if } T \text{ is bounded,} \\ \infty & \text{if } T \text{ is unbounded.} \end{cases}$$

Proof (i) To obtain a contradiction, suppose that $z_0 \in \mathbf{C}$, $|z_0| > R$ and that the sequence $(a_n z_0^n)_{n \in \mathbf{N}_0}$ is bounded. Since $R \geq 0$, $z_0 \neq 0$ and so, by Lemma 3.3.1, the set $\{z \in \mathbf{C} : |z| < |z_0|\} \subset S$. It follows that $|z_0| \leq R$, contradicting our assumption.
(ii) Obviously $S \subset T$. By (i), if $0 \leq R < \infty$ then $T \subset \{z \in \mathbf{C} : |z| \leq R\}$; also, if $R = \infty$ then $S = T = \mathbf{C}$. Hence S is bounded if, and only if, T is bounded. The formula for R follows immediately. $\square$

Formulae for the evaluation of the radius of convergence are evidently desirable and we now give some.

Theorem 3.3.4 *Let (a_n) be a sequence of non-zero complex numbers such that*

$$\lim_{n \to \infty} \left| \frac{a_{n+1}}{a_n} \right| \text{ exists (possibly equal to } \infty\text{)}.$$

Then the radius of convergence R of $\sum a_n z^n$ is given by the formula

$$\frac{1}{R} = \lim_{n \to \infty} \left| \frac{a_{n+1}}{a_n} \right|,$$

provided that we interpret $1/\infty$ as 0 and $1/0$ as ∞.

Proof By Theorem 3.1.15, $\sum a_n z^n$ converges absolutely or diverges according to whether $\lim_{n \to \infty} |z| \left| \frac{a_{n+1}}{a_n} \right|$ is less than or greater than 1. Thus

$$R = \sup \left\{ |z| : \sum a_n z^n \text{ is convergent} \right\} = 1 / \lim_{n \to \infty} \left| \frac{a_{n+1}}{a_n} \right|. \qquad \square$$

Useful though this formula is, it relies on the existence of $\lim_{n \to \infty} \left| \frac{a_{n+1}}{a_n} \right|$ and so cannot cope with, for example, $\sum z^{n!}$ where, for $n > 1$, a_{n+1}/a_n is either $1/0, 0/0$ or $0/1$. To remedy this, the nth root formula given below can be used.

Theorem 3.3.5 (Cauchy-Hadamard) *Let (a_n) be a sequence of complex numbers. Then the radius of convergence R of $\sum a_n z^n$ is given by*

$$\frac{1}{R} = \limsup_{n \to \infty} |a_n|^{1/n},$$

with the understanding that $1/\infty$ is 0 and $1/0$ is ∞.

Proof First suppose that $R = \infty$, so that $\sum a_n z^n$ is convergent for all $z \in \mathbf{C}$: thus given any particular $z \in \mathbf{C}$, certainly $|a_n z^n| \leq 1$ for all large enough n. Hence

$$\limsup_{n \to \infty} |a_n z^n|^{1/n} = |z| \limsup_{n \to \infty} |a_n|^{1/n} \leq 1,$$

and as this must hold for all $z \in \mathbf{C}$, $\limsup_{n \to \infty} |a_n|^{1/n} = 0$.

Next, suppose that $R = 0$, so that $\sum a_n z^n$ is divergent for all $z \in \mathbf{C}$, $z \neq 0$. Then

$$\limsup_{n \to \infty} |a_n z^n|^{1/n} = |z| \limsup_{n \to \infty} |a_n|^{1/n} \geq 1 \text{ for all } z \neq 0,$$

by Theorem 3.1.17. Thus $\limsup_{n \to \infty} |a_n|^{1/n} = \infty$.

Finally, suppose that $0 < R < \infty$. Since $\sum a_n z^n$ is convergent if $|z| < R$ and divergent if $|z| > R$, we see that

$$\limsup_{n\to\infty} \left|a_n z^n\right|^{1/n} \le 1 \text{ if } |z| < R, \limsup_{n\to\infty} \left|a_n z^n\right|^{1/n} \ge 1 \text{ if } |z| > R,$$

(by Theorem 3.1.17 again). Hence $|z| \limsup_{n\to\infty} |a_n|^{1/n}$ is ≤ 1 if $|z| < R$, and ≥ 1 if $|z| > R$. Thus $1/R = \limsup_{n\to\infty} |a_n|^{1/n}$. □

Example 3.3.6

(i) Use of Theorem 3.3.4 shows immediately that the radius of convergence of $\sum \frac{z^n}{n!}$ is ∞, for $\lim_{n\to\infty} \frac{n!}{(n+1)!} = 0$.
(ii) The series $\sum z^{n!}$ defies Theorem 3.3.4, as has already been remarked. However, since for this series a_n is either 0 or 1, it is plain that $\limsup_{n\to\infty} |a_n|^{1/n} = 1$, and so $R = 1$.

Now we begin to establish the connection between power series and analyticity.

Definition 3.3.7 Let G be an open subset of **C**. A function $f : G \to \mathbf{C}$ is said to be **representable by power series in** G if, for each $z_0 \in G$ and each $r > 0$ such that $B(z_0, r) \subset G$, there is a sequence $(a_n)_{n\ge 0}$ in **C** such that for all $z \in B(z_0, r)$,

$$f(z) = \sum_{0}^{\infty} a_n(z - z_0)^n.$$

In fact, $(a_n)_{n\ge 0}$ depends on z_0 but not on r, as the following theorem shows.

Theorem 3.3.8 *If f is representable by power series in G, then $f \in H(G)$ and f' is also representable by power series in G; in fact, if*

$$f(z) = \sum_{0}^{\infty} a_n(z - z_0)^n \text{ for all } z \in B(z_0, r),$$

then

$$f'(z) = \sum_{1}^{\infty} n a_n(z - z_0)^{n-1} \text{ for all } z \in B(z_0, r).$$

Proof Let $z_0 \in G$, let $B(z_0, r) \subset G$ and suppose that $f(z) = \sum_{0}^{\infty} a_n(z - z_0)^n$ for all $z \in B(z_0, r)$. With the obvious change of variable we may suppose that $z_0 = 0$. Let R be the radius of convergence of $\sum_{0}^{\infty} a_n z^n$; plainly $0 < r \le R$. For each $z \in B(0, r)$ put $g(z) = \sum_{n=1}^{\infty} n a_n z^{n-1}$. The radius of convergence of this series is

$$1/\limsup_{n\to\infty} |n a_n|^{1/n} = 1/\limsup_{n\to\infty} |a_n|^{1/n} = R,$$

since $\lim_{n\to\infty} n^{1/n} = 1$. Fix $w \in B(0, r)$ and choose ρ such that $|w| < \rho < r$. Then if $z \neq w$ and $|z| < \rho$,

$$\frac{f(z) - f(w)}{z - w} - g(w) = \sum_{n=1}^{\infty} a_n \left\{ \frac{z^n - w^n}{z - w} - nw^{n-1} \right\}. \qquad (3.3.1)$$

Note that $(z^n - w^n)/(z - w) - nw^{n-1}$ equals 0 if $n = 1$, while if $n \geq 2$ it equals

$$z^{n-1} + z^{n-2}w + \cdots + w^{n-1} - nw^{n-1} = (z - w) \sum_{k=1}^{n-1} kw^{k-1}z^{n-k-1}.$$

Moreover, if $|z| < \rho$,

$$\left| \sum_{k=1}^{n-1} kw^{k-1}z^{n-k-1} \right| \leq \sum_{k=1}^{n-1} k\rho^{k-1}\rho^{n-k-1} = \sum_{k=1}^{n-1} k\rho^{n-2} = \frac{1}{2}n(n-1)\rho^{n-2} < n^2\rho^{n-2}.$$

Hence if $z \neq w$ and $|z| < \rho$,

$$\left| \frac{f(z) - f(w)}{z - w} - g(w) \right| \leq |z - w| \sum_{n=2}^{\infty} n^2\rho^{n-2} |a_n|. \qquad (3.3.2)$$

Since $\rho < R$,

$$\limsup_{n\to\infty} \left| n^2\rho^{n-2}a_n \right|^{1/n} = \lim_{n\to\infty} \left(n^{2/n}\rho^{1-2/n} \right) \limsup_{n\to\infty} |a_n|^{1/n} = \rho/R < 1,$$

and so the series on the right-hand side of (3.3.2) converges. Thus the left-hand side of (3.3.2) tends to 0 as $z \to w$. It follows that $f'(w) = g(w)$, and the proof is complete. $\qquad \square$

Remark 3.3.9 Since f' satisfies the same hypotheses as f, the theorem can be applied to f'. It follows that f has derivatives of all orders, that each derivative is representable by power series in G and is thus analytic in G, and that for each $k \in \mathbf{N}$,

$$f^{(k)}(z) = \sum_{n=k}^{\infty} n(n - 1)\ldots(n - k + 1)a_n(z - z_0)^{n-k} \text{ for all } z \in B(z_0, r),$$

so that $f^{(k)}(z_0) = k! \, a_k$. Thus given any $z_0 \in G$, there is a unique sequence (a_n) such that $f(z) = \sum_0^{\infty} a_n(z - z_0)^n$ in $B(z_0, r) \subset G$: it is given by

$$a_n = f^{(n)}(z_0)/n!$$

We emphasise that this depends on z_0 but not on the particular r. The converse of Theorem 3.3.8, namely that if $f \in H(G)$ then f is representable by power series in

G, is true and as mentioned earlier, is one of our main objectives and will be proved in Theorem 3.6.10 to follow.

We now turn to special functions represented by power series, beginning with the exponential function.

Theorem 3.3.10 *The series $E(z) := \sum_0^\infty \frac{z^n}{n!}$ converges absolutely for all $z \in \mathbf{C}$, $E'(z) = E(z)$ for all $z \in \mathbf{C}$ and $E(0) = 1$. [The function $E : \mathbf{C} \to \mathbf{C}$ is called the (complex)* **exponential function.** *It is more frequently written as* $\exp : \mathbf{C} \to \mathbf{C}$, *but in this and much of Sect. 3.4, an economy of notation is obtained by using E in place of* $\exp$].

Proof We have already seen in Example 3.3.6 (i) that the radius of convergence of the series for $E(z)$ is ∞. Now use Theorem 3.3.8. $\square$

Theorem 3.3.11 *For all $w, z \in \mathbf{C}$ we have*

$$E(z)E(-z) = 1,\ E(z) \neq 0 \text{ and } E(z+w) = E(z)E(w).$$

Proof By Theorems 3.2.5 and 3.3.10, $(E(-z))' = -E(-z)$ and so

$$(E(z)E(-z))' = E(z)E(-z) - E(z)E(-z) = 0 \text{ for all } z \in \mathbf{C}.$$

Hence by Theorem 3.2.11, $E(z)E(-z)$ is constant. Since $E(0) = 1$, $E(z)E(-z) = 1$; thus for all $z \in \mathbf{C}$, $E(z) \neq 0$. Finally, if we fix $w \in \mathbf{C}$, take any $z \in \mathbf{C}$ and differentiate with respect to z, we find that

$$\{E(z+w)/E(z)\}' = \{E(z)E(z+w) - E(z+w)E(z)\}/E(z)^2 = 0.$$

Thus $E(z+w)/E(z)$ remains constant as z varies; taking $z = 0$ we see that $E(z+w)/E(z) = E(w)$. $\square$

Remark 3.3.12 From Theorem 3.3.11 it is easy to show, by inductive arguments, that

$$E(\sum_1^n z_k) = \prod_1^n E(z_k) \text{ for all } n \in \mathbf{N} \text{ and all } z_1, \ldots, z_n \in \mathbf{C},$$

and that

$$E(pz) = (E(z))^p \text{ for all } z \in \mathbf{C} \text{ and all } p \in \mathbf{Z}.$$

Note also that

$$e = E(1) = \sum_0^\infty \frac{1}{n!}.$$

Moreover, Theorem 3.3.11 shows that for all $z, z_0 \in \mathbf{C}$,

$$E(z) = E(z_0)E(z - z_0) = E(z_0) \sum_{n=0}^{\infty} \frac{(z - z_0)^n}{n!},$$

which proves that E is representable by power series in $\mathbf{C}$. Finally, we recall that the function $x \longmapsto E(x) : \mathbf{R} \to (0, \infty)$ is injective and has inverse $x \longmapsto \log x : (0, \infty) \to \mathbf{R}$.

Next we deal with the trigonometric functions.

Definition 3.3.13 The functions $\sin : \mathbf{C} \to \mathbf{C}$ and $\cos : \mathbf{C} \to \mathbf{C}$ are defined by

$$\sin z = \sum_{0}^{\infty} (-1)^n z^{2n+1} / (2n + 1)!, \cos z = \sum_{0}^{\infty} (-1)^n z^{2n} / (2n)! \ (z \in \mathbf{C}).$$

Use of Theorem 3.3.4 shows that both these series have infinite radius of convergence. Obviously $\cos 0 = 1$ and $\sin 0 = 0$; by Theorem 3.3.8,

$$(\cos z)' = - \sin z \text{ and } (\sin z)' = \cos z \text{ for all } z \in \mathbf{C}.$$

Before establishing further properties of these functions we recall some elementary facts about $x \longmapsto \sin x, x \longmapsto \cos x : \mathbf{R} \to \mathbf{R}$. First, $\cos 2 < 0$: for

$$\cos 2 = 1 - \frac{2^2}{2} + \frac{2^4}{4!} - \sum_{n=2}^{\infty} \left(\frac{2^{4n-2}}{(4n - 2)!} - \frac{2^{4n}}{(4n)!} \right),$$

and

$$\frac{2^{4n-2}}{(4n - 2)!} - \frac{2^{4n}}{(4n)!} = \frac{2^{4n}}{(4n)!} \{n(4n - 1) - 1\} > 0 \text{ for all } n \in \mathbf{N}.$$

Next, $\sin x > 0$ if $0 < x < 2$, since

$$\sin x = \sum_{n=1}^{\infty} \left(\frac{x^{4n-3}}{(4n - 3)!} - \frac{x^{4n-1}}{(4n - 1)!} \right),$$

and

$$\frac{x^{4n-3}}{(4n - 3)!} - \frac{x^{4n-1}}{(4n - 1)!} = \frac{x^{4n-3}}{(4n - 3)!} \left\{ 1 - \frac{x^2}{(4n - 1)(4n - 2)} \right\} > 0$$

if $0 < x < 2$ and $n \in \mathbf{N}$. Moreover, since $(\cos x)' = - \sin x, x \longmapsto \cos x$ is strictly decreasing in $[0, 2]$ and so has exactly one zero in $[0, 2]$. We denote this real number by $\pi/2$: $\cos \frac{\pi}{2} = 0$, $\sin \frac{\pi}{2} > 0$ and $3 < \pi < 4$, since with $x = 3/2$ the expansion

$$\cos x = 1 - \frac{x^2}{2} + \frac{x^4}{4!} - \frac{x^6}{6!} + \sum_{n=2}^{\infty} \left(\frac{x^{4n}}{(4n)!} - \frac{x^{4n+2}}{(4n + 2)!} \right)$$

shows that $\cos(3/2) > 0$. From the definitions of sin and cos we see directly that

$$\cos z = \cos(-z), \sin(-z) = -\sin z, \cos z + i \sin z = E(iz), \cos z - i \sin z = E(-iz);$$

hence

$$\cos z = (E(iz) + E(-iz))/2, \sin z = (E(iz) - E(-iz))/(2i).$$

Also, if $w, z \in \mathbf{C}$, then

$$2 \cos z \cos w = \frac{1}{2}(E(iz) + E(-iz))(E(iw) + E(-iw)) = \cos(z+w) + \cos(z-w)$$

and

$$-2 \sin z \sin w = \frac{1}{2}(E(iz) - E(-iz))(E(iw) - E(-iw)) = \cos(z+w) - \cos(z-w);$$

thus

$$\cos(z + w) = \cos z \cos w - \sin z \sin w. \tag{3.3.3}$$

Setting $w = -z$ in this, we see that

$$\cos^2 z + \sin^2 z = 1, \tag{3.3.4}$$

and so $\cos^2 \frac{\pi}{2} + \sin^2 \frac{\pi}{2} = 1$, which shows that $\sin \frac{\pi}{2} = 1$. Since $E\left(\frac{i\pi}{2}\right) = i$, $E(2\pi i) = E(4 \cdot \frac{\pi i}{2}) = i^4 = 1$; thus

$$E(z + 2\pi i) = E(z) \text{ for all } z \in \mathbf{C}. \tag{3.3.5}$$

Now differentiate (3.3.3) with respect to z, holding w fixed; we obtain

$$\sin(z + w) = \sin z \cos w + \cos z \sin w. \tag{3.3.6}$$

From (3.3.3) and (3.3.6) it follows that cos and sin are representable by power series: thus in the case of cos, given any $a \in \mathbf{C}$,

$$\cos z = \cos(z - a) \cos a - \sin(z - a) \sin a$$
$$= \cos a \sum_0^\infty \frac{(-1)^n (z - a)^{2n}}{(2n)!} - \sin a \sum_0^\infty \frac{(-1)^n (z - a)^{2n+1}}{(2n + 1)!}.$$

Other trigonometric functions can be defined as in the real case: thus

$$\tan z = \sin z / \cos z, \cot z = \cos z / \sin z, \sec z = 1 / \cos z, \operatorname{cosec} z = 1 / \sin z$$

whenever the denominators are non-zero.

Notice how power series can be used to handle limits. For example, if $z \neq 0$, then $\frac{\sin z}{z} = 1 - \frac{z^2}{3!} + \frac{z^4}{5!} - \dots$ This power series converges for all $z \in \mathbf{C}$ and so determines a continuous function: hence

$$\lim_{z \to 0} \frac{\sin z}{z} = 1.$$

The hyperbolic functions are defined by

$$\sinh z = \frac{1}{2}(E(z) - E(-z)), \cosh z = \frac{1}{2}(E(z) + E(-z)),$$

$$\tanh z = \sinh z / \cosh z, \coth z = \cosh z / \sinh z, \operatorname{cosech} z = 1/\sinh z,$$

$$\operatorname{sech} z = 1/\cosh z;$$

$\sinh$ and $\cosh$ are defined for all $z \in \mathbf{C}$, the others whenever the denominator is non-zero. Naturally there are relationships between these functions and the trigonometric functions. For example, it is plain that

$$\sinh z = -i \sin(iz) \text{ and } \cosh z = \cos(iz) \text{ for all } z \in \mathbf{C}.$$

Note that if $z = x + iy$, with x, y real, then

$$E(z) = E(x + iy) = E(x)E(iy) = E(x)(\cos y + i \sin y), |E(z)| = E(x),$$

and

$$\cos z = \cos(x + iy) = \cos x \cos(iy) - \sin x \sin(iy) = \cos x \cosh y - i \sin x \sinh y,$$

$$\sin z = \sin x \cosh y + i \cos x \sinh y.$$

Exercise 3.3.14

1. Let (a_n) and (b_n) be sequences of non-negative real numbers and suppose that $\lim_{n \to \infty} a_n = a$, where $0 < a < \infty$. Prove that

$$\limsup_{n \to \infty} a_n b_n = a \limsup_{n \to \infty} b_n.$$

2. Find the radius of convergence of each of the following power series:

$$\sum n^p z^n \ (p \in \mathbf{N}), \ \sum n! z^n, \ \sum \lambda^{n^2} z^n \ (|\lambda| < 1), \ \sum \{2 + (-1)^n\}^{-n} z^n.$$

3. Given that $\sum a_n z^n$ has radius of convergence R, find the radius of convergence of each of the following power series:

$$\sum n^3 a_n z^n, \ \sum a_n z^{3n}, \ \sum a_n^3 z^n.$$

4. Let $(a_n)_{n\geq 0}$ and $(b_n)_{n\geq 0}$ be sequences of complex numbers such that $\sum_0^\infty a_n$ and $\sum_0^\infty b_n$ are absolutely convergent, and put $c_n = \sum_{r=0}^n a_r b_{n-r}$ $(n \in \mathbf{N_0})$. Prove that $\sum_0^\infty c_n$ is absolutely convergent and that

$$\sum_0^\infty c_n = \left(\sum_0^\infty a_n\right)\left(\sum_0^\infty b_n\right).$$

($\sum c_n$ is called the **Cauchy product** of $\sum a_n$ and $\sum b_n$.)
Use this result to show that:

(i) $\sum_{n=0}^\infty \frac{(z+w)^n}{n!} = \left(\sum_{n=0}^\infty \frac{z^n}{n!}\right)\left(\sum_{n=0}^\infty \frac{w^n}{n!}\right)$ for all $w, z \in \mathbf{C}$;

(ii) $\sum_{n=1}^\infty n z^{n-1} = (1-z)^{-2}$ if $|z| < 1$.

5. Let $f(z) = \sum_{n=1}^\infty a_n z^n$ $(|z| \leq 1)$; suppose that $a_1 = 1$ and $\sum_{n=2}^\infty n|a_n| < 1$. Prove that f is continuous and injective on $\{z \in \mathbf{C} : |z| \leq 1\}$.

6. Prove that for all $x, y \in \mathbf{R}$,

$$|\cos(x+iy)|^2 = \sinh^2 y + \cos^2 x, \ |\sin(x+iy)|^2 = \sinh^2 y + \sin^2 x.$$

3.4 Arguments, Logarithms and the Winding Number

We have already observed that in the real case, the inverse of the exponential function is the logarithmic function. This makes it natural to study the corresponding situation in the complex case, but to do this it is essential to consider the **argument** function, and we now set about this.

Lemma 3.4.1 *Given any* $z \in \mathbf{C}\backslash\{0\}$, *there are unique* $r > 0$ *and* $w \in \mathbf{C}$, *with* $|w| = 1$, *such that* $z = rw$.

Proof Plainly we may write $z = |z|\,(z/|z|)$, so that $r := |z|$ and $w := z/|z|$ have the required properties. As for uniqueness, if $z = r_1 w_1$ and $z = r_2 w_2$, with $r_1, r_2 > 0$ and $|w_1| = |w_2| = 1$, then $|r_1 w_1| = |r_2 w_2|$ and so $r_1 = r_2$. Thus $w_1 = w_2$. $\qquad\square$

Theorem 3.4.2 *Let* $w \in \mathbf{C}$ *be such that* $|w| = 1$. *Then there exists* $\theta \in \mathbf{R}$ *such that* $-\pi < \theta \leq \pi$ *and* $w = E(i\theta)$.

Proof Put $w = u + iv$, with $u, v \in \mathbf{R} : u^2 + v^2 = 1$. First suppose that $u, v \geq 0$. Since $x \longmapsto \cos x$ is continuous and strictly decreasing on $[0, \pi/2]$ and $\cos 0 = 1$, $\cos(\pi/2) = 0$, there exists $\theta \in [0, \pi/2]$ such that $\cos\theta = u$; and evidently $v = \pm \sin\theta$. In fact, $v = \sin\theta$ since $v \geq 0$ and $\sin\theta \geq 0$. Hence $w = \cos\theta + i\sin\theta = E(i\theta)$.

If $u \leq 0$ and $v \geq 0$, then $u + iv = i(v - iu) = iE(i\theta')$ for some θ' in $[0, \pi/2]$, by the argument just given. But $iE(i\theta') = E(i\pi/2)E(i\theta') = E(i(\theta' + \pi/2))$. If $u \leq 0$ and $v < 0$, then $u + iv = -(-u - iv) = -E(i\theta_1) = E(i(-\pi + \theta_1))$, where

$\theta_1 \in (0, \pi/2]$ since $v < 0$. Finally, if $u \geq 0$ and $v \leq 0$, then $u + iv = -i(-v + iu) = -iE(i\theta_2) = E(i(\theta_2 - \pi/2))$, where $\theta_2 \in [0, \pi/2]$. $\qquad\square$

The next result is of considerable importance.

Theorem 3.4.3 $E(z) = 1$ if, and only if, $z = 2k\pi i$ for some $k \in \mathbf{Z}$.

Proof Since $E(2\pi i) = 1$, $E(2k\pi i) = (E(2\pi i))^k = 1$ for all $k \in \mathbf{Z}$. Conversely, suppose that $1 = E(z) = E(\operatorname{re} z)E(i \operatorname{im} z)$. Then since $E(\operatorname{re} z) \geq 0$ and $|E(i \operatorname{im} z)| = 1$, we must have $\operatorname{re} z = 0$ and $E(iy) = 1$, where $y = \operatorname{im} z$. There is a unique integer n such that $n \leq 2y/\pi < n + 1$; that is, such that $0 \leq y - \frac{1}{2}n\pi < \frac{1}{2}\pi$. Put $t = y - \frac{1}{2}n\pi$; we claim that $t = 0$. For suppose not. Then $0 < t < \frac{1}{2}\pi$ and $E(it) = E(iy)E(-in\pi/2) = i^{-n}$, so that $E(4it) = i^{-4n} = 1$. However, $E(it) = l + im$, where $l > 0$, $m > 0$, $l^2 + m^2 = 1$, and thus $E(4it) = (l + im)^4 = l^4 - 6l^2m^2 + m^4 + 4ilm(l^2 - m^2)$, which is real if, and only if, $l^2 = m^2 = \frac{1}{2}$. But then $E(4it) = -1$ and we have a contradiction. Thus $t = 0$; that is, $y = \frac{1}{2}n\pi$, and so $E(iy) = i^n = 1$. Hence $n = 4k$ for some integer k, and thus $y = 2k\pi$, so that $z = 2k\pi i$. $\qquad\square$

Corollary 3.4.4 *Let* $z \in \mathbf{C}\backslash\{0\}$. *Then there exist a unique* $r > 0$ *and a unique* $\theta \in (-\pi, \pi]$ *such that* $z = rE(i\theta)$. *In fact,* $r = |z|$; θ *is called the **principal argument** of* z *and is written as* arg z.

Proof By Lemma 3.4.1, there are unique numbers $r > 0$ and $w \in \mathbf{C}$, with $|w| = 1$, such that $z = rw$; and $r = |z|$. By Theorem 3.4.2, $w = E(i\theta)$ for some $\theta \in (-\pi, \pi]$. If $w = E(i\theta')$ for another $\theta' \in (-\pi, \pi]$, then $E(i\theta) = E(i\theta')$ and hence $E(i(\theta - \theta')) = 1$. By Theorem 3.4.3, $\theta - \theta'$ is a multiple of $2k\pi$; and since both θ and θ' belong to $(-\pi, \pi]$, $\theta = \theta'$. $\qquad\square$

The argument and modulus functions are very important; we give their continuity properties next.

Theorem 3.4.5 (a) *The map* $z \longmapsto |z|$ *of* $\mathbf{C}$ *to* $\mathbf{R}$ *is continuous.*
(b) *The map* $z \longmapsto$ arg z *of* $\mathbf{C}\backslash\{0\}$ *to* $(-\pi, \pi]$ *is continuous at each point of*

$$D(\pi) := \{z \in \mathbf{C} : z \neq |z| E(i\pi)\} = \{z \in \mathbf{C} : z \neq -|z|\}$$

and at no point of $\mathbf{C}\backslash D(\pi)$.

Proof (a) This follows directly from the triangle inequality $||w| - |z|| \leq |w - z|$.
(b) Let $z \in D(\pi)$ and suppose that arg is not continuous at z. Then there is a sequence (z_n) in $D(\pi)$ such that $z_n \to z$ but $t_n := \arg z_n \nrightarrow \arg z := t$. Hence there exist $\varepsilon > 0$ and a subsequence $(t_{m(n)})$ of (t_n) such that $|t_{m(n)} - t| > \varepsilon$ for all n. Let $s_n = t_{m(n)}$, $w_n = z_{m(n)}$; $w_n \to z$ and no subsequence of (s_n) converges to t. Since (s_n) is bounded (in fact, $s_n \in (-\pi, \pi]$), it has a convergent subsequence $(s_{p(n)})$, with $s_{p(n)} \to s \neq t$, $s \in [-\pi, \pi]$. As both the exponential map and $z \longmapsto |z|$ are continuous,

$$\left| w_{p(n)} \right| E(i s_{p(n)}) \to |z| E(is),$$

so that $|z| E(is) = z = |z| E(it)$, and hence $E(i(s - t)) = 1$. Since $|s - t| < 2\pi$, it follows from Theorem 3.4.3 that $s = t$. This contradiction proves that arg is continuous at z.

Finally, suppose that $z \neq 0$ and $z \in \mathbf{C} \backslash D(\pi)$. For each $n \in \mathbf{N}$ put $z_n = |z| E(i\pi(1 - n^{-1}))$, $w_n = |z| E(-i\pi(1 - n^{-1}))$. Then $z_n \to z$, $w_n \to z$, arg $z_n \to \pi$ and arg $w_n \to -\pi$, from which it follows that arg is not continuous at z. $\square$

Since $\mathbf{C} \backslash D(\pi) = \{ z \in \mathbf{C} : \text{im } z = 0, \text{re } z \leq 0 \}$, we see that arg is continuous everywhere on $\mathbf{C}$ except along the non-positive real axis: at 0, arg is not defined; and at points on the negative real axis its oscillation is 2π. Corresponding to the principal argument we define the principal logarithm, which also behaves well except on the non-positive real axis.

Theorem 3.4.6 *Given any $z \in \mathbf{C} \backslash \{0\}$, there is a unique $w \in \mathbf{C}$ with im $w \in (-\pi, \pi]$ such that $E(w) = z$: explicitly,*

$$w = \log |z| + i \arg z.$$

*This w is called the **principal logarithm** of z. It is denoted by $\log z$, the same symbol being used for the principal logarithm, i.e. the map $z \longmapsto w$, as for the real logarithm, given their natural identification when z is real and positive.*

Proof Let $z \in \mathbf{C} \backslash \{0\}$ and put $w = \log |z| + i \arg z$. Then

$$E(w) = E(\log |z|) E(i \arg z) = |z| E(i \arg z) = z.$$

If w' is such that $E(w') = z$ and im $w' \in (-\pi, \pi]$, then $E(w) = E(w')$ and so $E(w - w') = 1$. By Theorem 3.4.3, $w = w'$ since both im w and im w' belong to $(-\pi, \pi]$. $\square$

This Theorem implies that the exponential function maps $\{z \in \mathbf{C} : \text{im } z \in (-\pi, \pi]\}$ injectively onto $\mathbf{C} \backslash \{0\}$. The principal logarithm is the inverse of this map and the analyticity of its restriction to $D(\pi)$ is established next.

Theorem 3.4.7 *The function $z \longmapsto \log z : D(\pi) \to \mathbf{C}$ is in $H(D(\pi))$ and has derivative $z \longmapsto 1/z : D(\pi) \to \mathbf{C}$.*

Proof Put $f(z) = \log z$ $(z \in D(\pi))$. By Theorem 3.4.5, since $f(z) = \log |z| + i \arg z$, it follows that f is continuous. Since $E(f(z)) = z$ for all $z \in D(\pi)$, $E \in H(\mathbf{C})$ and for all $w \in \mathbf{C}$, $E'(w) = E(w) \neq 0$, we see from Theorem 3.2.6 that $f \in H(D(\pi))$ and $f'(z) = 1/z$ for all $z \in D(\pi)$. $\square$

Now we can define arbitrary powers of z.

Definition 3.4.8 Given any $z \in \mathbf{C}\backslash\{0\}$ and any $a \in \mathbf{C}$, we define z^a, the **principal** ath **power** of z, by $z^a = E(a \log z)$.

Remark 3.4.9 If we take $z = e$ we have $e^a = E(a)$.

Theorem 3.4.10 Let $a \in \mathbf{C}$. The function $z \longmapsto z^a : D(\pi) \rightarrow \mathbf{C}$ belongs to $H(D(\pi))$ and has derivative $z \longmapsto az^{a-1} : D(\pi) \rightarrow \mathbf{C}$.

Proof By Theorem 3.2.5, $z \longmapsto z^a$ is in $H(D(\pi))$. Also,

$$(E(a \log z))' = a \cdot \frac{1}{z} E(a \log z) = aE(- \log z)E(a \log z)$$

$$= aE((a - 1) \log z) = az^{a-1}. \qquad \square$$

We recall that the principal argument arg z of $z \neq 0$ is required to belong to $(-\pi, \pi]$. There is no particularly good reason for this other than definiteness, and in various circumstances it becomes inconvenient to have to put up with the associated discontinuity on the negative real axis. Evidently any real number which differs from arg z by a multiple of 2π is just as good a candidate for the position of the argument of z as arg z, and we now recognise this fact by defining the argument of z to be the set of all such candidates; we make the corresponding definitions of the logarithm and powers of z.

Definition 3.4.11 Let $z \in \mathbf{C}\backslash\{0\}$. The **argument** of z, Arg z, and the **logarithm** of z, Log z, are defined by

$$\text{Arg } z = \{\theta \in \mathbf{R} : z = |z| \, E(i\theta)\}, \text{Log } z = \{w \in \mathbf{C} : E(w) = z\}.$$

By Theorems 3.4.2 and 3.4.3,

$$\text{Arg } z = \{2k\pi + \arg z : k \in \mathbf{Z}\}, \text{Log } z = \{\log z + 2k\pi i : k \in \mathbf{Z}\}.$$

Out of the infinitely many members of Arg z and Log z we need a procedure for selecting desired members, just as the principal argument and principal logarithm were chosen, and we now give this procedure.

Definition 3.4.12 Let $\alpha \in \mathbf{R}$, $z \in \mathbf{C}\backslash\{0\}$. The unique element of Arg $z \cap (\alpha - 2\pi, \alpha]$ is called the α—**branch of the argument** of z and is denoted by $(\alpha - \arg)(z)$. Similarly, the α—**branch of the logarithm** of z is defined to be the unique element w of Log z with im $w \in (\alpha - 2\pi, \alpha]$, and is written $(\alpha - \log)(z)$.

Evidently $(\pi - \arg)(z)$ and $(\pi - \log)(z)$ are the principal argument and the principal logarithm of z respectively. The α—branches of the argument and the logarithm behave well everywhere in $\mathbf{C}$ except on $\{z \in \mathbf{C} : z = |z| \, E(i\alpha)\}$, the ray from the origin at an angle α with the positive real axis: we summarise the position below.

Theorem 3.4.13 *Let $\alpha \in \mathbf{R}$. Then $(\alpha - \arg) : \mathbf{C}\backslash\{0\} \to (\alpha - 2\pi, \alpha]$ is continuous at each point of*

$$D(\alpha) := \{z \in \mathbf{C} : z \neq |z|\, E(i\alpha)\}$$

and at no point of $\mathbf{C}\backslash D(\alpha)$. Moreover, the exponential function E maps $\{w \in \mathbf{C} : \operatorname{im} w \in (\alpha - 2\pi, \alpha]\}$ onto $\mathbf{C}\backslash\{0\}$ injectively; also $z \longmapsto (\alpha - \log)(z) : D(\alpha) \to \mathbf{C}$ is analytic and has derivative $z \longmapsto 1/z$.

Proof All these assertions follow from obvious modifications of the proofs of Theorems 3.4.5 and 3.4.7. □

Definition 3.4.14 Let $a \in \mathbf{C}$ and $z \in \mathbf{C}\backslash\{0\}$. The set $\{E(a(\log z + 2k\pi i)) : k \in \mathbf{Z}\}$ is called the ath **power** of z. Given any $\alpha \in \mathbf{R}$, the $\alpha-$branch of the ath power of z, written $(z^a)_\alpha$, is defined to be $E(a(\alpha - \log)(z))$.

Plainly the $\pi-$branch of the ath power of z is the principal ath **power** of z, z^a. By proceeding along lines similar to the proof of Theorem 3.4.10 we obtain:

Theorem 3.4.15 *Let $a \in \mathbf{C}$ and $\alpha \in \mathbf{R}$. The function $z \longmapsto (z^a)_\alpha : D(\alpha) \to \mathbf{C}$ is in $H(D(\alpha))$ and has derivative $z \longmapsto a(z^{a-1})_\alpha : D(\alpha) \to \mathbf{C}$.*

Example 3.4.16

(i) If $a \in \mathbf{Z}$, $E(a(\log(z + 2k\pi i))) = E(a \log z)$ for all $k \in \mathbf{Z}$, and so all branches of the ath **power** of z coincide with the principal branch, z^a.

(ii) Suppose we wish to find the $3\pi/2-$branch of the argument , logarithm and ith power of $(-1 - i)$. Then applying the definitions we see that

$$\left(\frac{3\pi}{2} - \arg\right)(-1 - i) = \frac{5\pi}{4}, \quad \left(\frac{3\pi}{2} - \log\right)(-1 - i) = \log\sqrt{2} + \frac{5\pi i}{4},$$

$$((-1 - i)^i)_{3\pi/2} = E\left(i\left(\frac{3\pi}{2} - \log\right)(-1 - i)\right)$$

$$= E\left(-\frac{5\pi}{4}\right)\left\{\cos(\log\sqrt{2}) + i\sin(\log\sqrt{2})\right\}.$$

By way of contrast, the principal argument, logarithm and ith power are

$$\arg(-1 - i) = -\frac{3\pi}{4}, \quad \log(-1 - i) = \log\sqrt{2} - \frac{3\pi i}{4},$$

$$(-1 - i)^i = E(i\log(-1 - i)) = E\left(\frac{3\pi}{4}\right)\left\{\cos(\log\sqrt{2}) + i\sin(\log\sqrt{2})\right\}.$$

It is convenient to be able to use different branches of the argument, logarithm and powers because of the desirability of keeping discontinuities or lack of analyticity away from particular parts of the complex plane. We shall see how useful this is once integration theorems have been developed in Sects. 3.5 and 3.6.

To conclude the present section we turn to the notion of the winding number, which provides a measure of the number of times a closed path in the plane winds about a given point. This is achieved by means of continuous arguments, which we give in a metric space setting for greater flexibility.

Definition 3.4.17 Let X be a metric space and let $f : X \to \mathbf{C}\backslash\{0\}$ be continuous. A continuous map $g : X \to \mathbf{C}$ is called a **continuous logarithm of** f (on X) if $\exp(g(x)) = f(x)$ for all $x \in X$ (or, equivalently, if $g(x) \in \mathrm{Log}\, f(x)$ for all $x \in X$). A continuous map $\theta : X \to \mathbf{R}$ is called a **continuous argument of** f (on X) if $f(x) = |f(x)| \exp(i\theta(x))$ for all $x \in X$ (or, equivalently, if $\theta(x) \in \mathrm{Arg}\, f(x)$ for all $x \in X$). In the special case in which X is a region in $\mathbf{C}$ such that $0 \notin X$, and $f : X \to X$ is the identity map ($f(x) = x$ for all $x \in X$), a continuous logarithm of the identity is called a **branch** of the logarithm (it is analytic by Theorem 3.2.6), and a continuous argument of the identity is called a **branch** of the argument.

Various relationships of interest are given in the following:

Theorem 3.4.18 *Let X be a metric space and let $f : X \to \mathbf{C}\backslash\{0\}$ be continuous. Then:*

(a) *if g is a continuous logarithm of f, im g is a continuous argument of f;*
(b) *if θ is a continuous argument of f, $\log |f| + i\theta$ is a continuous logarithm of f;*
(c) *if X is connected and g_1, g_2 are continuous logarithms of f, then $g_2 - g_1 = 2\pi i n$ for some $n \in \mathbf{Z}$; if θ_1, θ_2 are continuous arguments of f (and X is connected), then $\theta_2 - \theta_1 = 2\pi m$ for some $m \in \mathbf{Z}$;*
(d) *if X is connected and $x, y \in X$, then $g(x) - g(y) = \log |f(x)| - \log |f(y)| + i\{\theta(x) - \theta(y)\}$ for all continuous logarithms g and all continuous arguments θ of f.*

Proof (a) For all $x \in X$,

$$f(x) = \exp(g(x)) = \exp(\mathrm{re}\, g(x)) \exp(i\, \mathrm{im}\, g(x)) = |f(x)| \exp(i\, \mathrm{im}\, g(x)),$$

and so im g is a continuous argument of f.
(b) For all $x \in X$,

$$f(x) = |f(x)| \exp(i\theta(x)) = \exp(\log |f(x)| + i\theta(x)),$$

which proves the result.
(c) Since $f(x) = \exp(g_1(x)) = \exp(g_2(x))$ for all $x \in X$, it follows that $(g_1 - g_2)/(2\pi i)$ is a continuous integer-valued function on the connected set X (by Theorem 3.4.3). Thus by Theorems 2.4.11 and 2.4.8, $(g_1 - g_2)/(2\pi i)$ is constant, so that $g_1 - g_2 = 2\pi i n$ for some $n \in \mathbf{Z}$. From this and (b) we see that $\theta_2 - \theta_1 = 2\pi m$ for some $m \in \mathbf{Z}$.

(d) Let g, θ be respectively a continuous logarithm and a continuous argument of f. Then by (b), $\log|f|+i\theta$ is a continuous logarithm of f; and by (c), $g-\log|f|-i\theta = 2\pi in$ for some $n \in \mathbf{Z}$. The result follows.   □

At the level of generality we have used so far, continuous arguments need not exist. For example, let $X = \{z \in \mathbf{C} : |z| = 1\}$ and define $f : X \to \mathbf{C}\backslash\{0\}$ by $f(z) = z$ for all $z \in X$. Suppose that f has a continuous argument θ, and define $\gamma : [0, 2\pi] \to X$ by $\gamma(t) = e^{it}$; since $z = |z|\exp(i\theta(z)) = \exp(i\theta(z))$ when $z \in X$, we see that $e^{it} = \exp(i\theta(e^{it}))$ for all $t \in [0, 2\pi]$. Hence $t \longmapsto t$ and $t \longmapsto \theta(e^{it})$ are continuous arguments of γ, and so by Theorem 3.4.18 (c), $\theta(e^{it}) = t + 2\pi k$ for some $k \in \mathbf{Z}$. With $t = 0$ this gives $\theta(1) = 2\pi k$, but with $t = 2\pi$ we have $\theta(1) = 2\pi(1+k)$. This contradiction shows that no continuous argument can exist.

This difficulty vanishes if X is a closed, bounded interval $[a, b]$ in $\mathbf{R}$, as we now show.

Theorem 3.4.19 *Let $\gamma : [a, b] \to \mathbf{C}\backslash\{0\}$ be continuous. Then γ has a continuous logarithm.*

Proof Let D be any open ball in $\mathbf{C}$ which does not contain 0. We claim that there is an analytic branch of the logarithm on D: that is, there is a function g which is analytic on D and is such that $\exp(g(z)) = z$ for all $z \in D$. Evidently $D \subset \{z \in \mathbf{C} : z \neq |z|e^{i\alpha}\} = D(\alpha)$ for some $\alpha \in \mathbf{R}$. Since, by Theorem 3.4.13, the map $z \mapsto (\alpha - \log)(z)$ belongs to $H(D(\alpha))$, it suffices to choose g to be its restriction to D.

Since $[a, b]$ is compact, $|\gamma|$ attains its minimum, m say, on $[a, b]$; clearly $m > 0$. As γ is uniformly continuous on $[a, b]$ (by Theorem 2.3.30), there is a partition $\{t_0, t_1, \ldots, t_n\}$ of $[a, b]$, with $a = t_0 < t_1 < \cdots < t_n = b$, such that $|\gamma(t) - \gamma(t_j)| < m$ whenever $t \in [t_j, t_{j+1}]$ and $j = 0, 1, \ldots, n-1$. Hence for all $j \in \{0, 1, \ldots, n-1\}$, there is a $g_j \in H(B(\gamma(t_j), m))$ such that $\exp(g_j(\gamma(t))) = \gamma(t)$ for all $t \in [t_j, t_{j+1}]$, as we saw in the first part of the proof. Thus the restriction of γ to each sub-interval $[t_j, t_{j+1}]$ has a continuous logarithm, h_j say, defined by $h_j(t) = g_j(\gamma(t))$ for each $t \in [t_j, t_{j+1}]$.

Since $\gamma(t) = \exp(h_0(t))$ on $[t_0, t_1]$ and $\gamma(t) = \exp(h_1(t))$ on $[t_1, t_2]$, it follows that $\exp(h_0(t_1)) = \exp(h_1(t_1))$, so that by Theorem 3.4.3, $h_0(t_1) = h_1(t_1) + 2\pi ik$ for some $k \in \mathbf{Z}$. Since $h_1 + 2\pi ik$ is also a continuous logarithm of γ on $[t_1, t_2]$, the function $h : [t_0, t_2] \to \mathbf{C}$ defined by $h(t) = h_0(t)$ for $t_0 \leq t \leq t_1$, $h(t) = h_1(t) + 2\pi ik$ for $t_1 \leq t \leq t_2$, is thus a continuous logarithm of γ on $[t_0, t_2]$. Repetition of this process a further $n-2$ times gives a continuous logarithm of γ on $[a, b]$.   □

Theorem 3.4.20 *Let $\gamma : [a, b] \to \mathbf{C}$ be a path and suppose that $0 \notin \gamma^*$ $(= \gamma([a, b]))$. Let θ and ϕ be continuous arguments of γ. Then $\theta(b) - \theta(a) = \phi(b) - \phi(a)$; if γ is closed, $(\theta(b) - \theta(a))/(2\pi) \in \mathbf{Z}$.*

Proof By Theorem 3.4.18 (c), $\theta - \phi = 2\pi m$ for some $m \in \mathbf{Z}$, whence $\theta(b) - \theta(a) = \phi(b) - \phi(a)$. For each $t \in [a, b]$, $\exp(i\theta(t)) = \gamma(t)/|\gamma(t)|$; thus if γ is closed,

$$\exp(i\theta(b) - i\theta(a)) = \gamma(b)\,|\gamma(a)|\,/(\gamma(a)\,|\gamma(b)|) = 1.$$

By Theorem 3.4.3, $(\theta(b) - \theta(a))/(2\pi) \in \mathbf{Z}$.   □

We are now in a position to define the winding number.

Definition 3.4.21 Let $\gamma : [a, b] \to \mathbf{C}$ be a path and, for any $w \in \mathbf{C}$, denote by $\gamma + w$ the path $t \longmapsto \gamma(t) + w : [a, b] \to \mathbf{C}$. Let $z_0 \in \mathbf{C}\backslash\gamma^*$ and let θ be a continuous argument of $\gamma - z_0$. The **winding number of γ with respect to z_0** is defined to be $\{\theta(b) - \theta(a)\}/(2\pi)$, written $n(\gamma, z_0)$.

Note that for all $w \in \mathbf{C}, n(\gamma, z_0) = n(\gamma + w, z_0 + w)$; note also that if γ is closed, $n(\gamma, z_0)$ is an integer (by Theorem 3.4.20).

Example 3.4.22

(i) Let $p, q \in \mathbf{N}, \gamma : [0, 2\pi p] \to \mathbf{C}, \gamma(t) = re^{it}$ for some $r > 0, f : \mathbf{C} \to \mathbf{C},$ $f(z) = z^q$ and $\sigma = f \circ \gamma$. Then γ^*, a circle of centre 0 and radius r, is traversed p times, the sense of that traverse being with the bounded component of $\mathbf{C}\backslash\gamma^*$ on the left. Since $\sigma(t) = f(\gamma(t)) = r^q e^{iqt}$, the function $\theta : [0, 2\pi p] \to \mathbf{R}$ defined by $\theta(t) = qt$ is a continuous argument of σ, and hence $n(\sigma, 0) = \{\theta(2\pi p) - \theta(0)\}/(2\pi) = pq$. In particular, if $q = 1$ this shows that $n(\gamma, 0) = p$: the winding number gives a measure of the number of times the circle is traversed; the path γ winds p times around 0.

(ii) Let

$$\gamma_1(t) = 3 + 2t - i\,(-2 \leq t \leq -1), \gamma_2(t) = 1 + i\,(1 + 2t)\,(-1 \leq t \leq 0),$$
$$\gamma_3(t) = 1 - 2t + i\,(0 \leq t \leq 1), \gamma_4(t) = -1 + i\,(3 - 2t)\,(1 \leq t \leq 2),$$

and let $\gamma : [-2, 2] \to \mathbf{C}$ be the simple, closed polygonal path defined by

$$\gamma(t) = \gamma_k(t) \quad \text{for } k - 3 \leq t \leq k - 2, k = 1, 2, 3, 4.$$

The rectangle which is the track of γ and the sense of its traverse which, as in (i), has the bounded component of $\mathbf{C}\backslash\gamma^*$ on the left, is illustrated below.

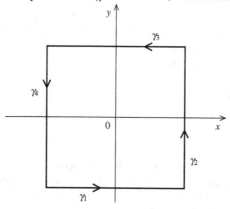

Evidently, γ winds once around the origin and intuitively one might expect that $n(\gamma, 0) = 1$. To verify this, let θ be a continuous argument of γ. Then

$$n(\gamma, 0) = \frac{\theta(2) - \theta(-2)}{2\pi} = \sum_{k=1}^{4} \frac{\theta(k-2) - \theta(k-3)}{2\pi} = \sum_{k=1}^{4} n(\gamma_k, 0).$$

We claim that for each k, $n(\gamma_k, 0) = 1/4$, and prove this for the case $k = 1$ only, as the other cases are similar. Since $\gamma_1^* \subset D(\pi)$, the principal argument of γ_1, that is $\arg \circ \gamma_1$, is a continuous argument of γ_1, and

$$\theta(-1) - \theta(-2) = \arg(\gamma_1(-1)) - \arg(\gamma_1(-2)) = \arg(1 - i) - \arg(-1 - i)$$

$$= -\frac{\pi}{4} - \left(-\frac{3\pi}{4}\right) = \frac{\pi}{2}.$$

Hence $n(\gamma_1, 0) = \{\theta(-1) - \theta(-2)\}/(2\pi) = 1/4$.

On the other hand, the path γ does not wind at all around the point $-2i$ and so we would expect that $n(\gamma, -2i) = 0$. To check that this is so, note that $(\gamma + 2i)^* \subset D(3\pi/2)$ and so $\left(\frac{3\pi}{2} - \arg\right) \circ (\gamma + 2i)$ is a continuous argument of $\gamma + 2i$. Hence

$$2\pi n(\gamma, -2i) = \left(\frac{3\pi}{2} - \arg\right)(-1 + i) - \left(\frac{3\pi}{2} - \arg\right)(-1 + i) = 0.$$

(iii) Generalising (ii) somewhat, we consider now a simple closed polygonal path whose track is a rectangle with vertices $\pm a \pm ib$ $(a, b > 0)$. Specifically, let

$$\gamma_1(t) = 2t + a + 2b - ib \, (-a - b \le t \le -b),$$
$$\gamma_2(t) = a + i(2t + b) \, (-b \le t \le 0),$$
$$\gamma_3(t) = a - 2t + ib \, (0 \le t \le a),$$
$$\gamma_4(t) = -a + i(2a + b - 2t) \, (a \le t \le a + b),$$

and let $\gamma : [-a - b, a + b] \to \mathbf{C}$ be defined by $\gamma(t) = \gamma_1(t)$ for $-a - b \le t \le -b$, etc. As in (ii), one might expect that $n(\gamma, 0) = 1$. Let θ be a continuous argument of γ. Since $\gamma_1^* \cup \gamma_2^* \subset D(\pi)$, the principal arguments of γ_1 and γ_2 are continuous arguments of γ_1 and γ_2 respectively, and

$$\theta(0) - \theta(-a - b) = \arg(\gamma_2(0)) - \arg(\gamma_1(-a - b))$$
$$= \arg(a + ib) - \arg(-a - ib)$$
$$= \arg(a + ib) - (\arg(a + ib) - \pi) = \pi.$$

Similarly, $\theta(a + b) - \theta(0) = \pi$ and so $n(\gamma, 0) = (\pi + \pi)/(2\pi) = 1$.

(iv) Let $R > 1$ and let $\gamma : [-R, R + \pi] \to \mathbf{C}$ be the simple closed path defined by

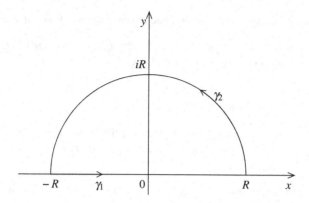

$$\gamma(t) = t \text{ if } -R \le t \le R, \ \gamma(t) = Re^{i(t-R)} \text{ if } R \le t \le R + \pi.$$

Then γ^* has the semicircular form shown above and the bounded component of $\mathbf{C}\backslash\gamma^*$ is on the left as it is traversed.

We claim that $n(\gamma, i) = 1$. To prove this, let γ_1, γ_2 be the restrictions of γ to $[-R, R]$, $[R, R + \pi]$ respectively. Since $\phi := \left(\frac{3\pi}{2} - \arg\right) \circ (\gamma_2 - i)$ is a continuous argument of $\gamma_2 - i$,

$$n(\gamma_2, i) = (\phi(R + \pi) - \phi(R))/(2\pi),$$

and since $-\frac{\pi}{2} < \phi(R) < \phi(R + \pi) < \frac{3\pi}{2}$, we see that $0 < n(\gamma_2, i) < 1$. Similarly, $\psi := \left(\frac{\pi}{2} - \arg\right) \circ (\gamma_1 - i)$ is a continuous argument of $\gamma_1 - i$, and

$$n(\gamma_1, i) = (\psi(R) - \psi(-R))/(2\pi);$$

hence, since $-\frac{3\pi}{2} < \psi(-R) < \psi(R) < \frac{\pi}{2}$, we obtain the inequality $0 < n(\gamma_1, i) < 1$. It follows that $n(\gamma, i) = 1$, since $n(\gamma, i) = n(\gamma_1, i) + n(\gamma_2, i) \in \mathbf{Z}$. Alternatively, appeal to the proof of Theorem 3.4.2 shows that $\phi(R) = \psi(R)$ and $\phi(R + \pi) = 2\pi + \psi(-R)$, equalities which establish the claim directly.

Our next result establishes the invariance of $n(\gamma, w)$ under translation, rotation and magnification.

Theorem 3.4.23 *Let* $\alpha, \beta \in \mathbf{C}$, $\alpha \ne 0$, $f(z) = \alpha z + \beta$ *(for all* $z \in \mathbf{C}$*); let* $\gamma :$ *$[a, b] \to \mathbf{C}$ be continuous and suppose that* $w \in \mathbf{C}\backslash\gamma^*$*. Then* $n(f \circ \gamma, f(w)) = n(\gamma, w)$*.*

Proof Let $\theta_0 \in \text{Arg}\,\alpha$ and let $\theta : [a, b] \to \mathbf{R}$ be a continuous argument of $\gamma - w$. Since $f(\gamma(t)) - f(w) = \alpha(\gamma(t) - w)$ for all $t \in [a, b]$, the map $t \longmapsto \theta(t) + \theta_0 :$ $[a, b] \to \mathbf{R}$ is a continuous argument of $f \circ \gamma - f(w)$, and thus

$$n(f \circ \gamma, f(w)) = \{\theta(b) + \theta_0 - \theta(a) - \theta_0\}/(2\pi) = n(\gamma, w). \qquad \square$$

As might be expected, constant paths have zero winding number.

Proposition 3.4.24 *Let $z \in \mathbf{C}$ and let $\gamma : [a, b] \to \mathbf{C}$ be such that $\gamma(t) = z$ for all $t \in [a, b]$. Then $n(\gamma, w) = 0$ for all $w \in \mathbf{C} \backslash \{z\}$.*

Proof Let $\theta \in \text{Arg}\,(z - w)$. Since $t \to \theta : [a, b] \to \mathbf{R}$ is a continuous argument of $\gamma - w$, $2\pi n(\gamma, w) = \theta - \theta = 0$. □

Our principal object now is to establish the invariance of the winding number under homotopies, and we begin with a few subsidiary results, some of independent interest.

Proposition 3.4.25 *Let $w \in \mathbf{C}$ and for $j = 1, \ldots, k$ let $\gamma_j : [a, b] \to \mathbf{C}$ be a path such that $0 \notin \gamma_j^*$. For each $t \in [a, b]$ let $\gamma(t) = w + \gamma_1(t)\gamma_2(t) \ldots \gamma_k(t)$. Then*

$$n(\gamma, w) = \sum_{j=1}^{k} n(\gamma_j, 0).$$

Proof For each j let θ_j be a continuous argument of γ_j. Then $\sum_{j=1}^{k} \theta_j$ is continuous and for each $t \in [a, b]$,

$$\gamma(t) - w = |\gamma(t) - w| \exp \left\{ i \sum_{j=1}^{k} \theta_j(t) \right\}.$$

Hence

$$2\pi n(\gamma, w) = \sum_{j=1}^{k} (\theta_j(b) - \theta_j(a)) = \sum_{j=1}^{k} 2\pi n(\gamma_j, 0). \qquad □$$

Theorem 3.4.26 *Let $\gamma, \sigma : [a, b] \to \mathbf{C}$ be closed paths, let $w \in \mathbf{C}$ and suppose that for all $t \in [a, b]$,*

$$|\gamma(t) - \sigma(t)| < |w - \gamma(t)|.$$

Then $n(\gamma, w) = n(\sigma, w)$.

Proof Evidently $w \notin \gamma^* \cup \sigma^*$. Define $\mu : [a, b] \to \mathbf{C}$ by $\mu(t) = (\sigma(t) - w)/(\gamma(t) - w)$, and note that $|1 - \mu(t)| = |\gamma(t) - \sigma(t)| / |\gamma(t) - w| < 1$ for all $t \in [a, b]$ and $\mu^* \subset \{z \in \mathbf{C} : z \neq -|z|\}$. Hence $\arg \mu$ is a continuous argument of μ and

$$2\pi n(\mu, 0) = \arg(\mu(b)) - \arg(\mu(a)) = 0.$$

Since $\sigma - w = \mu(\gamma - w)$, it follows from Proposition 3.4.25 that

$$n(\sigma, w) = n(\mu, 0) + n(\gamma, w) = n(\gamma, w). \qquad □$$

Theorem 3.4.27 *Let $\gamma : [a, b] \to \mathbf{C}$ be a closed path and suppose that w and z are points in the same component of $\mathbf{C}\backslash\gamma^*$. Then $n(\gamma, z) = n(\gamma, w)$. If w belongs to the unbounded component of $\mathbf{C}\backslash\gamma^*$, $n(\gamma, w) = 0$.*

Proof For the first part we merely have to prove that $n(\gamma, \cdot)$ is continuous on $\mathbf{C}\backslash\gamma^*$, since it is integer-valued. Let $z \in \mathbf{C}\backslash\gamma^*$ and put $\eta = \text{dist}(z, \gamma^*) > 0$. Let $w \in B(z, \eta)$, $w \neq z$, and set $\sigma = \gamma - w + z$. Then for all $t \in [a, b]$,

$$|\gamma(t) - \sigma(t)| = |w - z| < |\gamma(t) - z|,$$

and thus by Theorem 3.4.26,

$$n(\gamma, z) = n(\sigma, z) = n(\gamma - w + z, z) = n(\gamma, w).$$

This shows that $n(\gamma, \cdot)$ is continuous on $\mathbf{C}\backslash\gamma^*$ and the first part follows.

Since γ^* is compact, there exists $r > 0$ such that $\gamma^* \subset B(0, r)$. As $^c\overline{B(0, r)}$ is evidently connected, there is only one unbounded component of $\mathbf{C}\backslash\gamma^*$. Moreover, $(\gamma + 2r)^* \subset \{z \in \mathbf{C} : re z > 0\}$, and so arg $(\gamma + 2r)$ is a continuous argument of $\gamma + 2r$ and $n(\gamma, -2r) = 0$. Since $-2r$ lies in the unbounded component of $\mathbf{C}\backslash\gamma^*$, the result follows from the first part of the Theorem. □

We are now in a position to establish the homotopy-invariance of the winding number.

Theorem 3.4.28 *Let G be an open set in $\mathbf{C}$ and let $\gamma, \sigma : [0, 1] \to G$ be closed paths in G which are freely homotopic. Then $n(\gamma, w) = n(\sigma, w)$ for all $w \in \mathbf{C}\backslash G$.*

Proof Let $w \in \mathbf{C}\backslash G$. There is a continuous map $H : I \times I \to G$ (where $I = [0, 1]$) such that for all $s, t \in I$, $H(s, 0) = \gamma(s)$, $H(s, 1) = \sigma(s)$ and $H(0, t) = H(1, t)$. Since $H(I \times I)$ is compact, there exists $\varepsilon > 0$ such that for all $s, t \in I$, $|w - H(s, t)| > \varepsilon$. As H is uniformly continuous on $I \times I$, there exists $k \in \mathbf{N}$ such that

$$\left|H(s, t) - H(s', t')\right| < \varepsilon \text{ if } |s - s'| + |t - t'| \le 1/k.$$

Define closed paths $\mu_j : I \to G$ $(j = 0, 1, \ldots, k)$ by $\mu_j(s) = H(s, j/k)$; $\mu_0 = \gamma$, $\mu_k = \sigma$. For $j = 1, 2, \ldots, k$ and all $s \in I$,

$$\left|\mu_j(s) - \mu_{j-1}(s)\right| = |H(s, j/k) - H(s, (j - 1)/k)| < \varepsilon. \qquad (3.4.1)$$

Clearly $\left|w - \mu_j(s)\right| > \varepsilon$ for $j = 0, 1, \ldots, k$ and for all $s \in I$. From this, (3.4.1) and k applications of Theorem 3.4.26, it follows that w has the same winding number with respect to each of the closed paths $\gamma = \mu_0, \mu_1, \ldots, \mu_k = \sigma$. Hence $n(\gamma, w) = n(\sigma, w)$. □

Theorem 3.4.29 *Let (γ_k) be a sequence of closed paths in $\mathbf{C}$, each with parameter interval $[a, b]$, and suppose that (γ_k) converges uniformly on $[a, b]$ to a path $\gamma : [a, b] \to \mathbf{C}$. If $w \in \mathbf{C}\backslash\gamma^*$, then for all large enough k,*

$$n(\gamma_k, w) = n(\gamma, w).$$

Proof Let $w \in \mathbf{C}\backslash\gamma^*$. There exists $\varepsilon > 0$ such that $|\gamma(t) - w| > \varepsilon$ for all $t \in [a, b]$. Since (γ_k) converges uniformly on $[a, b]$ to γ, there exists $k_0 \in \mathbf{N}$ such that for all $k \geq k_0$ and all $t \in [a, b]$, $|\gamma_k(t) - \gamma(t)| < \varepsilon$. From this and Theorem 3.4.26, it follows that $n(\gamma_k, w) = n(\gamma, w)$ for all $k \geq k_0$. $\qquad\qquad\qquad\square$

The winding number can be used to clarify our ideas about the 'inside' and 'outside' of a closed path γ.

Definition 3.4.30 Let γ be a closed path in $\mathbf{C}$. We say that a point $z \in \mathbf{C}$ is **inside** γ if $z \notin \gamma^*$ and $n(\gamma, z) \neq 0$; and that z is **outside** γ if $z \notin \gamma^*$ and $n(\gamma, z) = 0$.

Remark 3.4.31 For simple closed paths this fits in well with our intuition. For example, suppose that $\gamma : [0, 2\pi] \to \mathbf{C}$ is given by $\gamma(t) = e^{it}$, so that γ^* is the unit circle. Then $\mathbf{C}\backslash\gamma^*$ has exactly two components, one bounded and the other unbounded. If $|z| > 1$, z is in the unbounded component and we know from Theorem 3.4.27 that $n(\gamma, z) = 0$; further, if $|z| < 1$, using Theorem 3.4.27 again, we see that $n(\gamma, z) = n(\gamma, 0) = 1$, the final equality following from Example 3.4.22 (i). Hence, the points inside γ correspond to those in the bounded component of $\mathbf{C}\backslash\gamma^*$, while those outside correspond to those in the unbounded component. Evidently, Theorem 3.4.23 extends this result to a larger class of circular paths, allowing tracks with arbitrary centre and radius. Although γ is so special a simple closed path, in a sense it is typical of simple closed paths: we shall see later that the Jordan curve theorem shows that given any simple closed path γ, $\mathbf{C}\backslash\gamma^*$ has exactly two connected components, one bounded, the other unbounded, and that $n(\gamma, z) = 0$ for all z in the unbounded component, while $n(\gamma, z) = \pm 1$ for all z in the bounded component.

To conclude this section we show how the winding number may be used to prove the two-dimensional version of a famous fixed-point theorem, due to Brouwer.

Theorem 3.4.32 *Let $D = \{z \in \mathbf{C} : |z - a| \leq r\}$, $r > 0$, so that the boundary ∂D is the track of the closed path $\gamma : [0, 1] \to \mathbf{C}$, $\gamma(t) = a + re^{2\pi i t}$. Let $f : D \to \mathbf{C}$ be continuous, let $w \in \mathbf{C}\backslash f(\partial D)$ and suppose that $n(f \circ \gamma, w) \neq 0$. Then there is a point z in D such that $f(z) = w$.*

Proof Suppose that no such point z exists, so that $w \notin f(D)$. Let $I = [0, 1]$, let $\gamma_0 : I \to \mathbf{C}$ be the constant path defined by $\gamma_0(t) = a$ ($t \in I$) and define $h : I \times I \to \mathbf{C}$ by $h(s, t) = (1 - t)\gamma(s) + t\gamma_0(s)$ ($s, t \in I$). Plainly h is continuous and $h(I \times I) \subset D$. Let $H : I \times I \to \mathbf{C}\backslash\{w\}$ be defined by $H(s, t) = f(h(s, t))$. Then H establishes that, regarded as closed paths in $\mathbf{C}\backslash\{w\}$, $f \circ \gamma$ and the constant path $f \circ \gamma_0$ are freely homotopic. The homotopy invariance of the winding number (Theorem 3.4.28) now shows that $n(f \circ \gamma, w) = n(f \circ \gamma_0, w)$; further, since $f \circ \gamma_0$ is a constant path, Proposition 3.4.24 tells us that $n(f \circ \gamma_0, w) = 0$. Hence $n(f \circ \gamma, w) = 0$, contrary to hypothesis. The proof is complete. $\qquad\qquad\qquad\square$

Theorem 3.4.33 *Let D be a closed ball in $\mathbf{C}$ and let $f : D \to \mathbf{C}$ be continuous and such that $f(z) = z$ for all $z \in \partial D$. Then $D \subset f(D)$.*

Proof Let $D = \{z \in \mathbf{C} : |z - a| \leq r\}$ and let ∂D be represented as the track of $\gamma : [0, 1] \to \mathbf{C}$, $\gamma(t) = a + re^{2\pi it}$. Let $w \in \overset{o}{D}$. Then $n(f \circ \gamma, w) = n(\gamma, w) = 1$, by Remark 3.4.31. Application of Theorem 3.4.32 now shows that there is a point $z \in D$ such that $f(z) = w$, and hence $D \subset f(D)$. $\qquad\square$

Corollary 3.4.34 *Let D be a closed ball in $\mathbf{C}$. Then there is no continuous map $f : D \to \partial D$ which leaves fixed every point of ∂D.*

In other words, ∂D is not a **retract** of D: a subset A of a metric space X is called a **retract** of X if there is a continuous map $r : X \to A$, called a **retraction**, such that $r(x) = x$ for all $x \in A$.

We can now prove the Brouwer fixed-point theorem in a two-dimensional form.

Theorem 3.4.35 *Let $D = \{z \in \mathbf{C} : |z - a| \leq r\}$ and let $f : D \to D$ be continuous. Then there is a point $z_0 \in D$ such that $f(z_0) = z_0$.*

Proof Suppose that f does not have a fixed point in D. Given any $z \in D$, let $g(z)$ be the point on ∂D nearer to z than to $f(z)$ on the line through z and $f(z)$. This gives a map $g : D \to \partial D$, the restriction of which to ∂D is the identity map. If we can prove that g is continuous, we shall have a contradiction, by Corollary 3.4.34. To do this, note that $g(z) = z + tu$, where $u = (z - f(z))/|z - f(z)|$ and $t \geq 0$ is so chosen that $|g(z) - a| = r$. Since $|u| = 1$,

$$t^2 + 2t\,\mathrm{re}((z - a)\bar{u}) - (r^2 - |z - a|^2) = 0,$$

and thus

$$t = -\mathrm{re}((z - a)\bar{u}) + \left[(\mathrm{re}((z - a)\bar{u}))^2 + (r^2 - |z - a|^2)\right]^{1/2}.$$

The continuity of g is now clear, and the Theorem follows. $\qquad\square$

Note that by identification of $\mathbf{C}$ with $\mathbf{R}^2$ we have Brouwer's theorem in the following form: any continuous map of a closed ball in $\mathbf{R}^2$ into itself has a fixed point.

Exercise 3.4.36

1. Let $\alpha \in \mathbf{C}$ and define

$$f(z) = 1 + \sum_{n=1}^{\infty} \frac{\alpha(\alpha - 1) \cdots (\alpha - n + 1)}{n!} z^n.$$

   Prove that if α is not a non-negative integer, then this power series has unit radius of convergence. Deduce that $f'(z) = \alpha f(z)/(1 + z)$ if $|z| < 1$. By considering ϕ', where $\phi(z) = f(z)/(1 + z)^\alpha$, show that $f(z) = (1 + z)^\alpha$ if $|z| < 1$.

2. Let $S^1 = \{z \in \mathbf{C} : |z| = 1\}$. Show that the map $t \longmapsto E(it) : (-\pi, \pi] \to S^1$ is bijective and continuous, but that its inverse is not continuous at $-1 \in S^1$. [This exercise highlights the hypothesis of a compact domain in Theorem 2.3.24 (ii).]

3. Find the α—branch of the argument, logarithm and ith power of $-1 - i$ in the cases $\alpha = \pi$ and $\alpha = 3\pi/2$.

4. Let α be an irrational real number. Prove that:
   (i) $\{m + n\alpha : m, n \in \mathbf{Z}\}$ is a countable dense subset of $\mathbf{R}$; (ii) the αth power of 1 is dense in $\{z \in \mathbf{C} : |z| = 1\}$.

5. (i) Prove that there exists $\lambda \in \mathbf{C}$ such that $|\lambda| = 1$ and for all $n \in \mathbf{N}$, $\lambda^n \neq 1$.
   (ii) Let λ be as in (i), let $f(z) = \sum_0^\infty a_n z^n$ have radius of convergence $R > 0$, let $A_n(z) = \frac{1}{n+1} \sum_{k=0}^n f(\lambda^k z)$ $(0 < |z| < R, n \in \mathbf{N}_0)$, and let

$$M(r) = \sup\{|f(z)| : |z| = r\} \quad \text{for } 0 < r < R.$$

   Prove that:
   (a) if $0 < |z| < R$, then $\lim_{n \to \infty} A_n(z) = a_0$;
   (b) if $0 < r < R$, then $|a_0| \leq M(r)$.
   (iii) By considering $f(z)/z^j$, show that $|a_j| \leq M(r)/r^j$ for all $j \in \mathbf{N}_0$ (Cauchy's inequalities).
   (iv) Deduce that if $R = \infty$ and there exists $M \in \mathbf{R}$ such that $|f(z)| \leq M$ for all $z \in \mathbf{C}$, then f is constant (Liouville's theorem).

6. Let $\gamma : [0, 2\pi] \to \mathbf{C}$ be given by

$$\gamma(t) = \cos t + \cos 2t - 1 + i(\sin t + \sin 2t) \ (0 \leq t \leq 2\pi).$$

   Find $n(\gamma, 0)$.

7. Let $\gamma : [a, b] \to \mathbf{C}$ be a path such that $0 \notin \gamma^*$. Suppose $c, d \in \mathbf{R}$, $c < d$, and $\psi : [c, d] \to [a, b]$ is a continuous map such that $\psi(c) = a$, $\psi(d) = b$. Show that

$$n(\gamma \circ \psi, 0) = n(\gamma, 0).$$

   Show also that, if the hypotheses are varied to require instead that $\psi(c) = b$, $\psi(d) = a$, then

$$n(\gamma \circ \psi, 0) = -n(\gamma, 0).$$

8. Use the homotopy invariance of the winding number to show that the circle S^1 is not simply connected.

3.5 Integration

We begin with a review of differentiation of complex-valued functions defined on an interval in $\mathbf{R}$ and an extension of the Riemann integral to cover continuous functions of this nature.

The notion of differentiability of a function $f : I \to \mathbf{C}$, where I is an interval in $\mathbf{R}$, has the expected definition: if $f(s) = u(s) + iv(s)$ $(s \in I)$, where $u(s)$ and $v(s)$ are real, then f is differentiable if, and only if, u and v are differentiable; also, in the event of differentiability, $f' = u' + iv'$. The usual rules for derivatives of sums, products and quotients of such functions apply, as do those relating to composition. Given the differentiability of f, if J is an interval in $\mathbf{R}$ and $\phi : J \to I$ is differentiable, then $f \circ \phi$ is differentiable and

$$(f \circ \phi)'(t) = f'(\phi(t))\phi'(t) \, (t \in J);$$

also, if G is an open set in $\mathbf{C}$, $f(I) \subset G$ and $g : G \to \mathbf{C}$ is analytic, then $g \circ f$ is differentiable and

$$(g \circ f)'(s) = g'(f(s))f'(s) \, (s \in I).$$

Proof of these assertions is elementary and is left to the reader.

Definition 3.5.1 Let $f : [a, b] \to \mathbf{C}$ be continuous. We define the **integral of** f **over** $[a, b]$ by

$$\int_a^b f = \int_a^b \operatorname{re} f + i \int_a^b \operatorname{im} f.$$

Note that $\operatorname{re} f$ and $\operatorname{im} f$ are continuous, real-valued functions on $[a, b]$ and hence are Riemann-integrable over $[a, b]$; thus the definition makes sense. For example,

$$\int_0^\pi e^{it} dt = \int_0^\pi \cos t \, dt + i \int_0^\pi \sin t \, dt = 2i.$$

Certain standard facts about Riemann integrals of real functions go over, with obvious proofs, to the complex case. For convenience, we collect these in the following theorem.

Theorem 3.5.2 *Let* $f, f_1, f_2 : [a, b] \to \mathbf{C}$ *be continuous and let* $\alpha_1, \alpha_2 \in \mathbf{C}$. *Then*

(i) $\int_a^b (\alpha_1 f_1 + \alpha_2 f_2) = \alpha_1 \int_a^b f_1 + \alpha_2 \int_a^b f_2$;

(ii) *if* $a \le c \le b$, *then*

$$\int_a^b f = \int_a^c f + \int_c^b f;$$

(iii) *if* $F(t) = \int_a^t f$ *for* $a \le t \le b$, *then* F *is differentiable and* $F'(t) = f(t)$ *for all* $t \in [a, b]$;

(iv) *if* $F : [a, b] \to \mathbf{C}$ *is differentiable and* $F'(t) = f(t)$ *for all* $t \in [a, b]$, *then*

$$\int_a^b f = F(b) - F(a);$$

(v) *if* $\phi : [c, d] \to \mathbf{R}$ *is continuously differentiable and* $\phi([c, d]) \subset [a, b]$, *then*

$$\int_{\phi(c)}^{\phi(d)} f = \int_{c}^{d} f(\phi(t))\phi'(t)dt;$$

(vi) *if (g_n) is a sequence of continuous, complex-valued functions on $[a, b]$ which converges uniformly on $[a, b]$ to g, then g is continuous on $[a, b]$ and*

$$\int_{a}^{b} g_n \rightarrow \int_{a}^{b} g \text{ as } n \rightarrow \infty.$$

Extended to the complex case, we require an important, but standard, inequality.

Theorem 3.5.3 *Let $f : [a, b] \rightarrow \mathbf{C}$ be continuous. Then $|f| : [a, b] \rightarrow \mathbf{R}$ is continuous and*

$$\left| \int_{a}^{b} f \right| \le \int_{a}^{b} |f|.$$

Proof Continuity of $|f|$ is immediate: given any $c \in [a, b]$, by the triangle inequality,

$$||f(t)| - |f(c)|| \le |f(t) - f(c)| \rightarrow 0 \text{ as } t \rightarrow c.$$

As for the integral inequality, put $\alpha = \int_{a}^{b} f = \alpha_1 + i\alpha_2$, with α_1, α_2 real. Then

$$|\alpha|^2 = \alpha_1^2 + \alpha_2^2 = \alpha_1 \int_{a}^{b} \operatorname{re} f + \alpha_2 \int_{a}^{b} \operatorname{im} f = \int_{a}^{b} (\alpha_1 \operatorname{re} f + \alpha_2 \operatorname{im} f)$$

$$\le \int_{a}^{b} (\alpha_1^2 + \alpha_2^2)^{1/2}[(\operatorname{re} f)^2 + (\operatorname{im} f)^2]^{1/2} = |\alpha| \int_{a}^{b} |f|,$$

using Cauchy's inequality. If $\alpha \ne 0$, this gives $|\alpha| \le \int_{a}^{b} |f|$, as required; if $\alpha = 0$, the result is obvious. $\square$

3.5.1 Integrals Along Contours

We next introduce the special kind of path along which we shall be integrating.

Definition 3.5.4 (i) *A continuously differentiable path in $\mathbf{C}$ is called an **arc**: thus a path $\gamma : [a, b] \rightarrow \mathbf{C}$ is an arc if it has a derivative γ' that is defined and continuous on $[a, b]$, the derivatives at a and b being one-sided.*

(ii) *A piecewise continuously differentiable path in $\mathbf{C}$ is termed a **contour** : a path $\gamma : [a, b] \rightarrow \mathbf{C}$ is a contour if there is a partition $\{t_0, t_1, \ldots, t_n\}$ of $[a, b]$ such that, for each $j \in \{1, 2, \ldots, n\}$, the restriction of γ to $[t_{j-1}, t_j]$ has a continuous derivative on $[t_{j-1}, t_j]$, so permitting the left and right derivatives at $t_1, \ldots, t_{n-1}$ to differ. Plainly, each arc is a contour, but a contour need not be an arc.*

(iii) *A closed path which is a contour is called a* **circuit**; further, a simple path which is a contour (circuit) is called a **simple contour (simple circuit)**.

The path $\gamma : [0, 1] \to \mathbf{C}$ defined by $\gamma(t) = e^{i\pi t}$ $(0 \leq t \leq 1)$ is an example of an arc; each polygonal path in $\mathbf{C}$ (see Definition 2.4.15) is a contour; the path $v : [0, 1] \to \mathbf{C}$ defined by $v(t) = e^{2\pi it}$ $(0 \leq t \leq 1/2)$, $v(t) = 4t - 3$ $(1/2 \leq t \leq 1)$ is a simple circuit.

Definition 3.5.5 Let $\gamma : [a, b] \to \mathbf{C}$ be a contour. The **length** of γ, written $l(\gamma)$, is defined by

$$l(\gamma) = \int_a^b \left|\gamma'(t)\right| dt = \int_a^b \{(\operatorname{re} \gamma')^2 + (\operatorname{im} \gamma')^2\}^{1/2}$$
$$= \sum_{j=1}^n \int_{t_{j-1}}^{t_j} \{(\operatorname{re} \gamma')^2 + (\operatorname{im} \gamma')^2\}^{1/2},$$

where $\{t_0, t_1, \ldots, t_n\}$ is a partition of $[a, b]$ such that the restriction of γ to each subinterval $[t_{j-1}, t_j]$ has a continuous derivative on $[t_{j-1}, t_j]$.

We leave it to the reader to show that the length of γ is independent of the choice of partition used to define γ.

To illustrate with examples, if $\gamma : [0, 1] \to \mathbf{C}$ is defined by $\gamma(t) = e^{2\pi it}$ $(0 \leq t \leq 1)$, so that γ^* is the unit circle, then

$$l(\gamma) = \int_0^1 2\pi(\sin^2 2\pi t + \cos^2 2\pi t)^{1/2} dt = 2\pi,$$

as expected; also, if $\mu : [a, b] \to \mathbf{C}$ is a polygonal path in $\mathbf{C}$, so that there exist $z_0, z_1, \ldots z_n \in \mathbf{C}$ and a partition $\{t_0, t_1, \ldots t_n\}$ of $[a, b]$ such that

$$\mu(t) = (t_j - t_{j-1})^{-1}\{(t_j - t)z_{j-1} + (t - t_{j-1})z_j\}$$

whenever $t_{j-1} \leq t \leq t_j$, then

$$l(\mu) = \sum_{j=1}^n \int_{t_{j-1}}^{t_j} \left|\mu'(t)\right| dt$$
$$= \sum_{j=1}^n \int_{t_{j-1}}^{t_j} (t_j - t_{j-1})^{-1} \left|z_j - z_{j-1}\right| dt$$
$$= \sum_{j=1}^n \left|z_j - z_{j-1}\right|.$$

Definition 3.5.6 Let $\gamma : [a, b] \to \mathbf{C}$ be a contour. The contour **opposite** to γ, written $-\gamma$, is defined by $(-\gamma)(t) = \gamma(a + b - t)$ $(a \leq t \leq b)$.

Clearly the track of $-\gamma$ is the same as that of γ, but it is traversed in the opposite sense. Naturally $l(-\gamma) = l(\gamma)$: to prove this formally, let $\{a = t_0, t_1, \ldots, t_n = b\}$ be a partition of $[a, b]$ which contains all the discontinuities of γ', so that $\{u_j = a + b - t_j : j = 0, 1, \ldots, n\}$ contains all those of $(-\gamma)'$, and $a = u_n < u_{n-1} < \cdots < u_0 = b$; then, appealing to Theorem 3.5.2 (v),

$$l(-\gamma) = \sum_1^n \int_{u_j}^{u_{j-1}} \left| (-\gamma)'(s) \right| ds = \sum_1^n \int_{u_j}^{u_{j-1}} \left| \gamma'(a+b-s) \right| ds$$

$$= \sum_1^n \int_{t_{j-1}}^{t_j} \left| \gamma'(t) \right| dt = l(\gamma). \tag{3.5.1}$$

Now we can define the integral of a continuous complex-valued function along a contour.

Definition 3.5.7 Let $\gamma : [a, b] \to \mathbf{C}$ be a contour and let $f : \gamma^* \to \mathbf{C}$ be continuous. The **integral of** f **over** γ, written $\int_\gamma f$ or $\int_\gamma f(z)dz$, is defined to be $\int_a^b f(\gamma(t))\gamma'(t)dt$. [The range $[a, b]$ of t is subdivided into subintervals corresponding to the discontinuities of γ' if γ' is not continuous on the whole of $[a, b]$.]

Example 3.5.8

(i) Let $w_1, w_2 \in \mathbf{C}$ and define $\gamma : [0, 1] \to \mathbf{C}$ by $\gamma(t) = (1 - t)w_1 + tw_2$; γ is a line segment joining w_1 to w_2. Then

$$\int_\gamma z dz = \int_0^1 \{(1 - t)w_1 + tw_2\}(w_2 - w_1)dt = \frac{1}{2}(w_2^2 - w_1^2).$$

(ii) Let $a \in \mathbf{C}$ and let r, θ be positive real numbers. Define $\gamma : [0, \theta] \to \mathbf{C}$ by $\gamma(t) = a + re^{it}$ and put $f(z) = (z - a)^{-1}$ $(z \neq a)$. Then

$$\int_\gamma (z - a)^{-1}dz = \int_0^\theta ire^{it}/(re^{it})dt = i\theta.$$

The particular case of this when $\theta = 2\pi$ is important: the contour γ is then called the **positively oriented circle with centre** a **and radius** r, and

$$\int_\gamma (z - a)^{-1}dz = 2\pi i. \tag{3.5.2}$$

Lemma 3.5.9 *Let* $\gamma : [a, b] \to \mathbf{C}$ *be a contour and let* $f, g : \gamma^* \to \mathbf{C}$ *be continuous. Then for all* $\alpha, \beta \in \mathbf{C}$,

$$\int_\gamma (\alpha f + \beta g) = \alpha \int_\gamma f + \beta \int_\gamma g;$$

and

$$\int_\gamma f = -\int_{-\gamma} f.$$

Proof The first part is obvious from the definition of the integral over a contour allied with Theorem 3.5.2 (i). For the second part, with the same notation as in the proof of (3.5.1), we have

$$\int_{-\gamma} f = \sum_1^n \int_{u_j}^{u_{j-1}} f(-\gamma(s))(-\gamma)'(s)ds$$

$$= -\sum_1^n \int_{u_j}^{u_{j-1}} f(\gamma(a+b-s))\gamma'(a+b-s)ds$$

$$= \sum_1^n \int_{t_j}^{t_{j-1}} f(\gamma(t))\gamma'(t)dt$$

$$= -\sum_1^n \int_{t_{j-1}}^{t_j} f(\gamma(t))\gamma'(t)dt = -\int_\gamma f. \qquad \square$$

A simple and frequently applied inequality, an outcome of Theorem 3.5.3, is given next.

Theorem 3.5.10 *Let $\gamma : [a, b] \to \mathbf{C}$ be a contour, let $f : \gamma^* \to \mathbf{C}$ be continuous and suppose that M is a constant such that for all $z \in \gamma^*$, $|f(z)| \le M$. Then*

$$\left| \int_\gamma f \right| \le Ml(\gamma). \tag{3.5.3}$$

More generally,

$$\left| \int_\gamma f \right| \le \int_a^b |f(\gamma(t))| \, |\gamma'(t)| \, dt. \tag{3.5.4}$$

[*The integral on the right-hand side of (3.5.4) is often written as $\int_\gamma |f| \, |dz|$.*]

Proof It is sufficient to assume that γ is an arc, as the general case proceeds by similar arguments on appropriate subintervals of $[a, b]$. Then

$$\left| \int_\gamma f \right| = \left| \int_a^b f(\gamma(t))\gamma'(t)dt \right| \le \int_a^b |f(\gamma(t))| \, |\gamma'(t)| \, dt$$

$$\le M \int_a^b |\gamma'(t)| \, dt = Ml(\gamma). \qquad \square$$

Definition 3.5.11 Let $\gamma : [a, b] \to \mathbf{C}$, $\gamma_1 : [a_1, b_1] \to \mathbf{C}$ be arcs. We say that γ_1 is a **reparametrisation** of γ if there is a continuously differentiable bijection ϕ

of $[a_1, b_1]$ onto $[a, b]$ with an everywhere strictly positive derivative ϕ', such that $\gamma_1 = \gamma \circ \phi$.

It is clear that γ and γ_1 have the same track, and we would expect them to have the same length. To give a formal proof: use of Theorem 3.5.2 (v) shows that

$$l(\gamma_1) = \int_{a_1}^{b_1} |\gamma_1'| = \int_{a_1}^{b_1} |\gamma' \circ \phi| \, |\phi'| = \int_{a}^{b} |\gamma'| = l(\gamma).$$

The commonest reparametrisation occurs when ϕ is defined by

$$\phi(t) = \{a(b_1 - t) + b(t - a_1)\}/(b_1 - a_1).$$

Thus, if $\gamma(t) = e^{it}$ $(0 \le t \le 1)$ and we want the parameter interval to be $[2, 8]$, then $\phi(t) = (t - 2)/6$ does the trick.

Lemma 3.5.12 *Let* $\gamma_1 : [a_1, b_1] \to \mathbf{C}$ *be a reparametrisation of an arc* $\gamma :$ *$[a, b] \to \mathbf{C}$. Then for all continuous maps* $f : \gamma^* \to \mathbf{C}$,

$$\int_{\gamma_1} f = \int_{\gamma} f.$$

[*Note that* $\gamma_1^* = \gamma^*$.]

Proof Let $\gamma_1 = \gamma \circ \phi$, where ϕ is a continuously differentiable bijection of $[a_1, b_1]$ onto $[a, b]$ and $\phi'(t) > 0$ whenever $a_1 \le t \le b_1$. Then

$$\int_{\gamma_1} f = \int_{a_1}^{b_1} f(\gamma_1(t))\gamma_1'(t)dt$$

$$= \int_{a_1}^{b_1} f(\gamma(\phi(t)))\gamma'(\phi(t))\phi'(t)dt$$

$$= \int_{a}^{b} f(\gamma(s))\gamma'(s)ds = \int_{\gamma} f. \qquad \square$$

Remark 3.5.13 Let $\mu : [a, b] \to \mathbf{C}$, $\nu : [c, d] \to \mathbf{C}$ be contours and suppose that there are partitions $\{s_0 = a, s_1, \ldots, s_m = b\}$ and $\{t_0 = c, t_1, \ldots, t_m = d\}$ of $[a, b]$ and $[c, d]$ respectively such that, for $1 \le j \le m$, the restrictions μ_j, ν_j of μ, ν to $[s_{j-1}, s_j]$, $[t_{j-1}, t_j]$, respectively, are arcs. Then the contour ν is said to be a **reparametrisation** of μ if, for $1 \le j \le m$, the arc ν_j is a reparametrisation of μ_j. In this event $\nu^* = \mu^*$; also, using Lemma 3.5.12, for all continuous $f : \mu^* \to \mathbf{C}$,

$$\int_{\mu} f = \sum_{j=1}^{m} \int_{\mu_j} f = \sum_{j=1}^{m} \int_{\nu_j} f = \int_{\nu} f.$$

As an illustration, let $\mu : [a, b] \to \mathbf{C}$ be a contour. It is elementary to check that $\mu \circ \phi : [c, d] \to \mathbf{C}$, where $c < d$ and

$$\phi(t) = \{(t - c)b + (d - t)a\}/(d - c),$$

is a reparametrisation of μ. It is called the **standard reparametrisation of μ relative to the interval** $[c, d]$.

Exercise 3.5.14

1. Sketch the tracks of the following contours:

    (i) $\gamma_1 : [-1, 1] \to \mathbf{C}, \gamma_1(t) = it \ (-1 \le t \le 1)$;

    (ii) $\gamma_2 : \left[-\frac{\pi}{2}, \frac{\pi}{2}\right] \to \mathbf{C}, \gamma_2(t) = e^{it} \ \left(-\frac{\pi}{2} \le t \le \frac{\pi}{2}\right)$;

    (iii) $\gamma_3 : \left[-\frac{\pi}{2}, \frac{\pi}{2}\right] \to \mathbf{C}, \gamma_3(t) = e^{-i(\pi+t)} \ \left(-\frac{\pi}{2} \le t \le \frac{\pi}{2}\right)$.

    For $k = 1, 2, 3$ evaluate $\int_{\gamma_k} |z| \, dz$.

2. Let a and b be positive real numbers. Let $\mu : [-1, 1] \to \mathbf{C}$ and $v : [0, \pi] \to \mathbf{C}$ be contours defined by

$$\mu(s) = \begin{cases} -sa + ib(1+s) & \text{if } -1 \le s \le 0, \\ -sa + ib(1-s) & \text{if } 0 \le s \le 1, \end{cases}$$

    and

$$v(s) = a \cos s + ib \sin s \text{ if } 0 \le s \le \pi.$$

    Evaluate $\int_\mu f(z)dz$ and $\int_v f(z)dz$ in each of the following cases:
    (i) $f(z) = \text{re } z$, (ii) $f(z) = \bar{z}$.

3. Show that $\int_\gamma z^{-1} dz = 2\pi i$ in each of the following cases:

    (i) $\gamma : [-\pi, \pi] \to \mathbf{C}$ is the circuit defined by

$$\gamma(t) = a \cos t + ib \sin t, -\pi \le t \le \pi,$$

    where a and b are positive real constants.
    (ii) $\gamma : [-1, 1] \to \mathbf{C}$ is the circuit defined by

$$\gamma(t) = \begin{cases} 3 + 4t - i, & -1 \le t \le -1/2, \\ 1 + i(1 + 4t), & -1/2 \le t \le 0, \\ 1 - 4t + i, & 0 \le t \le 1/2, \\ -1 + i(3 - 4t), & 1/2 \le t \le 1. \end{cases}$$

4. Let $\gamma : [0, 2\pi] \to \mathbf{C}$ be defined by $\gamma(t) = 2 \exp(it)$. Evaluate $\int_\gamma \frac{1}{z(z-1)} dz$.

3.6 Cauchy's Theorem

In this section we present basic integration theorems due to Cauchy upon which many important future developments rest.

Definition 3.6.1 Let G be an open subset of $\mathbf{C}$ and let $f : G \to \mathbf{C}$. A function $F : G \to \mathbf{C}$ is called a **primitive for** f **on** G if $F' = f$.

Our immediate concern is with the existence of a primitive for a given continuous function, and we begin with a reformulation of the problem.

Theorem 3.6.2 (The fundamental theorem of contour integration) *Let G be an open subset of $\mathbf{C}$ and let $f : G \to \mathbf{C}$ be continuous. Then the following three statements are equivalent:*

(i) *f has a primitive on G;*
(ii) *for every circuit γ in G, $\int_\gamma f = 0$;*
(iii) *for all contours γ_1, γ_2 in G with the same initial and terminal points,*

$$\int_{\gamma_1} f = \int_{\gamma_2} f.$$

Moreover, if f has a primitive F on G and γ is a contour in G with initial point z_1 and terminal point z_2, then

$$\int_\gamma f = F(z_2) - F(z_1).$$

Proof Suppose that (i) holds and let F be a primitive for f on G. Let $\gamma : [a, b] \to \mathbf{C}$ be a contour in G with $\gamma(a) = z_1$, $\gamma(b) = z_2$. An application of the fundamental theorem of integral calculus (Theorem 1.4.4) to the subintervals $[t_{j-1}, t_j]$ on which γ' is continuous shows that

$$\int_\gamma f = \sum_1^n \int_{t_{j-1}}^{t_j} F'(\gamma(t))\gamma'(t)dt = \sum_1^n \left\{F(\gamma(t_j)) - F(\gamma(t_{j-1}))\right\}$$

$$= F(\gamma(b)) - F(\gamma(a)) = F(z_2) - F(z_1).$$

This establishes the last part of the Theorem. Further, if γ is a circuit, so that $z_1 = z_2$, then $\int_\gamma f = 0$ and it follows that (i) implies (ii).

Next, suppose that (ii) holds and that γ_1, γ_2 are contours in G with the same initial and terminal points. Referring to Remark 3.5.13, let η_1, η_2 be the standard reparametrisations of γ_1, γ_2 relative to the intervals $[0, 1/2]$ and $[1/2, 1]$, respectively: then $\int_{\eta_k} f = \int_{\gamma_k} f$ ($k = 1, 2$). Define $\gamma : [0, 1] \to \mathbf{C}$ by

$$\gamma(t) = \begin{cases} \eta_1(t), & 0 \le t \le \frac{1}{2}, \\ (-\eta_2)(t) = \eta_2\left(\frac{3}{2} - t\right), & \frac{1}{2} < t \le 1. \end{cases}$$

Evidently γ is a circuit in G and, because assumption (ii) holds,

$$0 = \int_\gamma f = \int_{\eta_1} f - \int_{\eta_2} f = \int_{\gamma_1} f - \int_{\gamma_2} f.$$

Thus (ii) implies (iii).

Finally, suppose that (iii) holds and, for the moment, that G is connected. Fix $z_0 \in G$ and for any $z \in G$ define

$$F(z) = \int_\gamma f,$$

where γ is any contour in G joining z_0 to z. Note that such a contour exists because G is open and connected, and hence polygonally-connected (Theorem 2.4.23); moreover, since (iii) holds, $F(z)$ is independent of the particular contour γ. Let η be the standard reparametrisation of γ relative to the interval $[-1, 0]$: then $F(z) = \int_\eta f$. For $h \in \mathbf{C} \backslash \{0\}$, let $\mu : [0, 1] \to \mathbf{C}$ be defined by $\mu(t) = z + th$. Since G is open, $\mu^* \subset G$ whenever $|h|$ is small enough. Let $\nu : [-1, 1] \to \mathbf{C}$ be given by

$$\nu(t) = \begin{cases} \eta(t), & -1 \le t \le 0, \\ \mu(t), & 0 \le t \le 1. \end{cases}$$

Then, with $|h|$ small enough, ν is a contour in G joining z_0 to $z + h$ and

$$F(z + h) = \int_\nu f = \int_\eta f + \int_\mu f = F(z) + \int_\mu f.$$

Moreover, using Theorem 3.5.10,

$$\left| \frac{F(z + h) - F(z)}{h} - f(z) \right| = \left| \frac{1}{h} \int_\mu f(w)dw - f(z) \right|$$

$$= \left| \frac{1}{h} \int_\mu \{f(w) - f(z)\}dw \right|$$

$$\le \sup\{|f(w) - f(z)| : w \in \mu^*\}.$$

Now, f being continuous, this supremum tends to 0 as $|h| \to 0$. Hence $F'(z) = f(z)$, and F is a primitive for f on G.

If G is not connected, the argument given may be applied to each component of G since, by Theorem 2.4.27, every such component is open. Thus (iii) implies (i), and the proof is complete. □

Corollary 3.6.3 *Let $a \in \mathbf{C}$. Then*

$$\int_\gamma (z - a)^n dz = 0$$

for every circuit γ if $n = 0, 1, 2, \ldots$, and for those circuits γ such that $a \notin \gamma^*$ if $n = -2, -3, -4, \ldots$.

Proof Since $z \longmapsto (z - a)^n$ has $z \longmapsto (z - a)^{n+1}/(n + 1)$ as a primitive on $\mathbf{C}$ if $n = 0, 1, 2, \ldots$, and on $\mathbf{C}\backslash\{a\}$ if $n = -2, -3, -4, \ldots$, the result follows directly from Theorem 3.6.2. $\qquad\square$

Corollary 3.6.3 cannot be extended to include the case $n = -1$, as we see from (3.5.2).

Theorem 3.6.2 answers a natural question negatively. It is not the case that, if G is an arbitrary open set in $\mathbf{C}$, then every continuous function $f : G \to \mathbf{C}$ has a primitive. To see this, let $G = \mathbf{C}$, define f by $f(z) = \bar{z}$ ($z \in \mathbf{C}$) and let $\gamma : [0, 1] \to \mathbf{C}$ be the circuit given by $\gamma(t) = e^{2\pi it}$. Then

$$\int_\gamma f = \int_0^1 e^{-2\pi it} \cdot 2\pi i e^{2\pi it} dt = 2\pi i \neq 0.$$

More than this, replacing the requirement of continuity of f by analyticity does not help. This time, let $G = \left\{z \in \mathbf{C} : \frac{1}{2} < |z| < 2\right\}$, $f(z) = 1/z$ and let γ be as before. Then by (3.5.2),

$$\int_\gamma f = 2\pi i.$$

It turns out that determination of those conditions on G which ensure that all $f \in H(G)$ have a primitive involves some subtlety. This delicate matter we approach slowly, beginning with the supposition that G is convex. Recall that a subset S of $\mathbf{C}$ is called **convex** if $\lambda z_1 + (1 - \lambda)z_2 \in S$ whenever $z_1, z_2 \in S$ and $0 \leq \lambda \leq 1$. Plainly, every open ball in $\mathbf{C}$ is convex; also, no annulus $\{z \in \mathbf{C} : 0 < a \leq |z - z_0| \leq b\}$ is convex. The **convex hull** of a set $S \subset \mathbf{C}$ is the intersection of all convex sets which contain S; we denote this set by co S. An inductive argument (left to the reader) shows that co S consists of all points which can be expressed in the form $\sum_{k=1}^n \lambda_k z_k$ for some $n \in \mathbf{N}$, some $z_1, z_2, \ldots, z_n \in S$ and some non-negative real numbers $\lambda_1, \lambda_2, \ldots, \lambda_n$ with $\sum_{k=1}^n \lambda_k = 1$.

Definition 3.6.4 A closed polygonal path $\Delta : [u, v] \to \mathbf{C}$ such that a finite sequence $z_0, z_1, z_2, z_3 = z_0$ of complex numbers (its vertices) and a partition $u = t_0 < t_1 < t_2 < t_3 = v$ exist, and

$$\Delta(t) = (t_j - t_{j-1})^{-1}\left\{(t - t_{j-1})z_j + (t_j - t)z_{j-1}\right\}$$

whenever $t_{j-1} \le t \le t_j$ ($j = 1, 2, 3$), is described as a **triangular circuit**. It is said to be **degenerate** if the points z_0, z_1, z_2 are collinear.

Remark 3.6.5

(i) The vertices and sense of description of Δ are determined by the ordered triple $(z_0, z_1, z_2) \in \mathbf{C}^3$. Note that each ordered triple (w_0, w_1, w_2) determines the vertices and sense of description of some triangular circuit: for example Γ : $[0, 1] \to \mathbf{C}$ defined by

$$\Gamma(t) = \begin{cases} w_0 + 3t(w_1 - w_0), & 0 \le t \le 1/3, \\ w_1 + (3t - 1)(w_2 - w_1), & 1/3 \le t \le 2/3, \\ w_2 + (3t - 2)(w_0 - w_2), & 2/3 \le t \le 1. \end{cases}$$

(ii) Let $f : \Delta^* \to \mathbf{C}$ be continuous and, for $j = 1, 2, 3$, let v_j be the restriction of Δ to the interval $[t_{j-1}, t_j]$. Since the canonical line segment from z_{j-1} to z_j, that is, the map

$$t \longmapsto (1 - t)z_{j-1} + tz_j : [0, 1] \to \mathbf{C}$$

denoted by $[z_{j-1}, z_j]$ (as is its track), is the standard reparametrisation of v_j relative to the unit interval $[0, 1]$, using Remark 3.5.13 it follows that

$$\int_\Delta f = \sum_{j=1}^3 \int_{v_j} f = \sum_{j=1}^3 \int_{[z_{j-1}, z_j]} f.$$

(iii) The boundary of $\mathrm{co}\{z_0, z_1, z_2\}$ is the track of Δ.

Theorem 3.6.6 (Fundamental theorem of contour integration in a convex set) *Let G be a convex, open subset of $\mathbf{C}$ and let $f : G \to \mathbf{C}$ be continuous. Then the statement*

(iv)
$$\int_\Delta f = 0 \quad \text{for all triangular circuits } \Delta \text{ in } G$$

is equivalent to statements (i), (ii) *and* (iii) *of Theorem 3.6.2.*

Proof We use the convention that, for $w_1, w_2 \in \mathbf{C}$, the symbol $[w_1, w_2]$ may stand for the canonical line segment from w_1 to w_2 or its track, the meaning to be understood by context.

Since (ii) implies (iv), all we have to prove is that (iv) implies (i). Suppose that (iv) holds. Fix $z_0 \in G$ and define $F : G \to \mathbf{C}$ by

$$F(z) = \int_{[z_0, z]} f;$$

as G is convex, $[z_0, z] \subset G$. Let $h \ne 0$ be such that $z + h \in G$ and let Δ be a triangular circuit in G whose vertices and sense of description are determined by the triple $(z_0, z, z + h)$. Then, using Remark 3.6.5 (ii) and Lemma 3.5.9,

$$0 = \int_\Delta f = \int_{[z_0,z]} f + \int_{[z,z+h]} f + \int_{[z+h,z_0]} f = F(z) + \int_{[z,z+h]} f - F(z+h),$$

and thus

$$F(z+h) - F(z) = \int_{[z,z+h]} f.$$

From this last equality one may proceed, just as in the proof that (iii) implies (i) in Theorem 3.6.2, to show that (i) holds. $\qquad\square$

Using this result it will be shown that, if G is a convex open set of $\mathbf{C}$, then every $f \in H(G)$ has a primitive or, equivalently, $\int_\Delta f = 0$ whenever $f \in H(G)$ and Δ is a triangular circuit in G. First we give a preliminary lemma.

Lemma 3.6.7 *Let a, b, c be collinear complex numbers and let Δ be a degenerate triangular circuit in $\mathbf{C}$ whose vertices and sense of description are determined by the ordered triple (a, b, c). Let $f : \Delta^* \to \mathbf{C}$ be continuous. Then*

$$\int_\Delta f = \int_{[a,b]} f + \int_{[b,c]} f + \int_{[c,a]} f = 0.$$

(Here $[a, b]$ denotes the canonical line segment from a to b, etc.)

Proof Let a, b, c be distinct: otherwise the result is obvious. Plainly, the track of one of the line segments exhausts that of Δ. Three cases occur: $a \in [b, c]$, $b \in [c, a]$, $c \in [a, b]$. Cyclic interchange of a, b and c leads to the same result: for definiteness, suppose that $c \in [a, b] = \Delta^*$ and that $\theta \in (0, 1)$ is such that $c = a + \theta(b - a)$. Note that $b - a = \theta^{-1}(c - a) = (1 - \theta)^{-1}(b - c)$. Use of Theorem 3.5.2 (v) shows that

$$\int_{[a,b]} f = \int_0^1 f(a + t(b - a))(b - a)dt$$

$$= (b - a) \int_0^\theta f(a + t(b - a))dt + (b - a) \int_\theta^1 f(a + t(b - a))dt$$

$$= \theta^{-1}(c - a) \int_0^\theta f(a + \theta^{-1}t(c - a))dt$$

$$\quad + (1 - \theta)^{-1}(b - c) \int_\theta^1 f(c + (1 - \theta)^{-1}(t - \theta)(b - c))dt$$

$$= (c - a) \int_0^1 f(a + s(c - a))ds + (b - c) \int_0^1 f(c + s(b - c))ds$$

$$= \int_{[a,c]} f + \int_{[c,b]} f.$$

Hence, using Lemma 3.5.9 and Remark 3.6.5 (ii),

$$\int_\Delta f = \int_{[a,b]} f + \int_{[b,c]} f + \int_{[c,a]} f = 0. \qquad \square$$

Theorem 3.6.8 (Cauchy's theorem for a triangle) *Let G be an open subset of $\mathbf{C}$, let $f \in C(G, \mathbf{C})$, let $p \in G$ and suppose that f is analytic on $G\backslash\{p\}$. Let Δ be a triangular circuit such that $\mathrm{co}\,\Delta^*$, the convex hull of Δ^*, is contained in G. Then*

$$\int_\Delta f = 0.$$

Proof Let the vertices and sense of description of Δ be determined by the ordered triple $(a, b, c) \in \mathbf{C}^3$. Lemma 3.6.7 establishes the result when a, b, c are collinear: henceforth, we assume that this is not the case.

As a first step in the proof, suppose that $p \notin \mathrm{co}\,\Delta^*$. Let $a_1 = (b + c)/2$, $b_1 = (c+a)/2$ and $c_1 = (a+b)/2$, and let Δ^1, Δ^2, Δ^3 and Δ^4 be triangular circuits whose vertices and sense of description are determined by the ordered triples (a, c_1, b_1), (b, a_1, c_1), (c, b_1, a_1) and (a_1, b_1, c_1), respectively. Note that $l(\Delta^j) = l(\Delta)/2$ ($j = 1, 2, 3, 4$), where l denotes length. The configuration envisaged is illustrated below.

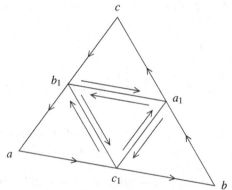

Adopting notation as for Lemma 3.6.7, from Remark 3.6.5 (ii), Lemma 3.6.7 and Lemma 3.5.9 it follows that

$$\begin{aligned}
\int_\Delta f &= \int_{[a,b]} f + \int_{[b,c]} f + \int_{[c,a]} f \\
&= \int_{[a,c_1]} f + \int_{[c_1,b]} f + \int_{[b,a_1]} f + \int_{[a_1,c]} f + \int_{[c,b_1]} f + \int_{[b_1,a]} f \\
&= \left(\int_{[a,c_1]} f + \int_{[c_1,b_1]} f + \int_{[b_1,a]} f \right) + \left(\int_{[b,a_1]} f + \int_{[a_1,c_1]} f + \int_{[c_1,b]} f \right) \\
&\quad + \left(\int_{[c,b_1]} f + \int_{[b_1,a_1]} f + \int_{[a_1,c]} f \right) + \left(\int_{[a_1,b_1]} f + \int_{[b_1,c_1]} f + \int_{[c_1,a_1]} f \right) \\
&= \sum_{j=1}^{4} \int_{\Delta^j} f.
\end{aligned}$$

Plainly, $\left|\int_{\Delta^j} f\right| \geq \left|\int_\Delta f\right|/4$ for some j. For definiteness, choose the least j for which this is true and relabel the corresponding triangular circuit as Δ_1: note that

$$co\,\Delta^* \supset co\,\Delta_1^*, \quad \left|\int_\Delta f\right| \leq 4\left|\int_{\Delta_1} f\right| \quad \text{and } l(\Delta_1) = 2^{-1}l(\Delta).$$

Evidently, with Δ_1 in place of Δ, the same process may be repeated to produce a further triangular circuit Δ_2, and so on. The procedure generates a sequence (Δ_n) of triangular circuits such that $co\,\Delta^* \supset co\,\Delta_1^* \supset co\,\Delta_2^* \supset \ldots$, with $l(\Delta_n) = 2^{-n}l(\Delta)$ and $\left|\int_\Delta f\right| \leq 4^n\left|\int_{\Delta_n} f\right|$ for all $n \in \mathbf{N}$. Now $\bigcap_{n=1}^\infty co\,\Delta_n^* = \{z_0\}$ for some $z_0 \in G$; moreover, since $z_0 \neq p$, f is differentiable at z_0. Let $\varepsilon > 0$. There exists $r > 0$ such that

$$\left|f(z) - f(z_0) - (z - z_0)f'(z_0)\right| \leq \varepsilon\,|z - z_0| \text{ if } |z - z_0| < r;$$

also there exists $n \in \mathbf{N}$ such that $|z - z_0| < r$ if $z \in co\,\Delta_n^*$. By Corollary 3.6.3,

$$\int_{\Delta_n} f = \int_{\Delta_n} \{f(z) - f(z_0) - (z - z_0)f'(z_0)\}dz,$$

and so, by Theorem 3.5.10,

$$\left|\int_{\Delta_n} f\right| \leq \varepsilon\left(\sup_{z\in\Delta_n^*} |z - z_0|\right)l(\Delta_n) \leq \varepsilon\,(l(\Delta_n))^2 = \varepsilon(2^{-n}l(\Delta))^2.$$

Hence $\left|\int_\Delta f\right| \leq \varepsilon(l(\Delta))^2$. As this holds for all $\varepsilon > 0$, it follows that $\int_\Delta f = 0$.

Next, suppose that p is a vertex of Δ^*, say $p = a$. Choose points $x \in [a, b]$, $y \in [a, c]$, both close to a, and note that $\int_\Delta f$ is the sum of the integrals of f over the triangles with vertices and sense of description determined by the triples (a, x, y), (y, x, b) and (y, b, c). Refer to the figure below.

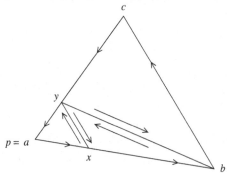

Using that part of the result already proved, it follows that each of the last two of these integrals is zero. Thus

$$\int_\Delta f = \int_{[a,x]} f + \int_{[x,y]} f + \int_{[y,a]} f,$$

and, since the lengths of the line segments $[a, x]$, $[x, y]$ and $[y, a]$ can be made arbitrarily small and f is bounded, it again follows that $\int_\Delta f = 0$.

Lastly, let p be an arbitrary point of $co\ \Delta^*$ and consider triangular circuits determined by the triples (a, b, p), (b, c, p) and (c, a, p), as portrayed in the figure below.

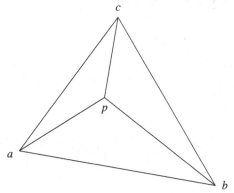

Application of the preceding result to the non-degenerate triangular circuits, and Lemma 3.6.7 to any that are degenerate, completes the proof. □

As an immediate consequence of Theorems 3.6.6 and 3.6.8 we have the following important form of Cauchy's theorem:

Theorem 3.6.9 (Cauchy's theorem in a convex set) *Let G be a convex open subset of* **C**, *let $f \in C(G, \mathbf{C})$, let $p \in G$ and suppose that f is analytic on $G\backslash\{p\}$. Then there is a function $F \in H(G)$ such that $F' = f$ and (equivalently) $\int_\gamma f = 0$ for all circuits γ in G.*

Used in conjunction with Theorem 3.3.8 the next theorem shows that in fact, $f \in H(G)$; that is, the point p is not really exceptional. This follows from

Theorem 3.6.10 *Let G be an open subset of* **C** *and let $f \in H(G)$. Then f is representable by power series in G.*

Proof Let $a \in G$, let $R > 0$ be such that $B(a, R) \subset G$, let $0 < r < R$ and let μ be the positively oriented circle with centre a and radius r. Let $z_0 \in G$ and define $g : G \to \mathbf{C}$ by

$$g(w) = \frac{f(w) - f(z_0)}{w - z_0} \ (w \neq z_0), \ g(z_0) = f'(z_0).$$

Plainly, g is continuous on G and analytic on $G\backslash\{z_0\}$. Since $B(a, R)$ is convex, we see from Theorem 3.6.9 that $\int_\mu g = 0$; thus, for all $z_0 \in G\backslash\mu^*$,

$$f(z_0) \int_\mu (w - z_0)^{-1} dw = \int_\mu (w - z_0)^{-1} f(w) dw.$$

We now use the fact, to be established by the lemma which follows, that $\int_\mu (w - z_0)^{-1} dw = 2\pi i$ if $z_0 \in B(a, r)$. Accepting this for the moment, we have that

$$f(z_0) = \frac{1}{2\pi i} \int_\mu (w - z_0)^{-1} f(w) dw \quad \text{if } z_0 \in B(a, r). \tag{3.6.1}$$

Let $z_0 \in B(a, r)$ and $s = r - |z_0 - a|$; evidently $B(z_0, s) \subset B(a, r)$. For all $z \in B(z_0, s)$,

$$f(z) = \frac{1}{2\pi i} \int_\mu (w - z_0)^{-1} \left\{ 1 - \left(\frac{z - z_0}{w - z_0} \right) \right\}^{-1} f(w) dw$$

$$= \frac{1}{2\pi i} \int_\mu (w - z_0)^{-1} f(w) \sum_{n=0}^{\infty} \left(\frac{z - z_0}{w - z_0} \right)^n dw.$$

Since

$$\left| f(w)(z - z_0)^n (w - z_0)^{-n-1} \right| \le M s^{-1} (|z - z_0| / s)^n,$$

where $M = \sup\{|f(w)| : w \in \mu^*\}$, the series $\sum_{n=0}^{\infty} (z - z_0)^n (w - z_0)^{-n-1} f(w)$ converges uniformly on μ^* (by the Weierstrass M-test) and thus can be integrated term by term (Theorem 3.5.2 (vi)) to give

$$f(z) = \sum_{n=0}^{\infty} a_n (z - z_0)^n \quad \text{for all } z \in B(z_0, s),$$

where, for all $n \in \mathbf{N}_0$,

$$a_n = \frac{1}{2\pi i} \int_\mu (w - z_0)^{-n-1} f(w) dw.$$

Appeal to Remark 3.3.9 shows that

$$a_n = f^{(n)}(z_0)/n! \quad \text{for all } n \in \mathbf{N}_0.$$

Thus, for all $z_0 \in B(a, r)$ and all $n \in \mathbf{N}_0$,

$$f^{(n)}(z_0) = \frac{n!}{2\pi i} \int_\mu (w - z_0)^{-n-1} f(w) dw, \tag{3.6.2}$$

a formula which includes (3.6.1) as a special case.

Setting $z_0 = a$, we see that for each $r \in (0, R)$,

$$f(z) = \sum_{n=0}^{\infty} c_n(z - a)^n \text{ for all } z \in B(a, r),$$

where the sequence $(c_n)_{n \in \mathbb{N}_0}$ is independent of r. Hence, for all $z \in B(a, R)$,

$$f(z) = \sum_{n=0}^{\infty} c_n(z - a)^n$$

where, for all $n \in \mathbb{N}_0$ and all $r \in (0, R)$,

$$c_n = \frac{f^{(n)}(a)}{n!} = \frac{1}{2\pi i} \int_{\mu} (w - a)^{-n-1} f(w) dw.$$

To complete the proof we need to justify our claim that $\int_{\mu} (w - z)^{-1} dw = 2\pi i$ if $z \in B(a, r)$. □

Lemma 3.6.11 Let $a \in \mathbb{C}$, $r > 0$ and define $\mu : [0, 2\pi] \to \mathbb{C}$ by $\mu(t) = a + re^{it}$ $(0 \le t \le 2\pi)$. Then

$$\int_{\mu} (w - z)^{-1} dw = \begin{cases} 2\pi i, & |z - a| < r, \\ 0, & |z - a| > r. \end{cases}$$

Proof Suppose first that $|z - a| > r$. Then $w \longmapsto (w - z)^{-1}$ is analytic on an open ball B containing μ^* but not z. Since B is convex, we conclude from Theorem 3.6.9 that $\int_{\mu} (w - z)^{-1} dw = 0$. On the other hand, if $|z - a| < r$, then

$$\int_{\mu} (w - z)^{-1} dw = \int_{\mu} (w - a)^{-1} \left\{ 1 - \left(\frac{z - a}{w - a} \right) \right\}^{-1} dw$$

$$= \int_{\mu} (w - a)^{-1} \sum_{n=0}^{\infty} \left(\frac{z - a}{w - a} \right)^n dw.$$

Since $\left| \frac{z-a}{w-a} \right| = \frac{|z-a|}{r} < 1$, the series is uniformly convergent on μ^* and thus, by Theorem 3.5.2 (vi), (3.5.2) and Corollary 3.6.3,

$$\int_{\mu} (w - z)^{-1} dw = \sum_{n=0}^{\infty} (z - a)^n \int_{\mu} (w - a)^{-n-1} dw = 2\pi i.$$

□

Remark 3.6.12 In view of Theorem 3.3.8, Theorem 3.6.10 shows that a function f is analytic on an open set G if, and only if, it is representable by power series in G.

Let us now take stock of the position. Suppose that G is an open subset of $\mathbf{C}$, $a \in G$, $r_0 = \text{dist}(a, \mathbf{C} \backslash G)$ ($= \infty$ if $G = \mathbf{C}$), μ is a positively oriented circle with centre a and radius $r \in (0, r_0)$, and $f \in H(G)$. Our results show that under these conditions,

$$f(z) = \sum_{n=0}^{\infty} \frac{f^{(n)}(a)}{n!}(z-a)^n \text{ for all } z \in B(a,r) \tag{3.6.3}$$

and that

$$f^{(n)}(a) = \frac{n!}{2\pi i} \int_{\mu} (w-a)^{-n-1} f(w) dw \text{ for all } n \in \mathbf{N}_0. \tag{3.6.4}$$

The series in (3.6.3) is called the **Taylor expansion of** f **about** a; it converges absolutely on $B(a, r_0)$ and uniformly on compact subsets of $B(a, r_0)$. The formulae (3.6.4) are well worth committing to memory; they are, however, special cases of an important and more general result to which we now proceed, namely Theorem 3.6.15 below.

Theorem 3.6.13 *Let γ be a contour in $\mathbf{C}$, let G be an open subset of $\mathbf{C}$, let $\phi :$ $\gamma^* \times G \to \mathbf{C}$ be continuous and suppose that for all $\zeta \in \gamma^*$, $\phi(\zeta, \cdot) \in H(G)$; put $f(z) = \int_{\gamma} \phi(\zeta, z) d\zeta$ $(z \in G)$. Then $f \in H(G)$ and for all $k \in \mathbf{N}$,*

$$f^{(k)}(z) = \int_{\gamma} \frac{\partial^k}{\partial z^k} \phi(\zeta, z) d\zeta \ (z \in G).$$

Proof Let $a \in G$, let $r \in (0, 1)$ be so small that $\overline{B(a, r)} \subset G$ and put $\mu(t) = a + re^{it}$ $(0 \le t \le 2\pi)$. Use of (3.6.2) shows that for all $(\zeta, z) \in \gamma^* \times B(a, r)$,

$$\phi(\zeta, z) = \frac{1}{2\pi i} \int_{\mu} (w-z)^{-1} \phi(\zeta, w) dw$$

and

$$\frac{\partial^k}{\partial z^k} \phi(\zeta, z) = \frac{k!}{2\pi i} \int_{\mu} (w-z)^{-k-1} \phi(\zeta, w) dw \ (k \in \mathbf{N}).$$

For simplicity, let ϕ_2 denote the first derivative of ϕ with respect to its second argument (so that $\phi_2(\zeta, z) = \partial \phi(\zeta, z)/\partial z$). We show that ϕ_2 is continuous. Let $\alpha \in \gamma^*$. It suffices to show that ϕ_2 is continuous at $(\alpha, a) \in \gamma^* \times G$. Let $\varepsilon > 0$. Since the map $(\zeta, w, z) \longmapsto (w-z)^{-2} \phi(\zeta, w)$ is uniformly continuous on the compact set $\gamma^* \times \mu^* \times \overline{B(a, r/2)}$, there exists $\delta \in (0, r/2)$ such that

$$\left| (w-z)^{-2} \phi(\zeta, w) - (w-a)^{-2} \phi(\alpha, w) \right| < \varepsilon$$

whenever $(\zeta, w, z) \in \gamma^* \times \mu^* \times \overline{B(a, r/2)}$ and $\{|\zeta - \alpha|^2 + |z - a|^2\}^{1/2} < \delta$. Use of this inequality in conjunction with the integral expression for ϕ_2 derived from (3.6.2) shows that

$$|\phi_2(\zeta, z) - \phi_2(\alpha, a)| \leq \frac{1}{2\pi} \cdot 2\pi r \cdot \varepsilon < \varepsilon$$

whenever $(\zeta, z) \in \gamma^* \times \overline{B(a, r/2)}$ and $\{|\zeta - \alpha|^2 + |z - a|^2\}^{1/2} < \delta$. Thus ϕ_2 is continuous at (α, a) and hence on $\gamma^* \times G$.

We now establish the result when $k = 1$. It is clearly enough to prove that

$$\int_\gamma (z - a)^{-1} \{\phi(\zeta, z) - \phi(\zeta, a) - (z - a)\phi_2(\zeta, a)\} \, d\zeta \to 0 \text{ as } z \to a.$$

We use the identity

$$(1 - t)^{-1} = 1 + t + (1 - t)^{-1} t^2 \, (t \neq 1)$$

with $t = (z - a)/(w - a)$; that is, $1 - t = (w - z)/(w - a)$. This shows that

$$2\pi i \phi(\zeta, z) = \int_\mu (w - a)^{-1} \left\{ 1 + \frac{z - a}{w - a} + \frac{(z - a)^2}{(w - a)(w - z)} \right\} \phi(\zeta, w) dw$$

$$= \int_\mu (w - a)^{-1} \phi(\zeta, w) dw + (z - a) \int_\mu (w - a)^{-2} \phi(\zeta, w) dw$$

$$+ (z - a)^2 \int_\mu (w - a)^{-2} (w - z)^{-1} \phi(\zeta, w) dw.$$

With the aid of the formulae from (3.6.2) quoted earlier, this gives

$$\phi(\zeta, z) - \phi(\zeta, a) - (z - a)\phi_2(\zeta, a) = \frac{(z - a)^2}{2\pi i} \int_\mu (w - a)^{-2} (w - z)^{-1} \phi(\zeta, w) dw$$

$$(3.6.5)$$

whenever $z \in B(a, r)$, $\zeta \in \gamma^*$. Put $M = \sup\{|\phi(\zeta, w)| : \zeta \in \gamma^*, w \in \mu^*\}$; $M < \infty$ since $\gamma^* \times \mu^*$ is compact. Thus from (3.6.5) and (3.5.3) we have that for all $z \in B(a, r)$ and all $\zeta \in \gamma^*$,

$$|\phi(\zeta, z) - \phi(\zeta, a) - (z - a)\phi_2(\zeta, a)| \leq M |z - a|^2 / \{r(r - |z - a|)\},$$

which shows, by (3.5.3) again, that

$$\left| \int_\gamma \left\{ \frac{\phi(\zeta, z) - \phi(\zeta, a)}{z - a} - \phi_2(\zeta, a) \right\} d\zeta \right| \leq \frac{M |z - a|}{r(r - |z - a|)} l(\gamma) \to 0 \text{ as } z \to a.$$

Hence $f'(a)$ exists and

$$f'(a) = \int_\gamma \phi_2(\zeta, a)d\zeta,$$

which proves the case $k = 1$.

Finally, for $k \in \mathbf{N}$, let $\phi_{2,k}$ denote the kth derivative of ϕ with respect to its second argument ($\phi_{2,1} \equiv \phi_2$). Let Γ be the set of natural numbers k such that $\phi_{2,k}$ is continuous and

$$f^{(k)}(z) = \int_\gamma \phi_{2,k}(\zeta, z)d\zeta \ (z \in G).$$

We have shown that $1 \in \Gamma$; moreover, since $\phi_{2,k}$ satisfies the same hypotheses as ϕ if $k \in \Gamma$, induction shows that $\Gamma = \mathbf{N}$. $\square$

It is now convenient to introduce the **index** of a point with respect to a circuit; as we shall soon see, this is nothing more than the winding number.

Definition 3.6.14 Let γ be a circuit in $\mathbf{C}$ and let $z \in \mathbf{C}\backslash\gamma^*$. The **index of z with respect to** γ is defined to be $\mathrm{ind}_\gamma(z)$, where

$$\mathrm{ind}_\gamma(z) = \frac{1}{2\pi i} \int_\gamma (\zeta - z)^{-1}d\zeta.$$

Before discussion of the meaning of $\mathrm{ind}_\gamma(z)$, we give the following important result.

Theorem 3.6.15 (Cauchy's integral formula in a convex open set) *Let γ be a circuit in a convex, open set $G \subset \mathbf{C}$ and let $f \in H(G)$; suppose that $z \in G\backslash\gamma^*$. Then*

$$f(z)\,\mathrm{ind}_\gamma(z) = \frac{1}{2\pi i} \int_\gamma (\zeta - z)^{-1}f(\zeta)d\zeta \qquad (3.6.6)$$

and

$$f^{(n)}(z)\,\mathrm{ind}_\gamma(z) = \frac{n!}{2\pi i} \int_\gamma (\zeta - z)^{-n-1}f(\zeta)d\zeta \ (n \in \mathbf{N}). \qquad (3.6.7)$$

Proof Fix $z \in G\backslash\gamma^*$ and define $g : G \to \mathbf{C}$ by

$$g(\zeta) = \{f(\zeta) - f(z)\}/(\zeta - z) \text{ for } \zeta \neq z, \ g(z) = f'(z).$$

By Theorem 3.6.9, $\int_\gamma g = 0$; (3.6.6) follows immediately. Next, define ϕ by

$$\phi(\zeta, z) = f(\zeta)/(\zeta - z) \text{ for } (\zeta, z) \in \gamma^* \times (G\backslash\gamma^*),$$

and apply Theorem 3.6.13 with G replaced by $G\backslash\gamma^*$. We see that the map

$$z \longmapsto \int_\gamma (\zeta - z)^{-1}f(\zeta)d\zeta = 2\pi i f(z)\,\mathrm{ind}_\gamma(z)$$

is in $H(G\backslash\gamma^*)$ and has derivatives of all orders:

$$2\pi i(f \cdot \mathrm{ind}_\gamma)^{(k)}(z) = k! \int_\gamma (\zeta - z)^{-k-1} f(\zeta)d\zeta \, (k \in \mathbf{N}). \qquad (3.6.8)$$

Put $\psi(\zeta, z) = (\zeta - z)^{-1}$ for $(\zeta, z) \in \gamma^* \times (\mathbf{C}\backslash\gamma^*)$; by Theorem 3.6.13, $\mathrm{ind}_\gamma \in H(\mathbf{C}\backslash\gamma^*)$ and

$$2\pi i(\mathrm{ind}_\gamma)^{(k)}(z) = k! \int_\gamma (\zeta - z)^{-k-1}d\zeta = 0(k \in \mathbf{N}),$$

the final step following from Theorem 3.6.2 since $\zeta \mapsto (\zeta - z)^{-k-1}$ has primitive $\zeta \mapsto -(\zeta - z)^{-k}/k$ on $\mathbf{C}\backslash\{z\}$. Thus (3.6.8) reduces to

$$2\pi i f^{(k)}(z) \, \mathrm{ind}_\gamma(z) = k! \int_\gamma (\zeta - z)^{-k-1} f(\zeta)d\zeta,$$

$\square$

which gives (3.6.7).

Note that formula (3.6.4), in which γ is the positively oriented circle with centre a and radius r, follows from (3.6.6) and (3.6.7) since in this case

$$\mathrm{ind}_\gamma(z) = 1 \text{ if } |z - a| < r, \mathrm{ind}_\gamma(z) = 0 \text{ if } |z - a| > r,$$

as we know from Lemma 3.6.11.

As mentioned earlier, if γ is a circuit and $z \in \mathbf{C}\backslash\gamma^*$, then $\mathrm{ind}_\gamma(z)$ coincides with the winding number $n(\gamma, z)$ of γ with respect to z. To prove this it is convenient to introduce the idea of an **analytic logarithm**, which is companion to that of a continuous logarithm mentioned in Sect. 3.4.

Definition 3.6.16 Let G be an open subset of $\mathbf{C}$ and let $f : G \to \mathbf{C}\backslash\{0\}$ be analytic. An analytic map $g : G \to \mathbf{C}$ is called an **analytic logarithm of** f **on** G if $g \in H(G)$ and $\exp(g(z)) = f(z)$ for all $z \in G$.

Theorem 3.6.17 *Let G be an open subset of $\mathbf{C}$, let $f \in H(G)$ and suppose that $0 \notin f(G)$. Then f has an analytic logarithm on G if, and only if, f'/f has a primitive on G.*

Proof Suppose f has an analytic logarithm g on G. Then $f = \exp g$, so that $f' = g' \exp g$; that is, $g' = f'/f$, and g is a primitive of f'/f. Conversely, if $g \in H(G)$ is such that $g' = f'/f$, then $[(\exp g)/f]' = 0$ on G. Hence $(\exp g)/f$ is constant, say equal to k_j, on each component G_j of G (by Theorem 3.2.11). Let c_j be so chosen that $\exp c_j = k_j$. Then $\exp(g(z) - c_j) = f(z)$ for all $z \in G_j$. Suppose the components G_j of G are indexed by a set J: then the function $\widetilde{g} : G \to \mathbf{C}$ defined by $\widetilde{g}(z) = g(z) - c_j$ for all $z \in G_j$ ($j \in J$), is an analytic logarithm of f on G. $\square$

Theorem 3.6.18 *Let G be a convex, open subset of $\mathbf{C}$ and let $f \in H(G)$ be such that $0 \notin f(G)$. Then f has an analytic logarithm on G.*

Proof Since $f'/f \in H(G)$, it follows from Theorem 3.6.9 that f'/f has a primitive in G. The result now follows from Theorem 3.6.17.   □

The final result preparatory to the identification of the index and the winding number for circuits is the following:

Theorem 3.6.19 *Let g be an analytic logarithm of f on an open set $G \subset \mathbf{C}$, and let $\gamma : [a, b] \to \mathbf{C}$ be a contour in G. Then*

$$\int_\gamma f'/f = g(\gamma(b)) - g(\gamma(a)).$$

Proof As in the proof of Theorem 3.6.17 we see that $g' = f'/f$. Thus

$$\int_\gamma f'/f = \int_a^b g'(\gamma(t))\gamma'(t)dt = g(\gamma(b)) - g(\gamma(a)).$$ □

The promised identification now follows.

Theorem 3.6.20 *Let $\gamma : [a, b] \to \mathbf{C}$ be a circuit and let $z_0 \in \mathbf{C} \backslash \gamma^*$. Then*

$$n(\gamma, z_0) = \mathrm{ind}_\gamma(z_0).$$

More generally, if $f \in H(G)$ for some open set $G \supset \gamma^*$ and $z_0 \in \mathbf{C} \backslash (f \circ \gamma)^*$, then

$$n(f \circ \gamma, z_0) = \mathrm{ind}_{f \circ \gamma}(z_0) = \frac{1}{2\pi i} \int_\gamma \frac{f'(z)}{f(z) - z_0}dz.$$

Proof By 'shrinking' G if necessary, we may and shall suppose that $f - z_0$ is never zero on G. For example, with $r = \mathrm{dist}(\gamma^*, f^{-1}\{z_0\}) > 0$, let $G' = \cup_{z \in \gamma^*} B(z, r/2)$ and note that G' is open, $\gamma^* \subset G'$ and $f - z_0 \neq 0$ on G'; G may now be replaced by G'. We claim that there is a partition $\{t_0, t_1, \ldots, t_k\}$ of $[a, b]$ such that there are open balls $B_1, \ldots, B_k$, each contained in G, with $\gamma(t) \in B_j$ whenever $t \in [t_{j-1}, t_j]$ $(j = 1, 2, \ldots, k)$. To establish this, let $\varepsilon = \mathrm{dist}(\gamma^*, \mathbf{C} \backslash G) > 0$. Since γ is continuous, and hence uniformly continuous on $[a, b]$, there exists $\delta > 0$ such that $|\gamma(t) - \gamma(t')| < \varepsilon$ if $|t - t'| < \delta$ and $t, t' \in [a, b]$. Let $\{t_0, t_1, \ldots, t_k\}$ be a partition of $[a, b]$ with $|t_j - t_{j-1}| < \delta$ for each $j \in \{1, 2, \ldots, k\}$, and put $B_j = B(\gamma(t_j), \varepsilon)$ $(\subset G)$ for $j = 1, 2, \ldots, k$. If $t \in [t_{j-1}, t_j]$, then $|t - t_j| < \delta$ and hence $|\gamma(t) - \gamma(t_j)| < \varepsilon$; thus $\gamma(t) \in B_j$, and our claim is justified.

By Theorem 3.6.18, $f - z_0$ has an analytic logarithm g_j on B_j; by Theorem 3.6.19,

$$\int_{\gamma_j} \frac{f'(z)}{f(z) - z_0}dz = g_j(\gamma(t_j)) - g_j(\gamma(t_{j-1})),$$

where γ_j is the restriction of γ to $[t_{j-1}, t_j]$. Let θ be a continuous argument of $f \circ \gamma - z_0$, and note that $g_j \circ \gamma$ is a continuous logarithm of $f \circ \gamma - z_0 : [t_{j-1}, t_j] \to \mathbf{C}$. Then

$$\frac{1}{2\pi i} \int_\gamma \frac{f'(z)}{f(z) - z_0} dz = \frac{1}{2\pi i} \sum_{j=1}^{k} \{g_j(\gamma(t_j)) - g_j(\gamma(t_{j-1}))\},$$

and by Theorem 3.4.18 (d) this equals

$$\frac{1}{2\pi i} \sum_{j=1}^{k} \{\log|f(\gamma(t_j)) - z_0| - \log|f(\gamma(t_{j-1})) - z_0| + i\theta(t_j) - i\theta(t_{j-1})\}$$

$$= \frac{1}{2\pi} \{\theta(b) - \theta(a)\} = n(f \circ \gamma, z_0). \qquad \square$$

Now that the identity between the winding number and the index has been established for circuits, the properties of the winding number derived in Sect. 3.4 also hold for the index. For convenience we summarise these properties in the following:

Theorem 3.6.21 (i) *Let γ be a circuit and let $G = \mathbf{C} \backslash \gamma^*$. Then ind_γ is an integer-valued function on G which is constant on each component of G and is zero on the unbounded component of G.*
(ii) *Let $\alpha, \beta \in \mathbf{C}$, $\alpha \neq 0$, let $f : \mathbf{C} \to \mathbf{C}$ be given by $f(z) = \alpha z + \beta$ $(z \in \mathbf{C})$, let γ be a circuit and suppose that $w \in \mathbf{C} \backslash \gamma^*$. Then*

$$\mathrm{ind}_\gamma(w) = \mathrm{ind}_{f \circ \gamma}(f(w)).$$

(iii) *Let $\widetilde{G}$ be an open set in $\mathbf{C}$ and let $\gamma, \sigma : [0, 1] \to \widetilde{G}$ be circuits in $\widetilde{G}$ which are freely homotopic. Then for all $w \in \mathbf{C} \backslash \widetilde{G}$, $\mathrm{ind}_\gamma(w) = \mathrm{ind}_\sigma(w)$.*

Proof Part (i) is just Theorems 3.4.27 and 3.4.20; (ii) is Theorem 3.4.23; and (iii) simply Theorem 3.4.28, all applied to circuits rather than to closed paths. Without reference to these earlier results, it should be noted that Definition 3.6.14 permits direct derivation of parts (i) and (ii). $\qquad \square$

Remark 3.6.22 (i) Use of Theorems 3.6.20 and 3.6.21 shows that if γ is either the rectangular circuit of Example 3.4.22, part (iii), or the semicircular circuit of part (iv), then $\mathrm{ind}_\gamma(z) = 1$ for all z in the bounded component of $\mathbf{C} \backslash \gamma^*$. In the remainder of this Remark, for use in another context, we show that the same result holds for variants of these circuits. For these variants, the result may be proved either by methods developed in Example 3.4.22 (a task left to the reader), by direct appeal to Definition 3.6.14, or via a choice of an appropriate homotopy.
(ii) Suppose $0 < \varepsilon < 1 < R < \infty$ and let $\sigma : [-R, R + \pi] \to \mathbf{C}$ be the simple circuit defined by

$$\sigma(s) = \begin{cases} s, & -R \le s \le -\varepsilon, \\ \varepsilon \exp\left(i\pi(1 - s/\varepsilon)/2\right), & -\varepsilon \le s \le \varepsilon, \\ s, & \varepsilon \le s \le R, \\ R \exp(i(s - R)), & R \le s \le R + \pi. \end{cases}$$

The figure below portrays σ^*.

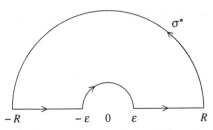

We seek to show that $\operatorname{ind}_\sigma(z) = 1$ for all z in the bounded component of $\mathbf{C}\backslash\sigma^*$ and, because of Theorem 3.6.21 (i), it suffices to prove that $\operatorname{ind}_\sigma(i) = 1$.

Loosely speaking, replacement of the smaller semicircle about the origin in the figure above by a line segment along the real axis transforms it into the track of the semicircular circuit of Example 3.4.22 (iv), a circuit $\gamma : [-R, R + \pi] \to \mathbf{C}$ defined by

$$\gamma(s) = \begin{cases} s, & -R \le s \le R, \\ R \exp(i(s - R)), & R \le s \le R + \pi. \end{cases}$$

Evidently, the map $H : [-R, R + \pi] \times [0, 1] \to \mathbf{C}\backslash\{i\}$ given by

$$H(s, t) = (1 - t)\sigma(s) + t\gamma(s)$$

establishes that γ and σ are homotopic and thus freely homotopic in $\mathbf{C}\backslash\{i\}$. Hence, by Theorem 3.6.21 (iii) and part (i) of this Remark, $\operatorname{ind}_\sigma(i) = \operatorname{ind}_\gamma(i) = 1$.

(iii) Let $r > 0$. Consider the map

$$H : [-1/2 - r, 1/2 + r] \times [0, 1] \to \mathbf{C}\backslash\{0\}$$

defined for the intervals specified and $0 \le t \le 1$ by

$$H(s, t) = \begin{cases} 2s + \dfrac{1}{2} + 2r - ire^{-it\pi/4}, & -\dfrac{1}{2} - r \le s \le -r, \\ \dfrac{1}{2} + i(2s + r)e^{-it\pi/4}, & -r \le s \le 0, \\ \dfrac{1}{2} - 2s + ire^{-it\pi/4}, & 0 \le s \le \dfrac{1}{2}, \\ -\dfrac{1}{2} + i(1 + r - 2s)e^{-it\pi/4}, & \dfrac{1}{2} \le s \le \dfrac{1}{2} + r. \end{cases}$$

The gluing lemma shows that H is continuous. Let $\gamma, \sigma : [-1/2 - r, 1/2 + r] \to \mathbf{C}$ be circuits defined by

$$\gamma(s) = H(s, 0), \sigma(s) = H(s, 1).$$

Evidently, γ^* is a rectangle with vertices $\pm\frac{1}{2} \pm ir$ and σ^* is a parallelogram with vertices $\pm\frac{1}{2} \pm re^{i\pi/4}$. The map H establishes that γ and σ are freely homotopic in $\mathbf{C}\setminus\{0\}$ and thus use of Theorems 3.6.20, 3.6.21 and Example 3.4.22 (iii) shows that

$$\text{ind}_\sigma(0) = \text{ind}_\gamma(0) = 1.$$

We shall now illustrate the significance of Theorem 3.6.15 by obtaining from it several results of great importance.

Theorem 3.6.23 *Let* $f \in H(B(a, R))$ *and put*

$$M_f(a, r) = \sup\{|f(z)| : |z - a| = r\}$$

for $0 < r < R$. *Then for all* $n \in \mathbf{N}_0$,

$$\left|f^{(n)}(a)\right| \leq n! M_f(a, r)r^{-n};$$

and if $f(z) = \sum_0^\infty a_n(z - a)^n$ $(z \in B(a, R))$, *then for all* $n \in \mathbf{N}_0$,

$$|a_n| \leq M_f(a, r)r^{-n}.$$

Proof Let $\gamma : [0, 2\pi] \to \mathbf{C}$ be defined by $\gamma(t) = a + re^{it}$. By Theorem 3.6.15, since $\text{ind}_\gamma(a) = 1$ we have

$$\left|f^{(n)}(a)\right| = \left|\frac{n!}{2\pi i}\int_\gamma (\zeta - a)^{-n-1} f(\zeta)d\zeta\right| \leq n! M_f(a, r)r^{-n},$$

by (3.5.3). The rest is now clear. $\qquad\square$

That these inequalities for $\left|f^{(n)}(a)\right|$ are best possible we see by considering the function f given by $f(z) = (z - a)^m$ for some $m \in \mathbf{N}$.

From these inequalities the famous theorem due to Liouville follows.

Theorem 3.6.24 (Liouville's theorem) *Every bounded function which is entire (that is, in* $H(\mathbf{C})$) *is constant.*

Proof Let $f \in H(\mathbf{C})$ and suppose there is a number M such that for all $z \in \mathbf{C}$, $|f(z)| \leq M$. By Theorem 3.6.10, there are constants a_n $(n \in \mathbf{N}_0)$ such that for all $z \in \mathbf{C}$, $f(z) = \sum_{n=0}^\infty a_n z^n$. If $n \geq 1$, Theorem 3.6.23 shows that for all $r > 0$, $|a_n| \leq Mr^{-n}$; and as $Mr^{-n} \to 0$ as $r \to \infty$, $a_n = 0$. Hence $f(z) = a_0$ for all $z \in \mathbf{C}$. $\qquad\square$

Theorem 3.6.25 *Let p be a polynomial of degree at least 1, with complex coefficients. Then there exists $z \in \mathbf{C}$ such that $p(z) = 0$.*

Proof Suppose p is never zero. Then $1/p \in H(\mathbf{C})$. Moreover, since $p(z) = a_0 + a_1 z + \cdots + a_n z^n$, where $a_n \neq 0$,

$$|p(z)| = |a_n z^n| \left| 1 + \frac{a_{n-1}}{a_n} z^{-1} + \cdots + \frac{a_0}{a_n} z^{-n} \right| \to \infty \text{ as } |z| \to \infty;$$

thus $1/|p(z)| \to 0$ as $|z| \to \infty$, and so there exists $N > 0$ such that $1/|p(z)| \leq 1$ if $|z| > N$. As $1/|p|$ is continuous on the compact set $\overline{B(0, N)}$, there exists $M > 0$ such that $1/|p(z)| \leq M$ for all z with $|z| \leq N$. Hence $1/|p(z)| \leq \max(M, 1)$ for all $z \in \mathbf{C} : 1/p$ is bounded on $\mathbf{C}$. Thus by Theorem 3.6.24, $1/p$ is constant, and so p is constant. Since $|p(z)| \to \infty$ as $|z| \to \infty$ we have a contradiction and the theorem is proved. $\square$

We now turn our attention to another remarkable property of analytic functions: two functions which are analytic in a region G and coincide in an open, non-empty subset of G (no matter how small!) must coincide throughout G. We lead up to this by means of the following theorem involving the zeros of an analytic function.

Theorem 3.6.26 *Let G be a region in $\mathbf{C}$, let $f \in H(G)$ and put $Z(f) = \{a \in G : f(a) = 0\}$. Then either $Z(f) = G$ or $Z(f)$ has no limit point in G; in the latter case, to each $a \in Z(f)$ there corresponds a unique positive integer m such that $f(z) = (z - a)^m g(z)$ for all $z \in G$, where $g \in H(G)$ and $g(a) \neq 0$. (The integer m is called the **order** or **multiplicity** of the zero which f has at a.)*

Proof Let L be the set of all limit points of $Z(f)$ in $G : L \subset Z(f)$ since f is continuous. Let $a \in Z(f)$ and let $r > 0$ be such that $B(a, r) \subset G$. Then $f(z) = \sum_0^\infty a_n (z - a)^n$ for all $z \in B(a, r)$. One of the following two possibilities must occur:

(i) $a_n = 0$ for all $n \in \mathbf{N}_0$; thus $B(a, r) \subset L$, $a \in \overset{o}{L}$;
(ii) there is a smallest integer m such that $a_m \neq 0$; $m \geq 1$ since $f(a) = a_0 = 0$.

In case (ii), put $g(z) = (z - a)^{-m} f(z)$ if $z \in G \setminus \{a\}$, $g(a) = a_m$. Then $f(z) = (z - a)^m g(z)$ for all $z \in G$ and g is analytic in $G \setminus \{a\}$. Moreover, for all $z \in B(a, r)$ we have $g(z) = \sum_{k=0}^\infty a_{m+k} (z - a)^k$ and so g is also analytic at a; thus $g \in H(G)$. Since $g(a) = a_m \neq 0$, there is a neighbourhood of a in which g has no zero, and it follows that f has an isolated zero at a.

If $a \in L$, case (i) must occur and so L must be open. Put $M = G \setminus L$. If $b \in M$, then there is a neighbourhood $V(\subset G)$ of b such that $Z(f) \cap (V \setminus \{b\}) = \emptyset$, and so $V \subset M$; thus $b \in \overset{o}{M}$. Hence M is open and, since G is connected, either $L = G$, in which case $Z(f) = G$, or $L = \emptyset$. $\square$

Theorem 3.6.27 *Let G be a region in $\mathbf{C}$ and let $f, g \in H(G)$. If $f(z) = g(z)$ for all z in some set which has a limit point in G, then $f(z) = g(z)$ for all $z \in G$.*

Proof Simply apply Theorem 3.6.26 to $f - g$. $\qquad\qquad\qquad\qquad\qquad\square$

As an immediate consequence of this we see that if f and g coincide on some open, non-empty subset of G, then they are identical on G.

Theorem 3.6.28 (The maximum modulus theorem) *Let G be a region in $\mathbf{C}$, let $f \in H(G)$ and suppose that f is non-constant on G. Then no point of G is a local maximum of the function $|f|$.*

Proof To obtain a contradiction, suppose that there exist $a \in G$ and $r' > 0$ such that

$$|f(z)| \leq |f(a)| \text{ whenever } z \in B(a, r').$$

Let $0 < r < r'$ and $\gamma(t) = a + re^{it}$ $(0 \leq t \leq 2\pi)$. Since $B(a, r')$ is open and convex, and $\gamma^* \subset B(a, r')$, application of Theorem 3.6.15 gives

$$f(a) \, \text{ind}_\gamma(a) = \frac{1}{2\pi i} \int_\gamma (z - a)^{-1} f(z) dz.$$

Use of Lemma 3.6.11 thus gives

$$f(a) = \frac{1}{2\pi} \int_0^{2\pi} f(a + re^{it}) dt,$$

and so

$$|f(a)| \leq \frac{1}{2\pi} \int_0^{2\pi} \left| f(a + re^{it}) \right| dt \leq \frac{1}{2\pi} \int_0^{2\pi} |f(a)| \, dt = |f(a)|.$$

It follows that

$$\int_0^{2\pi} \left[|f(a)| - \left| f(a + re^{it}) \right| \right] dt = 0,$$

and that, since the integrand is continuous and non-negative,

$$|f(a)| = \left| f(a + re^{it}) \right| (0 \leq t \leq 2\pi; 0 < r < r').$$

Hence $|f(z)| = |f(a)|$ for all $z \in B(a, r')$. By Exercise 3.2.12/5, the constancy of $|f|$ on $B(a, r')$ implies that of f on $B(a, r')$. In turn, through Theorem 3.6.27, this implies the constancy of f on G, contrary to hypothesis. $\qquad\qquad\square$

After these diversions we return to Cauchy's theorem, and attempt to answer the question as to whether, given a region G, we can determine those circuits γ in G such that $\int_\gamma f = 0$ for all $f \in H(G)$. To handle this we need a lemma and a new idea, that of a 'cycle'.

Lemma 3.6.29 *Let G be an open subset of* $\mathbf{C}$ *and let* $f \in H(G)$. *Define* $g : G \times G \to$ $\mathbf{C}$ *by*

$$g(\zeta, z) = \begin{cases} \frac{f(\zeta) - f(z)}{\zeta - z}, & \zeta \neq z, \\ f'(z), & \zeta = z. \end{cases}$$

Then g is continuous.

Proof Continuity of g is clear at all points $(\zeta, z) \in G \times G$ with $\zeta \neq z$. To discuss the 'diagonal' points, fix $a \in G$. Given any $\varepsilon > 0$, there exists $r > 0$ such that $B(a, r) \subset G$ and $|f'(w) - f'(a)| < \varepsilon$ for all $w \in B(a, r)$. If $\zeta, z \in B(a, r)$ and $\gamma : [0, 1] \to \mathbf{C}$ is defined by $\gamma(t) = (1 - t)z + t\zeta$, then $\gamma^* \subset B(a, r)$ and since

$$f(\zeta) - f(z) = \int_0^1 f'(\gamma(t))\gamma'(t)dt = (\zeta - z) \int_0^1 f'(\gamma(t))dt,$$

it follows that

$$g(\zeta, z) - g(a, a) = \int_0^1 \{f'(\gamma(t)) - f'(a)\}dt.$$

Thus $|g(\zeta, z) - g(a, a)| < \varepsilon$ for all $(\zeta, z) \in B((a, a), r) \subset B(a, r) \times B(a, r)$, which gives the required continuity.   □

Not only does this lemma play a key rôle in our proof of a general form of Cauchy's theorem, but it also leads to the following useful result.

Corollary 3.6.30 *Let G be an open subset of* $\mathbf{C}$, $a \in G$, $f \in H(G)$ *and* $f'(a) \neq 0$. *Then there is a neighbourhood U of a in G such that the restriction of f to U is injective; that is, f is locally injective at a.*

Proof The proof of Lemma 3.6.29 and the choice $\varepsilon = \frac{1}{2}|f'(a)|$ show that there is a neighbourhood U of a such that, if $\zeta, z \in U$ and $\zeta \neq z$, then

$$\left| \frac{f(\zeta) - f(z)}{\zeta - z} - f'(a) \right| < \frac{1}{2}|f'(a)|.$$

Thus

$$|f(\zeta) - f(z)| > \frac{1}{2}|f'(a)||\zeta - z|,$$

and so f is injective on U.   □

Definition 3.6.31 A finite sequence $\Gamma = (\gamma_j)_{j=1}^m$ of circuits is called a **cycle**; the **track of the cycle** Γ, denoted by Γ^*, is defined by $\Gamma^* := \bigcup_{k=1}^m \gamma_k^*$; the cycle **opposite** to Γ, written $-\Gamma$, is given by $-\Gamma = (-\gamma_j)_{j=1}^m$. Let $f : \Gamma^* \to \mathbf{C}$ be continuous. The **integral of** f **over** Γ, denoted by $\int_\Gamma f$, is defined by

$$\int_\Gamma f = \sum_{j=1}^m \int_{\gamma_j} f.$$

Plainly $\int_\Gamma f = -\int_{-\Gamma} f$. Cycles Γ_1 and Γ_2 are said to be **equivalent** if

$$\int_{\Gamma_1} f = \int_{\Gamma_2} f \ (f \in C(\Gamma_1^* \cup \Gamma_2^*, \mathbf{C}));$$

a cycle Γ is said to be **equivalent to zero** if

$$\int_\Gamma f = 0 \ (f \in C(\Gamma^*, \mathbf{C})).$$

Evidently, if $\Gamma = (\gamma_j)_{j=1}^m$ is a cycle and σ is a bijective mapping of $\{1, 2, \ldots, m\}$ onto itself (a rearrangement or permutation), then Γ and $\tilde{\Gamma} := (\gamma_{\sigma(j)})_{j=1}^m$ are equivalent; rearrangement generates an equivalent cycle. Given a cycle $\Gamma = (\gamma_j)_{j=1}^m$ and a point $z \in \mathbf{C} \backslash \Gamma^*$, the **index** of z relative to Γ is defined by

$$\text{ind}_\Gamma(z) = \frac{1}{2\pi i} \int_\Gamma (\zeta - z)^{-1} d\zeta = \sum_{j=1}^m \text{ind}_{\gamma_j}(z).$$

Note that if Γ_1 and Γ_2 are equivalent cycles and $z \in \mathbf{C} \backslash (\Gamma_1^* \cup \Gamma_2^*)$, then

$$\text{ind}_{\Gamma_1}(z) = \text{ind}_{\Gamma_2}(z).$$

We can now give a very general form of Cauchy's theorem.

Theorem 3.6.32 (The global version of Cauchy's theorem) *Let G be an open subset of $\mathbf{C}$, let $f \in H(G)$ and let Γ be a cycle such that $\Gamma^* \subset G$ and $\text{ind}_\Gamma(a) = 0$ for all $a \in \mathbf{C} \backslash G$. Then*

$$\text{ind}_\Gamma(z)f(z) = \frac{1}{2\pi i} \int_\Gamma (\zeta - z)^{-1} f(\zeta) d\zeta \text{ for all } z \in G \backslash \Gamma^*, \tag{3.6.9}$$

and

$$\int_\Gamma f = 0. \tag{3.6.10}$$

If Λ and Ψ are cycles in G (that is, $\Lambda^* \subset G$ and $\Psi^* \subset G$) such that $\text{ind}_\Lambda(a) = \text{ind}_\Psi(a)$ for all $a \in \mathbf{C} \backslash G$, then

$$\int_\Lambda f = \int_\Psi f. \tag{3.6.11}$$

Proof Define $g : G \times G \to \mathbf{C}$ by

$$g(\zeta, z) = \{f(\zeta) - f(z)\}/(\zeta - z) \ (\zeta \neq z), \ g(z, z) = f'(z).$$

By Lemma 3.6.29, g is continuous. Now define $h : G \to \mathbf{C}$ by

$$h(z) = \frac{1}{2\pi i} \int_\Gamma g(\zeta, z) d\zeta$$

and note that since for $z \in G \backslash \Gamma^*$ we have

$$h(z) = \frac{1}{2\pi i} \int_\Gamma \{f(\zeta) - f(z)\}(\zeta - z)^{-1} d\zeta,$$

(3.6.9) is equivalent to the statement that $h(z) = 0$. To prove this statement, first observe that since $g(\zeta, \cdot)$ is continuous on G and analytic on $G \backslash \{\zeta\}$, Theorems 3.6.9 and 3.6.10 show that $g(\zeta, \cdot)$ is analytic on G. By Theorem 3.6.13, extended in the obvious way to cycles rather than contours, h is analytic on G.

Put $G_1 = \{z \in \mathbf{C} : \mathrm{ind}_\Gamma(z) = 0\}$ and define $h_1 : G_1 \to \mathbf{C}$ by

$$h_1(z) = \frac{1}{2\pi i} \int_\Gamma (\zeta - z)^{-1} f(\zeta) d\zeta.$$

By Theorem 3.6.13 again, h_1 is analytic on G_1 (note that by Theorems 2.4.27 and 3.6.21 (i), G_1 is open). If $z \in G \cap G_1$, evidently $h(z) = h_1(z)$. Thus the function $\phi : G_1 \cup G \to \mathbf{C}$ defined by $\phi(z) = h(z) \ (z \in G)$, $\phi(z) = h_1(z) \ (z \in G_1)$ is analytic in $G_1 \cup G$.

By hypothesis, $\mathbf{C} \backslash G \subset G_1$, and so $G_1 \cup G = \mathbf{C}$ and ϕ is an entire function. Since $\mathrm{ind}_\Gamma(z) = 0$ on the unbounded component U of $\mathbf{C} \backslash \Gamma^*$, $U \subset G_1$; thus $\lim_{|z| \to \infty} \phi(z) = \lim_{|z| \to \infty} h_1(z) = 0$. Thus ϕ is bounded, and hence constant, by Liouville's theorem: the constant is clearly zero, and so $h(z) = 0$ for all $z \in G$, which proves (3.6.9).

To obtain (3.6.10), let $z_0 \in G \backslash \Gamma^*$ and define $F(z) = (z - z_0) f(z) \ (z \in G) : F$ is analytic in G. Thus by (3.6.9),

$$\int_\Gamma f = \int_\Gamma F(\zeta)(\zeta - z_0)^{-1} d\zeta = 2\pi i F(z_0) \, \mathrm{ind}_\Gamma(z_0) = 0.$$

Finally, suppose $\Lambda = (\lambda_j)_{j=1}^r$ and $\Psi = (\psi_j)_{j=1}^s$, the λ_j and ψ_j being circuits, and let $\Gamma = (\gamma_j)_{j=1}^{r+s}$ be the cycle defined by

$$\gamma_j = \lambda_j (1 \leq j \leq r), \ \gamma_j = -\psi_{j-r}(r + 1 \leq j \leq r + s).$$

Then

$$\mathrm{ind}_\Gamma(a) = \mathrm{ind}_\Lambda(a) - \mathrm{ind}_\Psi(a) = 0 \ (a \in \mathbf{C} \backslash G);$$

further, application of (3.6.10) to f shows that

$$0 = \int_\Gamma f = \int_\Lambda f - \int_\psi f,$$

and therefore

$$\int_\Lambda f = \int_\psi f.$$

$\square$

Remark 3.6.33 (i) Suppose that G is a **convex**, open subset of $\mathbf{C}$ and let γ be a circuit in G. If $a \in \mathbf{C} \backslash G$, Theorem 3.6.9 shows that $\mathrm{ind}_\gamma(a) = 0$; thus $\mathrm{ind}_\Gamma(a) = 0$ for all cycles Γ with $\Gamma^* \subset G$. Thus the basic hypothesis of Theorem 3.6.32 holds if G is convex.

(ii) Let G be an open subset of $\mathbf{C}$ and let γ be a circuit in G. Theorem 3.6.32 shows that if $\mathrm{ind}_\gamma(a) = 0$ for all $a \in \mathbf{C} \backslash G$ then $\int_\gamma f = 0$ for all $f \in H(G)$. The converse also holds: if $a \in \mathbf{C} \backslash G$ and $\mathrm{ind}_\gamma(a) \neq 0$, put $f(z) = (z - a)^{-1}$ $(z \in G)$, so that $f \in H(G)$ and $\int_\gamma f = 2\pi i \, \mathrm{ind}_\gamma(a) \neq 0$. Thus the circuits γ in G such that $\int_\gamma f = 0$ for all $f \in H(G)$ are exactly those for which $\mathrm{ind}_\gamma(z) = 0$ for all $z \in \mathbf{C} \backslash G$.

(iii) We have shown that for a given open subset G of $\mathbf{C}$, the following three statements are equivalent:

(1) $\mathrm{ind}_\gamma(z) = 0$ for all $z \in \mathbf{C} \backslash G$ and all circuits γ in G;
(2) $\int_\gamma f = 0$ for all $f \in H(G)$ and all circuits γ in G;
(3) every $f \in H(G)$ has a primitive on G.

It is desirable to formalise some of the ideas which appear in the above discussion.

Definition 3.6.34 Let G be an open subset of $\mathbf{C}$. A cycle Γ in G such that $\mathrm{ind}_\Gamma(z) = 0$ for all $z \in \mathbf{C} \backslash G$ is said to be **homologous to** 0 (in G). Cycles Γ_1 and Γ_2 in G such that

$$\mathrm{ind}_{\Gamma_1}(z) = \mathrm{ind}_{\Gamma_2}(z) \text{ for all } z \in \mathbf{C} \backslash G,$$

are said to be **homologous** (in G).

This definition enables us to present the conclusions (3.6.10) and (3.6.11) of Theorem 3.6.32 in the following form: if $f \in H(G)$, then

$$\int_\Gamma f = 0$$

for all cycles Γ in G which are homologous to 0; and

$$\int_{\Gamma_1} f = \int_{\Gamma_2} f$$

for all homologous cycles Γ_1 and Γ_2 in G.

The global version of Cauchy's theorem is thus called the **homology version** of the theorem. There is another version, the **homotopy version**, which is easy to prove from Theorem 3.6.32.

Theorem 3.6.35 (The homotopy version of Cauchy's theorem) *Let G be an open subset of* $\mathbf{C}$ *and let* $\gamma_0, \gamma_1 : [0, 1] \to \mathbf{C}$ *be circuits in G which are freely homotopic. Then*

$$\int_{\gamma_0} f = \int_{\gamma_1} f$$

for all $f \in H(G)$. *In particular, if* γ_0 *is freely homotopic to a constant path in G, then* $\int_{\gamma_0} f = 0$ *for all* $f \in H(G)$.

Proof Since γ_0 and γ_1 are freely homotopic, they are homologous, by Theorem 3.6.21 (iii). Thus by Theorem 3.6.32, $\int_{\gamma_0} f = \int_{\gamma_1} f$ for all $f \in H(G)$. The rest is now clear.   □

Corollary 3.6.36 *Let G be an open, simply-connected subset of* $\mathbf{C}$. *Then for all circuits* γ *in G and all* $f \in H(G)$, $\int_{\gamma} f = 0$.

Proof Since every circuit in a simply-connected subset is null-homotopic, the result follows immediately from Theorem 3.6.35.   □

We shall see later that if G is a *region* in $\mathbf{C}$ such that $\int_{\gamma} f = 0$ for every $f \in H(G)$ and every circuit γ in G, then G is simply-connected. In view of this, and Corollary 3.6.36, it follows that for a given region $G \subset \mathbf{C}$, statements 1, 2 and 3 of Remark 3.6.33 (iii) are each equivalent to the following statement:
4) G is simply-connected.

Exercise 3.6.37

1. (i) Sketch the track of the contour $\mu : \left[-\frac{\pi}{2}, \frac{\pi}{2}\right] \to \mathbf{C}$ defined by $\mu(t) = -t + i\left(\frac{\pi^2}{4} - t^2\right)$. Evaluate $\int_{\mu} \cos z\, dz$.

   (ii) Let $\nu : [0, 1] \to \mathbf{C}$ be defined by $\nu(0) = 0$, $\nu(t) = t^3 \exp(-2\pi i/t)$ if $0 < t \leq 1$. Show that ν is a contour and sketch its track. Evaluate $\int_{\nu} z^2 e^z\, dz$.

2. Let $\gamma : [0, 1] \to \mathbf{C}$ be defined by $\gamma(0) = 0$, $\gamma(1) = 0$, $\gamma(t) = t + it^3 \sin(\pi/t)$ if $0 < t \leq 1/2$ and $\gamma(t) = (1 - t) - i(1 - t)^3 \sin(\pi/(1 - t))$ if $1/2 \leq t < 1$. Show that γ is a circuit and sketch its track. Evaluate
   (i) $\int_{\gamma} \cos^3(z^2)\, dz$, (ii) $\int_{\gamma} \log(1 + z)\, dz$.

3. Let $\Delta : [u, v] \to \mathbf{C}$ be a triangular circuit with vertices z_1, z_2, z_3. Let

$$K = \left\{ \sum_{j=1}^{3} \alpha_j z_j : \alpha_j \geq 0 \ (j = 1, 2, 3), \sum_{j=1}^{3} \alpha_j = 1 \right\}.$$

Show that K is a compact, convex subset of $\mathbf{C}$ and that $\mathrm{co}\, \{z_1, z_2, z_3\} = \mathrm{co}\, \Delta^* = K$.

4. Let $a > 1$ and define $\gamma : [0, 2\pi] \to \mathbf{C}$ by $\gamma(t) = e^{it}$. Using Cauchy's integral formula evaluate

$$\int_\gamma (z^2 - 2az + 1)^{-1} dz,$$

and deduce that

$$\int_0^{2\pi} (a - \cos t)^{-1} dt = 2\pi (a^2 - 1)^{-1/2}.$$

5. Let $f \in H(\mathbf{C})$ be defined by

$$f(z) = \sum_{k=1}^\infty \frac{(iz)^{k-1}}{k!} = \begin{cases} (iz)^{-1}(\exp(iz) - 1) & \text{if } z \neq 0, \\ 1 & \text{if } z = 0. \end{cases}$$

Let $R > 0$ and let $\gamma : [-R, R + \pi] \to \mathbf{C}$ be the circuit given by

$$\gamma(s) = \begin{cases} s, & -R \leq s \leq R, \\ \exp(i(s - R)), & R \leq s \leq R + \pi. \end{cases}$$

By Cauchy's theorem in a convex set, plainly $\int_\gamma f = 0$. Exploit this fact to prove that

$$\left| \int_0^R \frac{\sin x}{x} dx - \frac{\pi}{2} \right| \leq \int_0^{\pi/2} \exp(-R \sin \theta) d\theta$$

$$\leq \int_0^{\pi/2} \exp(-2R\theta/\pi) d\theta \leq \frac{\pi}{R},$$

and deduce that

$$\int_0^\infty \frac{\sin x}{x} dx = \frac{\pi}{2}.$$

[Hint: if $0 \leq \theta \leq \pi/2$, then $2\theta/\pi \leq \sin \theta \leq \theta$.]

6. Let $\rho > 0$. Show that if $f \in H(\mathbf{C})$ and $|f(z)| \leq A |z|^m$ for all $z \in \mathbf{C}$ with $|z| > \rho$, where A and m are non-negative real constants, then f is a polynomial of degree at most m.

7. Let G be an open subset of $\mathbf{C}$, let $f : G \to \mathbf{C}$ be continuous and suppose that for every triangular circuit Δ such that co $\Delta^* \subset G$, $\int_\Delta f = 0$. Prove that f is analytic in G. (This is Morera's theorem .)

8. Let $R > 0$, $z_0 \in \mathbf{C}$ and suppose $f : B(z_0, R) \to \mathbf{C}$ is defined by

$$f(z) = \sum_{n=0}^\infty a_n (z - z_0)^n.$$

Prove that if $0 < r < R$, then

$$\int_{-\pi}^{\pi} \left| f(z_0 + re^{i\theta}) \right|^2 d\theta = 2\pi \sum_{n=0}^{\infty} |a_n|^2 r^{2n}.$$

Hence show that:
(a) Every bounded entire function is constant.
(b) If G is a region in $\mathbf{C}$, $g \in H(G)$ and $\overline{B(z_0, r)} \subset G$, then

$$|g(z_0)| \leq \sup \left\{ \left| g(z_0 + re^{i\theta}) \right| : 0 \leq \theta \leq 2\pi \right\},$$

with equality if, and only if, g is constant. [Thus $|g|$ has no local maximum at any point of G unless g is constant: this gives another proof of the Maximum Modulus Theorem (Theorem 3.6.28).]

9. Prove that

$$(1 - z - z^2)^{-1} = \sum_{n=0}^{\infty} a_n z^n \quad \left(z \in B(0, (\sqrt{5} - 1)/2 \right),$$

where the a_n are the Fibonacci numbers defined by

$$a_0 = a_1 = 1, \quad a_{n+1} = a_n + a_{n-1} (n \geq 1).$$

Further, show that

$$a_n = \frac{1}{\sqrt{5}} \left\{ \left(\frac{1 + \sqrt{5}}{2} \right)^{n+1} - \left(\frac{1 - \sqrt{5}}{2} \right)^{n+1} \right\} \quad (n \geq 0).$$

10. Compute the coefficients in the Taylor expansion of f about $z = 0$ as far as the term in z^7, in each of the following cases:
(a) $f(z) = \log(1 + z)$; (b) $f(z) = \sec z$.
What is the radius of convergence of each power series?

11. If each of the following functions were expanded as a Taylor series about the indicated points, what would be the radius of convergence? (Do not find the Taylor series.)
(a) $\frac{\sin z}{z^2 + 4}$, $z = 0$; (b) $\frac{z}{e^z + 1}$, $z = 0$; (c) $\frac{e^z}{z(z-1)}$, $z = 4i$.

12. Find the first four terms and the radius of convergence of the Taylor series about $z = 0$ for
(a) $f(z) = \frac{e^z}{(1-z)^2}$; (b) $f(z) = \frac{1}{1+\log(1+z)}$.

13. Let G be a region in $\mathbf{C}$, let $f, g \in H(G)$ and suppose that $fg = 0$ (that is, $f(z)g(z) = 0$ for all $z \in G$). Prove that either $f = 0$ or $g = 0 : H(G)$ is an integral domain.

14. Let $\gamma : [a, b] \to \mathbf{C}$ be a circuit and $z \in \mathbf{C} \backslash \gamma^*$. Let $F : [a, b] \to \mathbf{C}$ be defined by

$$F(t) = \int_a^t \frac{\gamma'(s)}{\gamma(s) - z} ds;$$

note that γ' is defined save possibly at a finite number of points of $[a, b]$. Show that, for all $t \in [a, b]$,

$$(\gamma(a) - z) \exp(F(t)) = (\gamma(t) - z),$$

and that therefore $\mathrm{ind}_\gamma(z) = (2\pi i)^{-1} F(b) \in \mathbf{Z}$.

15. Let $z_0, z_1 \in \mathbf{C}$, $z_0 \neq z_1$ and

$$S = S(z_0, z_1) = \{(1 - t)z_0 + t z_1 : 0 \leq t \leq 1\}.$$

Let $\gamma : [a, b] \to \mathbf{C}$ be a circuit such that $\gamma^* \cap S = \emptyset$. Show that, for all $w \in \mathbf{C} \backslash S$, $(w - z_1)(w - z_0)^{-1} \in D(\pi)$; that $w \longmapsto \log\{(w - z_1)(w - z_0)^{-1}\}$ is a primitive of $w \longmapsto (w - z_1)^{-1} - (w - z_0)^{-1}$ on $\mathbf{C} \backslash S$; and that

$$\mathrm{ind}_\gamma(z_1) = \mathrm{ind}_\gamma(z_0).$$

Deduce that $\mathrm{ind}_\gamma(\cdot)$ is constant on components of $\mathbf{C} \backslash \gamma^*$ (hint: use Theorem 2.4.23). Further, observing that if $|z| > r > \sup_{w \in \gamma^*} |w|$, then $w \longmapsto (w - z)^{-1}$ is analytic in $B(0, r)$, show that $\mathrm{ind}_\gamma(z) = 0$ if z lies in the unbounded component of $\mathbf{C} \backslash \gamma^*$.

16. Let $a \in \mathbf{C}$, $r > 0$ and $\gamma : [0, 2\pi] \to \mathbf{C}$ be defined by $\gamma(t) = a + r \exp(it)$. Use the preceding exercise to show that

$$\mathrm{ind}_\gamma(z) = \begin{cases} 1 & \text{if } |z - a| < r, \\ 0 & \text{if } |z - a| > r. \end{cases}$$

17. Let $0 < R < \infty$, $0 < \phi \leq \pi$. Sketch the circuits μ, ν defined by

$$\mu(s) = \begin{cases} -s \exp(i\phi) & \text{if } -R \leq s \leq 0, \\ s & \text{if } 0 \leq s \leq R, \\ R \exp(i(s - R)) & \text{if } R \leq s \leq R + \phi; \end{cases}$$

and

$$\nu(s) = \begin{cases} -s & \text{if } -R \leq s \leq 0, \\ s \exp(i\phi) & \text{if } 0 \leq s \leq R, \\ R \exp(i(\phi + s - R)) & \text{if } R \leq s \leq R + 2\pi - \phi. \end{cases}$$

Using Definition 3.6.14 directly, establish a cancellation of integrals over line segments common to μ and ν and show that, if $\gamma(t) = R \exp(it)$ $(0 \leq t \leq 2\pi)$, then

$$\text{ind}_\gamma(z) = \text{ind}_\mu(z) + \text{ind}_\nu(z) \ (z \in \mathbf{C}\backslash\{\mu^* \cup \nu^*\}).$$

Deduce that

$$\text{ind}_\mu(z) = \begin{cases} 1 & \text{if } |z| < R \text{ and } \arg z \in (0, \phi), \\ 0 & \text{if } |z| > R \text{ or } \arg z \in (-\pi, 0) \cup (\phi, \pi]. \end{cases}$$

[If $\phi = \pi$, then μ coincides with the semi-circular circuit of Example 3.4.22 (iv).]

3.7 Singularities

Let G be an open subset of $\mathbf{C}$, let $z_0 \in G$ and suppose that $f \in H(G\backslash\{z_0\})$. Our immediate aim is to describe the behaviour of f near z_0.

Lemma 3.7.1 *Let G be an open subset of $\mathbf{C}$ such that*

$$G \supset \{z \in \mathbf{C} : r_1 \le |z - z_0| \le r_2\}$$

for some r_1, r_2 with $0 < r_1 < r_2 < \infty$. For $k = 1, 2$ define $\nu_k : [0, 2\pi] \to \mathbf{C}$ by $\nu_k(t) = z_0 + r_k e^{it}$. Then for all $f \in H(G)$ and all z such that $r_1 < |z - z_0| < r_2$,

$$f(z) = \frac{1}{2\pi i} \int_{\nu_2} \frac{f(w)}{w - z} dw - \frac{1}{2\pi i} \int_{\nu_1} \frac{f(w)}{w - z} dw.$$

Proof Let Γ be the cycle $(\nu_2, -\nu_1)$. Appeal to Lemma 3.6.11 shows that if $a \in \mathbf{C}\backslash G$, then

$$\text{ind}_\Gamma(a) = \text{ind}_{\nu_2}(a) - \text{ind}_{\nu_1}(a) = 0.$$

Hence by Theorem 3.6.32, if $z \in G\backslash\Gamma^*$, then

$$\text{ind}_\Gamma(z) f(z) = \frac{1}{2\pi i} \int_\Gamma \frac{f(w)}{w - z} dw$$

$$= \frac{1}{2\pi i} \int_{\nu_2} \frac{f(w)}{w - z} dw - \frac{1}{2\pi i} \int_{\nu_1} \frac{f(w)}{w - z} dw.$$

Since, with $r_1 < |z - z_0| < r_2$, we have

$$\text{ind}_\Gamma(z) = \text{ind}_{\nu_2}(z) - \text{ind}_{\nu_1}(z) = 1 - 0 = 1,$$

the result follows.   □

Armed with this lemma we can now prove a result of first-rate importance.

Theorem 3.7.2 (Laurent's theorem) *Let G be an open subset of $\mathbf{C}$ such that $G \supset A = \{z \in \mathbf{C} : s_1 < |z - z_0| < s_2\}$, where $0 \le s_1 < s_2 \le \infty$, and let $f \in H(G)$. Then there is a unique sequence $(a_n)_{n \in \mathbf{Z}}$ of complex numbers such that the series $\sum_{n=0}^{\infty} a_n z^n$ is convergent when $|z| < s_2$, the series $\sum_{n=1}^{\infty} a_{-n} z^n$ is convergent when $|z| < s_1^{-1}$ (for all $z \in \mathbf{C}$ when $s_1 = 0$), and for all $z \in A$,*

$$f(z) = \sum_{n=0}^{\infty} a_n(z - z_0)^n + \sum_{n=1}^{\infty} a_{-n}(z - z_0)^{-n}. \tag{3.7.1}$$

Moreover, for all $n \in \mathbf{Z}$ and all $s \in (s_1, s_2)$,

$$a_n = \frac{1}{2\pi i} \int_{\gamma_s} \frac{f(w)}{(w - z_0)^{n+1}} dw,$$

where γ_s is the positively oriented circle with centre z_0 and radius s. The series in (3.7.1) converge absolutely in A and uniformly on every compact subset of A. The identity (3.7.1), commonly written

$$f(z) = \sum_{-\infty}^{\infty} a_n(z - z_0)^n,$$

*is called the **Laurent expansion of** f **in** A and the a_n are the **coefficients** associated with that expansion. If $s_1 = 0$, so that f is analytic in a deleted neighbourhood of z_0 (sets of the form $U \setminus \{z_0\}$, where U is a neighbourhood of z_0, are described as **deleted neighbourhoods** of z_0), then a_{-1} is called the **residue of** f **at** z_0, written $\mathrm{res}(f, z_0)$:*

$$\mathrm{res}(f, z_0) = \frac{1}{2\pi i} \int_{\gamma_s} f(w) \, dw \ (0 < s < s_2).$$

Proof Let K be a compact subset of A, so that $K \subset \{z \in \mathbf{C} : r' \le |z - z_0| \le r''\}$ for some r', r'' such that $s_1 < r' \le r'' < s_2$. Fix $\zeta \in A$ and choose r_1, r_2 so that $s_1 < r_1 < r', r'' < r_2 < s_2, r_1 < |\zeta - z_0| < r_2$; let ν_1, ν_2 be the positively oriented circles centred at z_0 with radii r_1, r_2 respectively. By Lemma 3.7.1,

$$f(\zeta) = \frac{1}{2\pi i} \int_{\nu_2} \frac{f(w)}{w - \zeta} dw - \frac{1}{2\pi i} \int_{\nu_1} \frac{f(w)}{w - \zeta} dw = g(\zeta) + h(\zeta), \text{ say,}$$

where

$$g(z) = \frac{1}{2\pi i} \int_{\nu_2} \frac{f(w)}{w - z} dw \ (z \in \mathbf{C} \setminus \nu_2^*)$$

and

$$h(z) = -\frac{1}{2\pi i} \int_{v_1} \frac{f(w)}{w - z} dw \ (z \in \mathbf{C} \backslash v_1^*).$$

By Theorem 3.6.13, g is analytic in $B(z_0, r_2)$ and so $g(z) = \sum_{n=0}^{\infty} a_n (z - z_0)^n$ for $|z - z_0| < r_2$, where $a_n = g^{(n)}(z_0)/n!$; also, further appeal to Theorem 3.6.13 shows that

$$a_n = \frac{1}{2\pi i} \int_{v_2} \frac{f(w)}{(w - z_0)^{n+1}} dw.$$

Since $w \longmapsto (w - z_0)^{-(n+1)} f(w) : A \to \mathbf{C}$ is analytic, equality (3.6.11) of Theorem 3.6.32 shows that v_2 may be replaced by γ_s for any $s \in (s_1, s_2)$. By Lemma 3.3.1 the series for g converges uniformly on $\{z \in \mathbf{C} : |z - z_0| \leq r'' < r_2\}$ and is therefore uniformly convergent on K.

In view of Theorem 3.6.13, h is analytic in $\{z \in \mathbf{C} : |z - z_0| > r_1\}$. Write

$$h(z) = \frac{1}{2\pi i} \int_{v_1} (z - z_0)^{-1} f(w) \sum_{n=0}^{\infty} \left(\frac{w - z_0}{z - z_0} \right)^n dw$$

and note that if $z \in \mathbf{C} \backslash \overline{B}(z_0, r_1)$,

$$\left| \frac{w - z_0}{z - z_0} \right| = \frac{r_1}{|z - z_0|} < 1.$$

The series in the integrand thus converges uniformly on v_1^* and term-by-term integration is permissible, giving

$$h(z) = \sum_{n=1}^{\infty} a_{-n} (z - z_0)^{-n} \ (|z - z_0| > r_1),$$

where

$$a_{-n} = \frac{1}{2\pi i} \int_{v_1} (w - z_0)^{n-1} f(w) dw \ (n \in \mathbf{N}).$$

By Theorem 3.6.32 again, v_1 may be replaced by γ_s for any $s \in (s_1, s_2)$. The series $\sum_1^{\infty} a_{-n} u^n$ is absolutely convergent for $|u| < r_1^{-1}$ and is thus uniformly convergent on $\{u \in \mathbf{C} : |u| \leq (r')^{-1}\}$. Hence the series for h is uniformly convergent on $\{z \in \mathbf{C} : |z - z_0| \geq r'\}$, and so is uniformly convergent on K.

All that remains is to prove that the sequence $(a_n)_{n \in \mathbf{Z}}$ is unique. Suppose that for all $z \in A$,

$$f(z) = \sum_{n=0}^{\infty} b_n (z - z_0)^n + \sum_{n=1}^{\infty} b_{-n} (z - z_0)^{-n},$$

that $\sum_{n=0}^{\infty} b_n u^n$ is convergent if $|u| < s_2$ and that $\sum_{n=1}^{\infty} b_{-n} u^n$ is convergent if $|u| < s_1^{-1}$. Let $r \in (s_1, s_2)$. Then for each $k \in \mathbf{Z}$,

$$a_k = \frac{1}{2\pi i} \int_{\gamma_r} (w - z_0)^{-k-1} f(w) dw = \frac{1}{2\pi r^k} \int_0^{2\pi} e^{-ik\theta} f(z_0 + re^{i\theta}) d\theta$$

$$= \frac{1}{2\pi r^k} \int_0^{2\pi} e^{-ik\theta} \sum_{n=0}^{\infty} b_n \left(re^{i\theta}\right)^n d\theta + \frac{1}{2\pi r^k} \int_0^{2\pi} e^{-ik\theta} \sum_{n=1}^{\infty} b_{-n} \left(re^{i\theta}\right)^{-n} d\theta.$$

The series here converge absolutely for $|z - z_0| = r$, and so are uniformly convergent (in θ) on $[0, 2\pi]$; thus

$$a_k = \frac{1}{2\pi} \sum_{n=0}^{\infty} b_n r^{n-k} \int_0^{2\pi} e^{i(n-k)\theta} d\theta + \frac{1}{2\pi} \sum_{n=1}^{\infty} b_{-n} r^{-n-k} \int_0^{2\pi} e^{-i(n+k)\theta} d\theta = b_k.$$

$\square$

Corollary 3.7.3 *Let* $A = \{z \in \mathbf{C} : s_1 < |z - z_0| < s_2\}$, *where* $0 \le s_1 < s_2 \le \infty$, *and let* $f \in H(A)$. *Then there are functions* $g \in H(\{z \in \mathbf{C} : |z - z_0| < s_2\})$ *and* $h \in H(\{z \in \mathbf{C} : |z - z_0| > s_1\})$ *such that, for all* $z \in A$,

$$f(z) = g(z) + h(z).$$

This decomposition is unique if we require that $h(z) \to 0$ *as* $|z| \to \infty$.

Proof Let $f(z) = \sum_{-\infty}^{\infty} a_n (z - z_0)^n$ be the Laurent expansion of f in the annulus A. Put $g(z) = \sum_0^{\infty} a_n (z - z_0)^n$, $h(z) = \sum_1^{\infty} a_{-n} (z - z_0)^{-n}$; in A, $f = g + h$, $g \in H(\{z \in \mathbf{C} : |z - z_0| < s_2\})$, $h \in H(\{z \in \mathbf{C} : |z - z_0| > s_1\})$ and $h(z) \to 0$ as $|z| \to \infty$. Let $f = g_1 + h_1$ be another such decomposition and let ϕ be the function equal to $g - g_1$ when $|z - z_0| < s_2$ and to $h_1 - h$ when $|z - z_0| > s_1$: note that $g - g_1 = h_1 - h$ in A. Then $\phi \in H(\mathbf{C})$ and $\lim_{|z| \to \infty} \phi(z) = 0$. By Liouville's theorem, $\phi = 0$. $\square$

Definition 3.7.4 Let G be an open subset of $\mathbf{C}$ and $z_0 \in G$. If $f \in H(G \backslash \{z_0\})$, then z_0 is said to be an **isolated singularity** of f.

Suppose G, z_0 and f are as in Definition 3.7.4. By Laurent's theorem, there is a unique sequence $(a_n)_{n \in \mathbf{Z}}$ such that if $0 < r \le \infty$ and the annulus $\{z \in \mathbf{C} : 0 < |z - z_0| < r\} \subset G$, then the Laurent expansion of f in the annulus is

$$f(z) = \sum_{-\infty}^{\infty} a_n (z - z_0)^n.$$

Define $h : \mathbf{C} \backslash \{z_0\} \to \mathbf{C}$ by $h(z) = \sum_{n=1}^{\infty} a_{-n} (z - z_0)^{-n}$; h is called the **principal part** of f at z_0. Evidently $h \in H(\mathbf{C} \backslash \{z_0\})$ and $h(z) \to 0$ as $|z| \to \infty$. Further, if Γ

is a cycle and $z_0 \in \mathbf{C}\backslash\Gamma^*$, then the series determining h converges uniformly on the compact set $\{(z - z_0)^{-1} : z \in \Gamma^*\}$ and, therefore,

$$\frac{1}{2\pi i} \int_\Gamma h = \frac{1}{2\pi i} \int_\Gamma \sum_{n=1}^\infty a_{-n}(z - z_0)^{-n} dz = \frac{1}{2\pi i} \sum_{n=1}^\infty a_{-n} \int_\Gamma (z - z_0)^{-n} dz$$

$$= a_{-1} \operatorname{ind}_\Gamma(z_0) = \operatorname{res}(f, z_0) \operatorname{ind}_\Gamma(z_0),$$

since those terms in the principal part of order greater than one have primitives on $\mathbf{C}\backslash\{z_0\}$ and Corollary 3.6.3 applies. The sequence $(a_{-n})_{n\in\mathbf{N}}$ is used to classify the singularity of f at z_0 as one of three types:

(i) f has a **removable singularity at** z_0 if $a_{-n} = 0$ for all $n \in \mathbf{N}$; that is, if $h = 0$;
(ii) f has a **pole of order** $m \ge 1$ **at** z_0 if $a_{-m} \ne 0$ and $a_{-n} = 0$ for all $n > m$;
(iii) f has an **essential singularity at** z_0 if for infinitely many $n \in \mathbf{N}$, $a_{-n} \ne 0$.

Example 3.7.5

(a) Let $f(z) = \frac{\sin z}{z}$ $(z \ne 0)$. Plainly $f \in H(\mathbf{C}\backslash\{0\})$. Since

$$f(z) = 1 - \frac{z^2}{3!} + \frac{z^4}{5!} - \cdots (z \ne 0)$$

and no negative powers of z appear, we see from the uniqueness of the Laurent expansion that f has a removable singularity at 0. Since $a_0 = 1$, the singularity can be removed by defining $f(0) = 1$; the extended f is thus in $H(\mathbf{C})$. This process of extension can be followed whenever a given function has a removable singularity.
(b) Let $f(z) = \frac{\sin z}{z^2}$ $(z \ne 0)$. Here f has an isolated singularity at 0; and since

$$f(z) = z^{-1} - \frac{z}{3!} + \frac{z^3}{5!} - \cdots (z \ne 0),$$

f has a pole of order 1 at 0.
(c) Suppose $f(z) = e^{1/z}$ $(z \ne 0)$. Since $e^{1/z} = \sum_{n=0}^\infty \frac{z^{-n}}{n!}$, there is an essential singularity at $z = 0$.
(d) When $f(z) = 1/\sin(z^{-1})$ $(z \ne 0, z \ne 1/k\pi$ for all $k \in \mathbf{Z}\backslash\{0\})$, the function f cannot be classified at 0 in any of the three ways given above, for it does not have an **isolated** singularity at 0: it has singularities at $1/k\pi$ for all $k \in \mathbf{Z}\backslash\{0\}$.

How can we tell the nature of an isolated singularity without going to the bother of determining the Laurent expansion? The next two lemmas help a great deal in this connection.

Lemma 3.7.6 *Let G be an open subset of* $\mathbf{C}$, $z_0 \in G$ *and* $f \in H(G\backslash\{z_0\})$. *Then the following statements are equivalent:*

(i) *f has a removable singularity at z_0;*
(ii) $\lim_{z \to z_0} f(z)$ *exists in* **C**;
(iii) *f is bounded on* $B(z_0, r) \setminus \{z_0\}$ *for some open ball* $B(z_0, r) \subset G$;
(iv) $\lim_{z \to z_0} (z - z_0) f(z) = 0$.

Proof Evidently (i) $\Rightarrow$ (ii) $\Longrightarrow$ (iii) $\Longrightarrow$ (iv). To prove that (iv) implies (i), put $h(z) = (z - z_0)^2 f(z)$ if $z \in G \setminus \{z_0\}$, and put $h(z_0) = 0$. Then

$$h'(z_0) = \lim_{z \to z_0} (z - z_0) f(z) = 0,$$

by hypothesis; also, h is differentiable at each point of $G \setminus \{z_0\}$. Hence $h \in H(G)$ and, being representable by power series in G, there is a unique sequence $(a_n)_{n \in \mathbb{N}_0}$ such that for any positive real number r such that $B(z_0, r) \subset G$,

$$h(z) = \sum_{n=0}^{\infty} a_n (z - z_0)^n$$

whenever $z \in B(z_0, r)$. By Remark 3.3.9, $a_0 = a_1 = 0$. Define $f(z_0) = a_2$; then $f(z) = \sum_0^{\infty} a_{n+2}(z - z_0)^n$ for all $z \in B(z_0, r)$. $\square$

Lemma 3.7.6 enables us to detect the presence of a removable singularity; the next Lemma does the same for poles.

Lemma 3.7.7 *Let G be an open subset of* **C**, *$z_0 \in G$ and $f \in H(G \setminus \{z_0\})$. Then f has a pole of order m at z_0 if, and only if, $\lim_{z \to z_0} (z - z_0)^m f(z) = \lambda$ for some non-zero $\lambda \in$ **C**. In this case, $\lim_{z \to z_0} |f(z)| = \infty$.*

Proof Suppose that f has a pole of order m at z_0, that $r \in (0, \infty)$ is such that $A = \{z \in \mathbf{C} : 0 < |z - z_0| < r\} \subset G$, and that the Laurent expansion of f in A is

$$f(z) = \sum_{n=0}^{\infty} a_n (z - z_0)^n + \sum_{n=1}^{m} a_{-n}(z - z_0)^{-n}, a_{-m} \neq 0.$$

Then $(z - z_0)^m f(z) = \sum_0^{\infty} b_n (z - z_0)^n$ for all $z \in A$, where $b_{m+p} = a_p$ for $p \geq -m$. By Lemma 3.7.6, $(z - z_0)^m f(z)$ has a removable singularity at z_0; also $\lim_{z \to z_0} (z - z_0)^m f(z) = b_0 = a_{-m} \neq 0$, and so there exists $r' \in (0, r)$ such that $|(z - z_0)^m f(z)| \geq \frac{1}{2} |a_{-m}|$ if $0 < |z - z_0| < r'$. Thus $\lim_{z \to z_0} |f(z)| = \infty$.

Conversely, if $\lim_{z \to z_0} (z - z_0)^m f(z) = \lambda$ for some $\lambda \in \mathbf{C} \setminus \{0\}$, then by Lemma 3.7.6, $(z - z_0)^m f(z)$ has a removable singularity at z_0 and, in a punctured open ball centred at z_0 and contained in G, is given by $\sum_0^{\infty} b_n (z - z_0)^n$ with $b_0 \neq 0$. Division by $(z - z_0)^m$ shows that f has a pole of order m at z. $\square$

The last two lemmas show that if f has an isolated singularity at z_0, then $\lim_{z \to z_0} f(z)$ exists in **C** if the singularity is removable, while if it is a pole of order m, then $\lim_{z \to z_0} |f(z)| = \infty$. The behaviour near z_0 is much more exotic if the singularity is an essential one, as the next result shows.

Theorem 3.7.8 (The Casorati-Weierstrass theorem) *Let G be an open subset of* $\mathbf{C}$, *let* $z_0 \in G$ *and suppose that* $f \in H(G\backslash\{z_0\})$. *Then* f *has an essential singularity at* z_0 *if, and only if, for all* $r > 0$ *such that* $B(z_0, r) \subset G$, $f(B(z_0, r)\backslash\{z_0\})$ *is dense in* $\mathbf{C}$.

Proof Lemmas 3.7.6 and 3.7.7 establish 'if'. Regarding 'only if', suppose that f has an essential singularity at z_0 and that the conclusion is false. Then there exist $w \in \mathbf{C}$, $\rho > 0$ and $r > 0$ such that $B(w, \rho) \cap f(B(z_0, r)\backslash\{z_0\}) = \varnothing$, that is, $|f(z) - w| \geq \rho$ whenever $0 < |z - z_0| < r$. Let $A = \{z \in \mathbf{C} : 0 < |z - z_0| < r\}$ and define $g \in H(A)$ by $g(z) = (f(z)-w)^{-1}$. Evidently g has an isolated singularity at z_0, g is zero-free in A and is bounded by ρ^{-1}. Hence, by Lemma 3.7.6, it has a removable singularity at z_0. Setting $g(z_0) = \lim_{z\to z_0} g(z)$, the extended function, also denoted by g, lies in $H(B(z_0, r))$. Note that, for all $z \in A$,

$$f(z) = w + \frac{1}{g(z)},$$

and therefore if $g(z_0) \neq 0$ then $1/g \in H(B(z_0, r))$ and f has a removable singularity at z_0, contrary to hypothesis. Thus $g(z_0) = 0$. It follows, using Theorem 3.6.26, that there is a positive integer m such that $g(z) = (z - z_0)^m h(z)$ for all $z \in B(z_0, r)$, where $h \in H(B(z_0, r))$ and $h(z_0) \neq 0$. Clearly, h is zero-free in $B(z_0, r)$: put $k = 1/h$. Then $k \in H(B(z_0, r))$ and

$$f(z) - w = (z - z_0)^{-m} k(z) \text{ for all } z \in A.$$

But $k(z) = \sum_0^\infty b_n(z - z_0)^n$ for all $z \in B(z_0, r)$, and $b_0 \neq 0$. Hence f has a pole of order m at z_0, again contrary to hypothesis. The result follows.   □

This result can be strengthened: in fact, $f(B(z_0, r)\backslash\{z_0\})$ is either all of $\mathbf{C}$ or all of $\mathbf{C}$ except for one point. This is Picard's theorem (see [16]).

We now turn our attention to the so-called **residue calculus**, which has far-reaching implications. The next lemma prepares the way for a significant extension of Cauchy's theorem.

Lemma 3.7.9 *Let G be an open subset of* $\mathbf{C}$ *and S be a subset of G which has no limit point in G. Then $G\backslash S$ is open and S is at most countable. Further, if $f \in H(G\backslash S)$, then f has an isolated singularity at each point of S.*

Proof Let $z \in G\backslash S$. Then there exists $r > 0$ such that $B(z, r) \subset G$ and $B(z, r) \cap S = \varnothing$; otherwise, z is a limit point of S in G, contradicting the hypothesis. Hence $G\backslash S$ is open. That S is at most countable is clear from the fact that each compact subset of G contains only finitely many points of S (otherwise S would have a limit point in G) and G has a compact exhaustion: specifically, setting aside the trivial case in which $G = \mathbf{C}$, if $n \in \mathbf{N}$ and

$$K_n = \{z \in \mathbf{C} : |z| \leq n\} \cap \{z \in G : \operatorname{dist}(z, \mathbf{C}\backslash G) \geq n^{-1}\},$$

then each K_n is compact, $K_n \subset \overset{o}{K}_{n+1}$ and $\cup_{n=1}^{\infty} K_n = G$.

Lastly, let $f \in H(G \backslash S)$ and $z_0 \in S$. Put $U = (G \backslash S) \cup \{z_0\}$. Then U is open and $U \backslash \{z_0\} = G \backslash S$, so that f has an isolated singularity at z_0. $\square$

Theorem 3.7.10 (The residue theorem) *Let G be an open set of $\mathbf{C}$, S be a subset of G with no limit point in G, $f \in H(G \backslash S)$ and Γ be a cycle in $G \backslash S$ which is homologous to zero in G. Then $\{w \in S : \mathrm{ind}_\Gamma(w) \neq 0\}$ is finite and*

$$\int_\Gamma f(z) dz = 2\pi i \sum_{w \in S} \mathrm{res}(f, w) \, \mathrm{ind}_\Gamma(w). \tag{3.7.2}$$

Proof Let $U = \{z \in \mathbf{C} : \mathrm{ind}_\Gamma(z) = 0\}$. The set U is open, since it is a union of components of $\mathbf{C} \backslash \Gamma^*$; it contains the unbounded component of $\mathbf{C} \backslash \Gamma^*$; and, by hypothesis, $\mathbf{C} \backslash G \subset U$. Put $K = \mathbf{C} \backslash U$ and $T = \{z \in S : \mathrm{ind}_\Gamma(z) \neq 0\}$. Then K is a bounded, closed and therefore compact subset of G and $T \subset K$. If T had infinitely many points, then it would have a limit point in K and so S would have a limit point in G. It follows that T is finite and that the right-hand side of (3.7.2) is a finite sum.

Considering the simplest case first, suppose $T = \emptyset$ so that $\mathrm{ind}_\Gamma(w) = 0$ for all $w \in S$. Then, since $G \backslash S$ is open, Γ is homologous to zero in $G \backslash S$ and $f \in H(G \backslash S)$, appeal to Theorem 3.6.32 shows that

$$\int_\Gamma f = 0 = 2\pi i \sum_{w \in S} \mathrm{res}(f, w) \, \mathrm{ind}_\Gamma(w),$$

and so (3.7.2) holds.

Now suppose that $T \neq \emptyset$ and that $w_1, \ldots, w_k$ are its distinct elements. For $1 \leq j \leq k$, let h_j be the principal part of f at w_j: observations immediately following Definition 3.7.4 show that $h_j \in H\left(\mathbf{C} \backslash \{w_j\}\right)$ and

$$\int_\Gamma h_j = 2\pi i \, \mathrm{res}(f, w_j) \, \mathrm{ind}_\Gamma(w_j).$$

Let $\widetilde{G} = G \backslash (S \backslash T)$; by Lemma 3.7.9, $\widetilde{G}$ is open. The function $f - \sum_{j=1}^k h_j$ defined on $G \backslash S$ has a removable singularity at each point of T and hence has an analytic extension, g say, defined on $\widetilde{G}$. Since Γ is homologous to zero in $\widetilde{G}$ and $g \in H(\widetilde{G})$, further appeal to Theorem 3.6.32 establishes that

$$0 = \int_\Gamma g = \int_\Gamma f - \sum_{j=1}^k \int_\Gamma h_j.$$

Hence

$$\int_\Gamma f = \sum_{j=1}^k \int_\Gamma h_j = \sum_{j=1}^k 2\pi i \operatorname{res}(f, w_j) \operatorname{ind}_\Gamma(w_j)$$

$$= 2\pi i \sum_{w \in S} \operatorname{res}(f, w) \operatorname{ind}_\Gamma(w),$$

as required.   □

In the rest of this section we concentrate on applications of the residue theorem. Naturally, routine procedures for the determination of the residue at an isolated singularity will be sought. Generally speaking, at poles such procedures are available; at essential singularities they are not. This accounts for the predominance of applications of the residue theorem exclusively involving poles.

An instance in which the hypotheses of the residue theorem arise naturally appears in Theorem 3.6.26 in connection with the zeros of an analytic function. Recapitulating, if G is a region in $\mathbf{C}$, $f \in H(G)$ and f is not identically zero in G, then $Z(f) := \{z \in G : f(z) = 0\}$ has no limit point in G. Further, if $a \in Z(f)$ and $m(a)$ denotes the multiplicity of this zero of f, then there is a neighbourhood V of a and a zero-free map $h \in H(V)$ such that, for all $z \in V$, $f(z) = (z - a)^{m(a)} h(z)$. Hence, for all $z \in V \setminus \{a\}$,

$$\frac{f'(z)}{f(z)} = \frac{m(a)}{z - a} + \frac{h'(z)}{h(z)}.$$

For $z \in G \setminus Z(f)$, put $\psi(z) = f'(z)/f(z)$. Then $\psi \in H(G \setminus Z(f))$, ψ has a pole of order 1 at a, and $\operatorname{res}(\psi, a) = m(a)$. It follows from the residue theorem that, if γ is a circuit in $G \setminus Z(f)$ which is homologous to zero in G, then $\{a \in Z(f) : \operatorname{ind}_\gamma(a) \neq 0\}$ is finite and

$$\frac{1}{2\pi i} \int_\gamma \frac{f'(z)}{f(z)} dz = \sum_{a \in Z(f)} \operatorname{res}(\psi, a) \operatorname{ind}_\gamma(a) = \sum_{a \in Z(f)} m(a) \operatorname{ind}_\gamma(a);$$

moreover, use of Theorem 3.6.20 shows that

$$\frac{1}{2\pi i} \int_\gamma \frac{f'(z)}{f(z)} dz = \operatorname{ind}_{f \circ \gamma}(0) = n(f \circ \gamma, 0).$$

Hence

$$\operatorname{ind}_{f \circ \gamma}(0) = \sum_{a \in Z(f)} m(a) \operatorname{ind}_\gamma(a),$$

an equality which may be used in conjunction with a suitable choice of γ to determine the number of zeros of f, counted according to multiplicity, in certain subsets of G. Assume γ chosen so that $\operatorname{ind}_\gamma(z) \in \{0, 1\}$ for all $z \in \mathbf{C} \setminus \gamma^*$, and let $\Omega = \{z \in \mathbf{C} : \operatorname{ind}_\gamma(z) = 1\}$ so that, in the terminology of Definition 3.4.30, Ω is the set of points "inside" γ. Then, since γ is homologous to zero in G, $\Omega \subset G$; also N_f, the number of zeros of f in Ω, is given by

$$N_f = \sum_{a \in Z(f) \cap \Omega} m(a) = \sum_{a \in Z(f)} m(a)\, \mathrm{ind}_\gamma(a) = \mathrm{ind}_{f \circ \gamma}(0).$$

These remarks establish parts (i) and (ii) of the following theorem.

Theorem 3.7.11 *Let G be a region in $\mathbf{C}$, $f \in H(G)$ and $Z(f) \neq G$. Let γ be a circuit in $G \backslash Z(f)$ which is homologous to zero in G. Then*

(i) *the set $\{a \in Z(f) : \mathrm{ind}_\gamma(a) \neq 0\}$ is finite and*

$$\mathrm{ind}_{f \circ \gamma}(0) = \frac{1}{2\pi i} \int_\gamma \frac{f'(z)}{f(z)} dz = \sum_{a \in Z(f)} m(a)\, \mathrm{ind}_\gamma(0),$$

where $m(a)$ is the multiplicity of the zero of f at a;

(ii) *if $\mathrm{ind}_\gamma(z) \in \{0, 1\}$ for all $z \notin \gamma^*$, $\Omega := \{z \in \mathbf{C} : \mathrm{ind}_\gamma(z) = 1\}$ and N_f denotes the number of zeros of f in Ω, counted according to multiplicity, then $\Omega \subset G$ and $N_f = \mathrm{ind}_{f \circ \gamma}(0)$; and*

(iii) *(Rouché's theorem) if in addition $g \in H(G)$ and $|f(z) - g(z)| < |f(z)|$ for all $z \in \gamma^*$, then $N_g = N_f$.*

Proof Given earlier remarks, it remains to prove Rouché's theorem. Since $|f(z) - g(z)| < |f(z)|$ for all $z \in \gamma^*$, evidently $Z(g) \cap \gamma^* = \emptyset$. Hence (i) and (ii) hold with g in place of f. It follows from Theorems 3.4.26 and 3.6.20 that

$$N_g = \mathrm{ind}_{g \circ \gamma}(0) = n(g \circ \gamma, 0) = n(f \circ \gamma, 0) = \mathrm{ind}_{f \circ \gamma}(0) = N_f. \qquad \square$$

Example 3.7.12 We shall use Rouché's theorem to prove that one root of the equation $z^4 + z^3 + 1 = 0$ lies in the first quadrant.

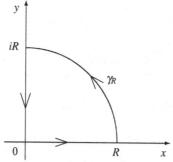

The idea is to apply Theorem 3.7.11 with $G = \mathbf{C}$, γ_R a simple, positively oriented circuit with the track indicated above, $f(z) = z^4 + 1$, $g(z) = z^4 + z^3 + 1$; the choice of f is dictated by the need to have a function known to have a zero in the first quadrant and easy to compare with g. Evidently, f has a single zero at $e^{i\pi/4}$ in the first quadrant and has no zeros on γ_R^* if $R > 1$. Moreover, routine arguments similar to those of Example 3.4.22 (iv) show that $\mathrm{ind}_{\gamma_R}(a) = 1$ for all a in the bounded component B of $\mathbf{C}\backslash\gamma_R^*$: thus $\Omega = B$, as $\mathrm{ind}_{\gamma_R}(a) = 0$ for all a in the unbounded component of $\mathbf{C}\backslash\gamma_R^*$.

Finally, consider $f(z) - g(z) = -z^3$. On the non-negative real axis, where $z = x \geq 0$,

$$\left|-z^3\right| = x^3 < 1 + x^4;$$

on the non-negative imaginary axis, where $z = iy$, $y \geq 0$,

$$\left|-z^3\right| = y^3 < y^4 + 1;$$

and on $|z| = R$, $\left|-z^3\right| < |f(z)|$ for large enough R. Thus all the conditions of Rouché's theorem hold, and so $N_g = N_f$.

Theorem 3.7.13 (The fundamental theorem of algebra) *Let $p(z) = a_0 + a_1 z + \cdots + a_n z^n$, where each a_k is in $\mathbf{C}$, $n \geq 1$ and $a_n \neq 0$. Then p has exactly n roots, allowing for multiplicities.*

Proof Put $f(z) = a_n z^n$ and let γ be a positively oriented circle with centre 0 and radius r so large that $|p(z) - f(z)| < |f(z)|$ for all $z \in \gamma^*$. Apply Theorem 3.7.11 (iii) with $G = \mathbf{C}$: then $N_p = N_f = n$. $\square$

Theorem 3.7.14 (The open mapping theorem) *Let G be a region in $\mathbf{C}$, let $f \in H(G)$ and suppose that f is not constant on G. Then $f(U)$ is open in $\mathbf{C}$ whenever U is an open subset of G (that is, f is an open mapping), and $f(G)$ is a region.*

Proof Let U be an open subset of G and let $a \in U$. To show that $f(U)$ is open, it is enough to establish that $f(a)$ is an interior point of $f(U)$. This is achieved below by proving that there exists $\lambda > 0$ such that $B(f(a), \lambda) \subset f(U)$. By Theorem 3.6.26, there are a unique $m \in \mathbf{N}$ and a function $g \in H(G)$, with $g(a) \neq 0$, such that

$$f(z) - f(a) = (z - a)^m g(z), z \in G.$$

Choose $r > 0$ so that $\overline{B(a, r)} \subset U$ and $g(z) \neq 0$ if $z \in \overline{B(a, r)}$. Define $\gamma :$ $[0, 2\pi] \to \mathbf{C}$ by $\gamma(t) = a + re^{it}$. If $z \in \gamma^*$, then plainly $f(z) \neq f(a)$. Put

$$\lambda = \inf\{|f(z) - f(a)| : z \in \gamma^*\};$$

since γ^* is compact, $\lambda > 0$. Let $w \in B(f(a), \lambda)$. If $z \in \gamma^*$, then

$$|(f(z) - f(a)) - (f(z) - w)| = |f(a) - w| < \lambda \leq |f(z) - f(a)|.$$

Rouché's theorem, Theorem 3.7.11 (iii), is now applicable and shows that, counting according to multiplicity, $f - w$ has exactly m zeros in $B(a, r)$. Thus w is the image under f of at least one point in $B(a, r)$:

$$B(f(a), \lambda) \subset f(B(a, r)) \subset f(U).$$

Hence $f(U)$ is open.

Finally, by the above, $f(G)$ is open; by Theorem 2.4.11, it is connected. It follows that $f(G)$ is a region. □

Corollary 3.7.15 (The inverse function theorem) *Let G be a region in $\mathbf{C}$, let $f \in H(G)$ and suppose that f is injective. Then*

(i) *f' is never zero in G;*
(ii) *$f^{-1} \in H(f(G))$ and*

$$(f^{-1})'(f(a)) = (f'(a))^{-1}, a \in G.$$

Proof (i) To obtain a contradiction, suppose that $a \in G$ and that $f'(a) = 0$. Retracing the steps in the proof of the last theorem and adopting the same notation, let U be any neighbourhood of a in G and observe that $m > 1$, since $f'(a) = 0$. Choose $r > 0$ so that it satisfies the following additional condition:

$$f'(z) \neq 0 \text{ if } z \in B(a, r)\backslash\{a\}.$$

Such a choice of r is possible because a cannot be a limit point of zeros of f': for if a was such a limit point then, by Theorem 3.6.26, f' would be identically zero in G and f could not be injective. Let $w \in B(f(a), \lambda)\backslash\{f(a)\}$. The revised choice of r ensures that the m zeros of $f - w$ in $B(a, r)$ are distinct: if z were such a zero of order at least 2, then $f'(z) = 0$, which contradicts our choice of r. It follows that $m > 1$ is incompatible with the hypothesis that f is injective. Hence (i) is established.

(ii) Theorem 3.7.14 and Lemma 2.1.33 together establish the continuity of f^{-1}: if V is open in G, then $(f^{-1})^{-1}(V) = f(V)$ is open in $f(G)$. Theorem 3.2.6 then deals with its analyticity and provides the formula for its derivative. □

Corollary 3.7.15 gives a necessary condition for an analytic map to be injective, namely, that its derivative is never zero. This condition is not sufficient, as the example of the exponential function shows:

$$z \longmapsto \exp(z) : \mathbf{C} \to \mathbf{C}\backslash\{0\}$$

is not injective.

These facts notwithstanding, if at a point z a map f is analytic and its derivative is non-zero, then f is locally injective at z in the sense of being injective in a neighbourhood of z. Although this result was proved earlier (see Corollary 3.6.30), because of its interest we restate it below and establish it as an outcome of the proof of the open mapping theorem. Note that the exponential function is an example of an analytic non-injective map which is nevertheless locally injective at each point of its domain of definition.

Corollary 3.7.16 *Let V be an open set in $\mathbf{C}$, $a \in V$, $f \in H(V)$ and $f'(a) \neq 0$. Then there exists a neighbourhood $\mathcal{O}$ of a such that the restriction of f to $\mathcal{O}$ is injective.*

Proof We again adopt the notation and retrace the steps of the proof of Theorem 3.7.14, taking G to be that component of V containing a. Using the hypothesis $f'(a) \neq 0$, it follows that $m = 1$. Thus each $w \in B(f(a), \lambda)$ is the image under f of precisely one point in $B(a, r)$. Let

$$\mathcal{O} = B(a, r) \cap f^{-1}(B(f(a), \lambda)).$$

It is plain that $\mathcal{O}$ is a neighbourhood of a and f is injective on $\mathcal{O}$.   □

Definition 3.7.17 Let G be an open set in $\mathbf{C}$ and let f and f_n ($n \in \mathbf{N}$) be complex-valued mappings defined on G. The sequence (f_n) is said to **converge uniformly to** f **on compact subsets of** G if, for each compact set $K \subset G$ and each $\varepsilon > 0$, there exists $N \in \mathbf{N}$ (depending on K and ε) such that, for all $z \in K$ and all $n \geq N$,

$$|f_n(z) - f(z)| < \varepsilon.$$

Equivalently, we require that for each compact set $K \subset G$,

$$\lim_{n \to \infty} \left(\sup_{z \in K} |f_n(z) - f(z)| \right) = 0.$$

Theorem 3.7.18 *Let G be an open set in $\mathbf{C}$ and let (f_n) be a sequence in $H(G)$ which converges uniformly to a mapping $f : G \to \mathbf{C}$ on compact subsets of G. Then $f \in H(G)$ and, for each $k \in \mathbf{N}$, the sequence $\left(f_n^{(k)} \right)_{n \in \mathbf{N}}$ converges uniformly to $f^{(k)}$ on compact subsets of G.*

Proof Plainly f is continuous on G, since convergence is uniform on each closed ball in G. Let Δ be a triangular circuit in G such that co $\Delta^* \subset G$. Using Cauchy's theorem 3.6.8 and the compactness of Δ^*, we see that

$$\int_\Delta f(z)dz = \lim_{n \to \infty} \int_\Delta f_n(z)dz = 0.$$

Thus, by Morera's theorem (Exercise 3.6.37/7), $f \in H(G)$.

Let K be a compact subset of G,

$$v = \frac{1}{2}\text{dist}(K, {}^c G) \quad \text{and} \quad L = \{z \in \mathbf{C} : \text{dist}(z, K) \leq v\}.$$

Then $v > 0$, L is compact and $K \subset L \subset G$. Fix $k \in \mathbf{N}$ and r, $0 < r < v$. Then, by Theorem 3.6.23, for each $n \in \mathbf{N}$ and each $a \in K$,

$$\left| f_n^{(k)}(a) - f^{(k)}(a) \right| \leq \frac{k!}{r^k} \sup_{z \in L} |f_n(z) - f(z)|.$$

Hence, for all $n \in \mathbf{N}$,

$$\sup_{a \in K} \left| f_n^{(k)}(a) - f^{(k)}(a) \right| \leq \frac{k!}{r^k} \sup_{z \in L} |f_n(z) - f(z)|.$$

Since (f_n) converges uniformly to f on L, it follows that $\left(f_n^{(k)} \right)$ converges uniformly to $f^{(k)}$ on K. $\qquad\square$

Theorem 3.7.19 (Hurwitz's theorem) *Let G be an open set in $\mathbf{C}$, let $f_n \in H(G)$ ($n \in \mathbf{N}$) and suppose that (f_n) converges uniformly to f on compact subsets of G. Suppose also that $\overline{B(z_0, r)} \subset G$ and that f is never zero on $\partial B(z_0, r)$. Then there exists $N \in \mathbf{N}$ such that if $n \geq N$, then f_n and f have the same number of zeros in $B(z_0, r)$.*

Proof Let $\min\{|f(z)| : |z - z_0| = r\} = \mu$; $\mu > 0$ since $|f|$ is continuous and never zero on the compact set $\partial B(z_0, r)$. For large enough $n \in \mathbf{N}$,

$$|f(z) - f_n(z)| < \mu \leq |f(z)| \quad \text{if } |z - z_0| = r.$$

Now use Rouché's theorem. $\qquad\square$

Theorem 3.7.20 *Let G be a region in $\mathbf{C}$, let $f_n \in H(G)$ ($n \in \mathbf{N}$) and suppose that (f_n) converges uniformly to f on compact subsets of G. Suppose that for all $n \in \mathbf{N}$, $0 \notin f_n(G)$. Then either f is the zero function on G, or $0 \notin f(G)$.*

Proof Suppose that $f(z_0) = 0$ for some $z_0 \in G$ and that f is not the zero function on G. By Theorem 3.6.26, there exists $r > 0$ such that $\overline{B(z_0, r)} \subset G$ and f is never zero on $\partial B(z_0, r)$. By Theorem 3.7.19, for all large enough $n \in \mathbf{N}$, f_n must have a zero in $B(z_0, r)$, contrary to hypothesis. $\qquad\square$

Theorem 3.7.21 *Let G be a region in $\mathbf{C}$, let $f_n \in H(G)$ ($n \in \mathbf{N}$) and suppose that (f_n) converges uniformly to f on compact subsets of G. If each f_n is injective, then f is either injective or constant on G.*

Proof Let $z_0 \in G$ and put $g_n(z) = f_n(z) - f_n(z_0)$ $(n \in \mathbf{N}, z \in G)$. Then g_n is analytic in $G\backslash\{z_0\}$ and (g_n) converges uniformly to $f - f(z_0)$ on compact subsets of the connected set $G\backslash\{z_0\}$. Since g_n is never zero on $G\backslash\{z_0\}$, Theorem 3.7.20 implies that $f - f(z_0)$ is either the zero function or never zero on $G\backslash\{z_0\}$. As z_0 is an arbitrary point of G, the result follows.   □

A particularly striking use of the residue theorem is to evaluate definite integrals. To help with this it is desirable to be able to calculate residues easily; the following lemma gives a procedure for doing this.

Lemma 3.7.22 *Let $f \in H(B(z_0, \rho)\backslash\{z_0\})$ have a pole of order m at z_0. Then*

$$\mathrm{res}(f, z_0) = \frac{1}{(m-1)!} \lim_{z \to z_0} \left[\frac{d^{m-1}}{dz^{m-1}} \{(z-z_0)^m f(z)\} \right].$$

*In particular, if there is a pole of order 1 at z_0 (a **simple** pole), then*

$$\mathrm{res}(f, z_0) = \lim_{z \to z_0} \{(z-z_0)f(z)\}.$$

Proof The Laurent expansion of f about z_0 is

$$f(z) = a_{-m}(z-z_0)^{-m} + \cdots + a_{-1}(z-z_0)^{-1} + \sum_{n=0}^{\infty} a_n(z-z_0)^n \quad (0 < |z-z_0| < \rho).$$

Thus

$$(z-z_0)^m f(z) = a_{-m} + \cdots + a_{-1}(z-z_0)^{m-1} + \sum_{n=0}^{\infty} a_n(z-z_0)^{m+n},$$

and hence

$$\frac{d^{m-1}}{dz^{m-1}} \{(z-z_0)^m f(z)\} = (m-1)!a_{-1} + (z-z_0)h(z),$$

where h is analytic in $B(z_0, \rho)$. The result follows if we let $z \to z_0$.   □

The first definite integrals we consider are those of the type

$$I = \int_0^{2\pi} R(\cos\theta, \sin\theta)d\theta,$$

where $(x, y) \mapsto R(x, y)$ is a rational function bounded on the circle $x^2 + y^2 = 1$. The idea is to let $\gamma : [0, 2\pi] \to \mathbf{C}$ be the circle $\gamma(\theta) = e^{i\theta}$ and to note that

$$I = \int_\gamma R\left(\frac{1}{2}(z + z^{-1}), \frac{1}{2i}(z - z^{-1})\right)(iz)^{-1}dz.$$

This integral can now be evaluated using the residue theorem. The procedure is illustrated by the following example.

Example 3.7.23 Evaluate

$$I = \int_0^{2\pi} \frac{1}{1 - 2p \cos\theta + p^2} d\theta, \text{ where } 0 < p < 1.$$

Proceeding as suggested above we see that

$$I = \int_\gamma \frac{1}{i(1 - pz)(z - p)} dz.$$

By Remark 3.4.31, the only pole of the integrand inside γ is the simple pole at $z = p$ and $n(\gamma, p) = 1$; by Lemma 3.7.22, the residue at p is

$$\lim_{z \to p} \frac{1}{i(1 - pz)} = \frac{1}{i(1 - p^2)}.$$

Thus, since $\mathrm{ind}_\gamma(p) = n(\gamma, p)$, use of the residue theorem shows that

$$I = 2\pi i \cdot \frac{1}{i(1 - p^2)} = \frac{2\pi}{1 - p^2}.$$

Of course, this integral may be evaluated by traditional real-variable methods, but the above treatment is admirably short and simple.

More complicated examples, typically involving an infinite interval of integration, require limiting processes.

Example 3.7.24

(i) Evaluate

$$I = \int_0^\infty \frac{1}{1 + x^6} dx.$$

Let $R > 1$ and $\gamma : [-R, R + \pi] \to \mathbf{C}$ be the circuit defined by

$$\gamma(s) = \begin{cases} s, & -R \le s \le R, \\ R \exp(i(s - R)), & R \le s \le R + \pi. \end{cases}$$

This circuit was introduced and its track illustrated in Example 3.4.22 (iv); experience shows that it is suited to the evaluation of a number of improper Riemann integrals. The sets

$$G_1 := \{z \in \mathbf{C} : |z| < R, \operatorname{im} z > 0\}$$

and

$$G_2 := \{z \in \mathbf{C} : |z| > R \text{ or im } z < 0\}$$

are the components of $\mathbf{C}\backslash\gamma^*$, G_2 being the unbounded component. By Theorem 3.6.21, $\text{ind}_\gamma(z) = 0$ for all $z \in G_2$; by Remark 3.6.22 (i), $\text{ind}_\gamma(z) = 1$ for all $z \in G_1$. Evidently, γ is formed of a pair of arcs, its restrictions to $[-R, R]$ and $[R, R + \pi]$, respectively. Let μ denote the first of these, and let ν denote the standard reparametrisation of the second relative to the interval $[0, \pi]$:

$$\mu(s) = s (-R \le s \le R); \quad \nu(\theta) = Re^{i\theta} (0 \le \theta \le \pi).$$

Then

$$\int_\gamma f = \int_\mu f + \int_\nu f$$

whenever $f : \gamma^* \to \mathbf{C}$ is continuous.

Let $f(z) = (1 + z^6)^{-1} (z^6 \ne -1)$. The function f has six simple poles, all roots of -1; those in G_1 are $z_1 = e^{i\pi/6}$, $z_2 = e^{i\pi/2}$, $z_3 = e^{5\pi i/6}$, application of Lemma 3.7.22 shows that

$$\text{res}(f, z_k) = \lim_{z \to z_k} \frac{z - z_k}{1 + z^6} = \lim_{z \to z_k} \left\{ \frac{1 + z^6 - (1 + z_k^6)}{z - z_k} \right\}^{-1} = 1/(6z_k^5) = -z_k/6.$$

Hence by the residue theorem,

$$J := \int_\gamma f = 2\pi i \sum_{k=1}^6 \text{ind}_\gamma(z_k) \text{res}(f, z_k) = \frac{-2\pi i}{6} \sum_{k=1}^3 z_k$$

$$= \frac{\pi}{3} \left(1 + 2\sin\frac{\pi}{6}\right) = \frac{2\pi}{3}.$$

However,

$$J = \int_\mu f + \int_\nu f,$$

and

$$\int_\mu f = \int_{-R}^R \frac{1}{1 + x^6} dx = 2 \int_0^R \frac{1}{1 + x^6} dx,$$

while

$$\left| \int_\nu f \right| = \left| \int_0^\pi \frac{Rie^{i\theta}}{1 + R^6 e^{6i\theta}} d\theta \right| \le \int_0^\pi \frac{R}{R^6 - 1} d\theta \to 0$$

as $R \to \infty$. Hence

$$\lim_{R\to\infty}\int_0^R \frac{1}{1+x^6}dx = \frac{\pi}{3}, \text{ and so } I = \frac{\pi}{3}.$$

(ii) Evaluate

$$I = \int_0^\infty \frac{\cos mx}{x^2+a^2}dz \ (m>0, a>0).$$

With $R > \max\{1, a\}$, take γ to be the circuit of (i) above. For $z \neq \{-ia, ia\}$ let $f(z) = e^{imz}/(z^2+a^2)$; this choice is to be preferred when compared with the more obvious one of $(\cos mz)/(z^2+a^2)$. At each of the points $\pm ia$ the function f has a simple pole; ia lies inside γ, $-ia$ lies outside and, by Lemma 3.7.22 ,

$$\text{res}(f, ia) = \frac{e^{-ima}}{2ia}.$$

Thus by the residue theorem,

$$J := \int_\gamma f = \frac{\pi}{a}e^{-ma}.$$

Also $J = \int_\mu f + \int_\nu f$, and

$$\int_\mu f = \int_{-R}^R \frac{\exp(imx)}{x^2+a^2}dx = 2\int_0^R \frac{\cos mx}{x^2+a^2}dx,$$

while

$$\left|\int_\nu f\right| = \left|\int_0^\pi \frac{\exp(im\,Re^{i\theta})}{R^2e^{2i\theta}+a^2}i\,Re^{i\theta}\,d\theta\right|$$

$$\leq \int_0^\pi \frac{R}{R^2-a^2}e^{-mR\sin\theta}d\theta$$

$$\leq \int_0^\pi \frac{R}{R^2-a^2}d\theta \to 0 \text{ as } R\to\infty.$$

Hence

$$\frac{\pi}{a}e^{-ma} = J = 2\lim_{R\to\infty}\int_0^R \frac{\cos mx}{x^2+a^2}dx,$$

so that

$$\int_0^\infty \frac{\cos mx}{x^2+a^2}dx = \frac{\pi}{2a}e^{-ma}.$$

(iii) Evaluate

$$I = \int_0^\infty \frac{x \sin mx}{x^2 + a^2} dx \ (a > 0, m > 0).$$

This looks very similar to the last example, but the inequality work needed is a little more tricky.

Put $f(z) = z e^{imz}/(z^2 + a^2)$ $(z \neq \pm ia)$ and take γ to be as in (ii) above. The function f has a simple pole at ia inside γ and $\mathrm{res}(f, ia) = \frac{1}{2} e^{-ma}$; the other pole at $-ia$ is outside γ. Thus by the residue theorem,

$$J := \int_\gamma f = \pi i e^{-ma}.$$

As before,

$$J = \int_\mu f + \int_\nu f,$$

and

$$\int_\mu f = \int_{-R}^R \frac{x e^{imx}}{x^2 + a^2} dx = 2i \int_0^R \frac{x \sin mx}{x^2 + a^2} dx,$$

while

$$\left| \int_\nu f \right| = \left| \int_0^\pi \frac{i R^2 e^{2i\theta} \exp(im \, Re^{i\theta})}{R^2 e^{2i\theta} + a^2} d\theta \right| \le \int_0^\pi \frac{R^2}{R^2 - a^2} e^{-mR \sin \theta} d\theta.$$

This time it is not enough to estimate $e^{-mR \sin \theta}$ from above by 1; instead, we observe that

$$\int_0^\pi e^{-mR \sin \theta} d\theta = 2 \int_0^{\pi/2} e^{-mR \sin \theta} d\theta$$

and make use of the inequality

$$\frac{2}{\pi} \le \frac{\sin \theta}{\theta} \le 1 \text{ if } 0 < \theta \le \frac{\pi}{2}.$$

This shows that

$$\left| \int_\nu f \right| \le \frac{2R^2}{R^2 - a^2} \int_0^{\pi/2} \exp\left(\frac{-2mR}{\pi} \theta \right) d\theta = \frac{\pi R}{m(R^2 - a^2)} (1 - e^{-mR}) \to 0$$

as $R \to \infty$. Hence

$$\lim_{R \to \infty} \int_0^R \frac{x \sin mx}{x^2 + a^2} dx = \frac{\pi}{2} e^{-ma};$$

that is, $I = \frac{\pi}{2}e^{-ma}$.

This result leads easily to the evaluation of $\int_0^\infty \frac{\sin x}{x}dx$: by Remark 1.6.13, this exists as an improper Riemann integral. Use of the inequality $|(\sin x)/x| \le 1$ $(x > 0)$ shows that, for any $R > 0$,

$$\left| \int_0^R \frac{x \sin x}{x^2 + a^2}dx - \int_0^R \frac{\sin x}{x}dx \right| = a^2 \left| \int_0^R \frac{\sin x}{x(x^2 + a^2)}dx \right| \le a^2 \int_0^R \frac{1}{x^2 + a^2}dx$$

$$= a \tan^{-1}(R/a) \le \pi a/2.$$

Hence

$$\left| \frac{\pi}{2}e^{-a} - \int_0^\infty \frac{\sin x}{x}dx \right| \le \pi a/2$$

for each $a > 0$, from which it follows, on letting $a \to 0$, that

$$\int_0^\infty \frac{\sin x}{x}dx = \frac{\pi}{2}.$$

(iv) Evaluate

$$I = \int_0^\infty \frac{\sin mx}{x(x^2 + a^2)}dx \, (a > 0, m > 0).$$

This problem admits a choice of method. One such is to use the function $z \longmapsto (e^{imz} - 1)/\{z(z^2 + a^2)\}$, which has a removable singularity at 0, in conjunction with the circuit γ adopted in parts (i), (ii) and (iii). Another is to use the slightly simpler function $z \longmapsto e^{imz}/\{z(z^2 + a^2)\}$, with a pole at 0, in combination with a slightly more complicated circuit. For variety, we choose the latter method. For $z \ne 0$, $z \ne \pm ia$ let $f(z) = e^{imz}/\{z(z^2 + a^2)\}$. Since γ cannot be used with f because $0 \in \gamma^*$ and f has a pole at this point, a semicircular detour about the origin is introduced: in place of γ we use the circuit $\sigma : [-R, R + \pi] \to \mathbf{C}$ defined by

$$\sigma(s) = \begin{cases} s, & -R \le s \le -\varepsilon, \\ \varepsilon \exp\left(\frac{i\pi}{2}(1 - s/\varepsilon)\right), & -\varepsilon \le s \le \varepsilon, \\ s, & \varepsilon \le s \le R, \\ R \exp(i(s - R)), & R \le s \le R + \pi, \end{cases}$$

where $0 < \varepsilon < r := \frac{1}{2}\min\{1, a\}$ and $\max\{1, a\} < R < \infty$. This circuit was introduced in Remark 3.6.22 (ii) where its track is portrayed. By Remark 3.6.22 (ii), ia lies inside σ and $\mathrm{ind}_\sigma(ia) = 1$; evidently, $-ia$ and 0 lie outside σ. Since f has a simple pole at each of the points $\pm ia$ and 0, and since $\mathrm{res}(f, ia) = -e^{-ma}/(2a^2)$, appeal to the residue theorem shows that

$$\int_\sigma f = -i\pi e^{-ma}/a^2.$$

Suppose that $\sigma_1, \sigma_2, \sigma_3, \sigma_4$ are the restrictions of σ to $[-R, -\varepsilon], [-\varepsilon, \varepsilon], [\varepsilon, R],$ $[R, R + \pi]$ respectively. Then

$$\int_\sigma f = \sum_{k=1}^{4} \int_{\sigma_k} f.$$

Proceeding as in earlier parts of this example it follows easily that

$$\int_{\sigma_1} f + \int_{\sigma_3} f = 2i \int_\varepsilon^R \frac{\sin mx}{x(x^2 + a^2)} dx.$$

Since f has a simple pole at $z = 0$ and $\mathrm{res}(f, 0) = a^{-2}$, use of Corollary 3.7.3 shows that $f(z) = a^{-2} z^{-1} + g(z)$ $(0 < |z| < a)$, where $g \in H(B(0, a))$. Let $M = \sup\{|g(z)| : |z| \le r\}$. Then

$$\int_{\sigma_2} f = a^{-2} \int_{\sigma_2} z^{-1} dz + \int_{\sigma_2} g$$

$$= -\frac{i\pi}{2a^2\varepsilon} \int_{-\varepsilon}^\varepsilon ds + \int_{\sigma_2} g$$

$$= -\frac{i\pi}{a^2} + \int_{\sigma_2} g$$

and

$$\left| \int_{\sigma_2} g \right| \le M\pi\varepsilon \to 0 \text{ as } \varepsilon \to 0.$$

Hence

$$\int_{\sigma_2} f \to -i\pi/a^2 \text{ as } \varepsilon \to 0,$$

and so

$$2i \int_0^R \frac{\sin mx}{x(x^2 + a^2)} dx - \frac{i\pi}{a^2} + \int_{\sigma_4} f = -\frac{i\pi}{a^2} e^{-ma}$$

for all $R > \max\{1, a\}$. But

$$\left| \int_{\sigma_4} f \right| \le \int_0^\pi \frac{e^{-mR\sin\theta}}{R^2 - a^2} d\theta \le \pi(R^2 - a^2)^{-1}$$

$$\to 0 \text{ as } R \to \infty.$$

We conclude that

$$I = \frac{\pi}{2a^2}\left(1 - e^{-ma}\right).$$

(v) Evaluate

$$I = \int_0^\infty \frac{x^{\lambda-1}}{1+x}dx, \quad \text{where } 0 < \lambda < 1.$$

The existence of I as an improper Riemann integral follows from Exercise 1.6.15/4(a), based on Example 1.6.8. However, its evaluation presents difficulties not previously encountered: no branch of $z^{\lambda-1}/(1+z)$ is analytic in any neighbourhood of the origin and there is a pole on the negative real axis. To resolve these difficulties, the method adopted here involves the use of a pair of circuits whose tracks have line-segment overlap; the circuits are a trifle more complicated than those used earlier.

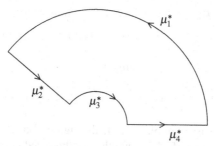

Let $0 < \varepsilon < 1 < R < \infty$. Let μ_k ($k = 1, 2, 3, 4$) be arcs defined by

$$\mu_1(s) = R\exp(3s\pi i/4), \mu_2(s) = \{R(1-s) + \varepsilon s\}\exp(3\pi i/4),$$
$$\mu_3(s) = \varepsilon \exp\{3(1-s)\pi i/4\}, \mu_4(s) = \varepsilon(1-s) + Rs$$

for each $s \in [0, 1]$, and let $\mu : [0, 4] \to \mathbf{C}$ be the simple circuit given by

$$\mu(s) = \mu_k(s - k + 1), k - 1 \le s \le k, k \in \{1, 2, 3, 4\}.$$

Clearly, $\mathbf{C}\backslash\mu^*$ has a single bounded component; μ^* is illustrated above. Let $f(z) = (z^{\lambda-1})_\pi/(1+z)$ ($z \in D(\pi)$). Since $f \in H(D(\pi))$ and μ is homologous to 0 in $D(\pi)$, change of variable and appeal to the global version of Cauchy's theorem shows that

$$\sum_{k=1}^4 \int_{\mu_k} f = \int_\mu f = 0.$$

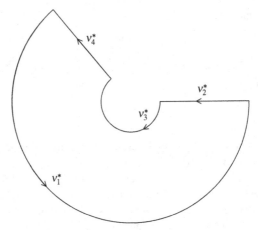

For each $s \in [0, 1]$ let

$$v_1(s) = R \exp\{(3 + 5s)\pi i/4\}, \; v_2(s) = R(1 - s) + \varepsilon s,$$
$$v_3(s) = \varepsilon \exp\{(8 - 5s)\pi i/4\}, \; v_4(s) = \{\varepsilon(1 - s) + Rs\} \exp(3\pi \, i/4),$$

and let $v : [0, 4] \rightarrow \mathbf{C}$ be defined by

$$v(s) = v_k(s - k + 1), k - 1 \leq s \leq k, k \in \{1, 2, 3, 4\}.$$

Evidently, v is a simple circuit; the figure above depicts v^*, its track. The sense of traverse of v^* is such that the bounded component of $\mathbf{C}\backslash v^*$ is on its left. We show that, whenever z lies in this bounded component, $\mathrm{ind}_v(z) = 1$. Identifying index and winding number, the method used is modelled on that given in Example 3.4.22 (iv).
Let ρ be the restriction of v to $[1, 4]$, let $w_1 = R + i$ and $w_2 = 2^{-1/2}R(-1 + i) + i$. Since $\phi := \left(\frac{3\pi}{2} - \arg\right) \circ (\rho + i)$ and $\psi := \left(\frac{\pi}{2} - \arg\right) \circ (v_1 + i)$ are continuous arguments of $\rho + i$ and $v_1 + i$ respectively,

$$n(v, -i) = n(\rho, -i) + n(v_1, -i)$$
$$= \frac{1}{2\pi} \{\phi(4) - \phi(1) + \psi(1) - \psi(0)\} = 1,$$

because

$$\phi(1) = \left(\frac{3\pi}{2} - \arg\right)(w_1) = \left(\frac{\pi}{2} - \arg\right)(w_1) = \psi(1)$$

and

$$\phi(4) = \left(\frac{3\pi}{2} - \arg\right)(w_2) = 2\pi + \left(\frac{\pi}{2} - \arg\right)(w_2) = 2\pi + \psi(0).$$

Use of Theorems 3.6.20 and 3.4.27 shows that, for all z in the bounded component of $\mathbf{C}\setminus\nu^*$,

$$\text{ind}_\nu(z) = n(\nu, z) = 1.$$

Let $g(z) = (z^{\lambda-1})_{5\pi/2}/(1+z)$ $(z \in D(5\pi/2)\setminus\{-1\})$. Since $\text{ind}_\nu(-1) = 1$ and $\text{res}(g, -1) = e^{i\pi(\lambda-1)}$, use of Theorem 3.5.2 (v) and the residue theorem shows that

$$\sum_{k=1}^{4} \int_{\nu_k} g = \int_\nu g = 2\pi i e^{i\pi(\lambda-1)}.$$

Note that μ_2 and ν_4 are opposite arcs and that, for all $z \in \mu_2^* = \nu_4^*$, $(\pi - \arg)(z) = 3\pi/4 = \left(\frac{5\pi}{2} - \arg\right)(z)$ and $f(z) = g(z)$. In defining g, the choice of the $\frac{5\pi}{2}$–branch of $z^{\lambda-1}$ is motivated by the fact that it coincides with the π–branch on $\mu_2^* = \nu_4^*$ and that therefore $\int_{\nu_4} g = -\int_{\mu_2} f$. It follows, adding $\int_\mu f$ to $\int_\nu g$, that

$$\left(\int_{\mu_1} f + \int_{\nu_1} g\right) + \left(\int_{\mu_4} f + \int_{\nu_2} g\right) + \left(\int_{\mu_3} f + \int_{\nu_3} g\right) = 2\pi i e^{i\pi(\lambda-1)}.$$

Now, elementary manipulation shows that

$$\int_{\mu_4} f + \int_{\nu_2} g = \left(1 - e^{2\lambda\pi i}\right)\int_\varepsilon^R \frac{x^{\lambda-1}}{1+x}dx,$$

and that a positive real number M exists such that for $0 < \varepsilon < 1/2$ and $R > 2$,

$$\left|\int_{\mu_1} f\right| + \left|\int_{\nu_1} g\right| \le MR^{\lambda-1}, \quad \left|\int_{\mu_3} f\right| + \left|\int_{\nu_3} g\right| \le M\varepsilon^\lambda.$$

Hence, letting $\varepsilon \to 0$ and $R \to \infty$, we see that

$$(1 - e^{2\lambda\pi i})\int_0^\infty \frac{x^{\lambda-1}}{1+x}dx = -2\pi i e^{i\pi\lambda},$$

and conclude that

$$I = \frac{\pi}{\sin\lambda\pi}.$$

3.7.1 Partial Fraction Decompositions

To conclude this section we indicate how the residue calculus may be used to derive expansions for various functions, expansions which enable us to determine the sums of certain celebrated series. We begin with the standard partial fraction expansion for a rational function.

Let $r(z) = p(z)/q(z)$ be a rational function (the quotient of two polynomials without a common zero). Without loss of generality we may assume that the degree of p is less than that of q, for the general case may be reduced to this by long division. This implies that $\lim_{|z| \to \infty} r(z) = 0$. Let

$$q(z) = c_0(z - \alpha_1)^{m_1} \ldots (z - \alpha_s)^{m_s},$$

where $\alpha_1, \ldots, \alpha_s$ are the distinct poles of r. Suppose that the principal part of r at α_k is

$$h_k(z) = \sum_{l=1}^{m_k} b_l^{(k)}(z - \alpha_k)^{-l} \quad (k = 1, \ldots, s).$$

Then

$$(m_k - l)! b_l^{(k)} = \lim_{z \to \alpha_k} \frac{d^{m_k-l}}{dz^{m_k-l}} \left\{ (z - \alpha_k)^{m_k} r(z) \right\} \quad (l = 1, \ldots, m_k).$$

The function $r - h_k$ has a removable singularity at α_k; clearly h_k is analytic at each α_j with $j \neq k$. Hence $g := r - \sum_{k=1}^{s} h_k$ may be extended to be analytic in $\mathbf{C}$. We claim that g is bounded. For since $r(z) \to 0$ and $h_k(z) \to 0$ as $|z| \to \infty$, $g(z) \to 0$ as $|z| \to \infty$. Thus by Liouville's theorem, g is constant, the constant being 0 as $g(z) \to 0$ as $|z| \to \infty$. Hence

$$r(z) = \sum_{k=1}^{s} \sum_{l=1}^{m_k} b_l^{(k)}/(z - \alpha_k)^l,$$

an equality which exhibits the rational function r as a sum of its principal parts, the usual partial fraction expansion for r in $\mathbf{C}$.

It is natural to ask whether similar expansions can be obtained for non-rational functions, functions with possibly an infinite number of isolated singularities. There are general theorems to this effect: see the treatment of Mittag-Leffler expansions given in [16]. Here we merely give a specific example to illustrate what may be achieved.

Let $S = \mathbf{Z} \setminus \{0\}$ and let $f : \mathbf{C} \setminus S \to \mathbf{C}$ be defined by

$$f(z) = z^{-2}(\pi z \cot \pi z - 1) \text{ if } z \in \mathbf{C} \setminus \mathbf{Z}, \ f(0) = -\pi^2/3.$$

Then $f \in H(\mathbf{C} \backslash S)$; also, S is the set of isolated singularities of f and each of these is a simple pole.

Let $\gamma : [-2, 2] \to \mathbf{C}$ be the circuit defined by

$$\gamma(s) = \begin{cases} 3 + 2s - i, & -2 \le s \le -1, \\ 1 + i(1 + 2s), & -1 \le s \le 0, \\ 1 - 2s + i, & 0 \le s \le 1, \\ -1 + i(3 - 2s), & 1 \le s \le 2, \end{cases}$$

and, for $k \in \mathbf{N}$, let $\gamma_k = (k + 1/2)\gamma$. The track of γ_k is the square whose vertices are the points $(k + 1/2)(\pm 1 \pm i)$. Example 3.4.22 (ii) shows that $n(\gamma, 0) = 1$ and $n(\gamma, 2i) = 0$, and from Theorems 3.4.23, 3.4.27 and 3.6.20 it follows that

$$\mathrm{ind}_{\gamma_k}(z) = n(\gamma_k, z) = \begin{cases} 1 & \text{if } \max\{|\mathrm{re}\, z|, |\mathrm{im}\, z|\} < k + 1/2, \\ 0 & \text{if } \max\{|\mathrm{re}\, z|, |\mathrm{im}\, z|\} > k + 1/2. \end{cases}$$

If $z \in \gamma_k^*$ and $|\mathrm{re}\, z| = k + 1/2$ then, for some $t \in \mathbf{R}$, $z = \pm(k + 1/2) + it$ and

$$|\cot \pi z|^2 = \frac{\cos^2(k + 1/2)\pi + \sinh^2 \pi t}{\sin^2(k + 1/2)\pi + \sinh^2 \pi t} = \tanh^2 \pi t \le 1;$$

moreover, if $z \in \gamma_k^*$ and $|\mathrm{im}\, z| = k + 1/2$ then, for some $t \in \mathbf{R}$, $z = t \pm (k + 1/2)i$ and

$$|\cot \pi z|^2 = \frac{\cos^2 \pi t + \sinh^2(k + 1/2)\pi}{\sin^2 \pi t + \sinh^2(k + 1/2)\pi} \le \frac{1 + \sinh^2(k + 1/2)\pi}{\sinh^2(k + 1/2)\pi} = \coth^2(k + 1/2)\pi$$

$$\le \coth^2(\pi/2)$$

(coth is a decreasing function on $(0, \infty)$ bounded below by 1). Thus for all $z \in \gamma_k^*$ and all $k \in \mathbf{N}$,

$$|f(z)| \le k^{-1} \left(\pi \coth \frac{\pi}{2} + 1 \right).$$

Now let K be a non-empty, bounded subset of $\mathbf{C} \backslash S$ and choose $M \in \mathbf{N}$ so large that $K \subset B(0, M)$. Let $z \in K$ and put $F(w) = (w - z)^{-1} f(w)$ $(w \in \mathbf{C} \backslash S, w \ne z)$. At the point z and at each point of S, F has a simple pole; moreover

$$\mathrm{res}(F, z) = \lim_{w \to z} f(w) = f(z),$$

and for each $k \in S$,

$$\mathrm{res}(F, k) = \lim_{w \to k} (w - k)F(w) = \lim_{w \to k} \frac{(w - k)}{w^2(w - z)} \left(\frac{\pi w \cos \pi w}{\sin \pi w} - 1 \right)$$

$$= \lim_{w \to k} \frac{(w-k)}{w^2(w-z)} \left(\frac{\pi w \cos \pi k \cos \pi w}{\sin \pi (w-k)} - 1 \right)$$

$$= \lim_{w \to k} \frac{\pi (w-k)}{\sin \pi (w-k)} \cdot \frac{\cos \pi k \cos \pi w}{w(w-z)} = \frac{1}{k(k-z)}.$$

By the residue theorem, for all $m > M$,

$$\left| f(z) + \sum_{k=1}^{m} \frac{1}{k(k-z)} + \sum_{k=-m}^{-1} \frac{1}{k(k-z)} \right| = \left| \frac{1}{2\pi i} \int_{\gamma_m} F \right| = \left| \frac{1}{2\pi i} \int_{\gamma_m} \frac{f(w)}{w-z} dw \right|$$

$$\leq \frac{\left(\pi \coth \frac{\pi}{2} + 1 \right) (8m + 4)}{2\pi m (m - M)}.$$

Thus, given any $\varepsilon > 0$, there exists $m_0 \in \mathbf{N}$ such that for all $z \in K$,

$$\left| f(z) + \sum_{k=1}^{m} \frac{2}{k^2 - z^2} \right| < \varepsilon \text{ if } m \geq m_0.$$

It follows that the sequence $\left(-2 \sum_{k=1}^{m} \frac{1}{k^2-z^2} \right)$ converges uniformly to f on any bounded subset of $\mathbf{C}$ which does not contain a pole of f. In particular,

$$f(z) = \frac{\pi \cot \pi z}{z} - \frac{1}{z^2} = \sum_{k=1}^{\infty} \frac{2}{z^2 - k^2} \text{ for all } z \in \mathbf{C} \backslash \mathbf{Z},$$

and

$$f(0) = -\frac{\pi^2}{3} = -\sum_{k=1}^{\infty} \frac{2}{k^2}$$

so that

$$\sum_{k=1}^{\infty} \frac{1}{k^2} = \frac{\pi^2}{6}.$$

Exercise 3.7.25

1. Describe the kind of singularity at 0 for each of the following functions:
   (a) $z^{-3} \sin^2 z$, (b) $z^{-2} - \operatorname{cosec}^2 z$, (c) $\sin(z) \sin(1/z)$.
2. Find res$(f; \alpha)$ in the following cases:
   (a) $f(z) = (z^4 + z^2 + 1)^{-1}$, $\alpha = e^{i\pi/3}$; (b) $f(z) = z^{-2} \cot z$, $\alpha = 0$;
   (c) $f(z) = (z^2 + 1)^{-2} \exp(iz)$, $\alpha = i$; (d) $f(z) = z^3 \exp(1/z)$, $\alpha = 0$.
3. Find the Laurent expansions which represent the function

$$f(z) = (1+z^2)^{-2}(z \in \mathbf{C}\backslash\{-i, i\})$$

in the following annuli:

(a) $\{z \in \mathbf{C} : 0 < |z| < 1\}$, (b) $\{z \in \mathbf{C} : |z| > 1\}$, (c) $\{z \in \mathbf{C} : 0 < |z - i| < 2\}$.
What kind of singularity does f have at i?

Find $\mathrm{res}(f; i)$ and use contour integration to evaluate

$$\int_0^\infty (1+t^2)^{-2} dt.$$

4. Use the method of residues to show that

(i) $\displaystyle\int_0^{2\pi} \frac{1}{5+3\cos\theta} d\theta = \frac{\pi}{2}$,

(ii) $\displaystyle\int_0^{2\pi} \frac{1}{(a+b\cos\theta)^2} d\theta = \frac{2\pi a}{(a^2-b^2)^{3/2}}$ $(0 < b < a)$,

(iii) $\displaystyle\int_0^{2\pi} \frac{\cos 4\theta}{1-2p\cos\theta+p^2} d\theta = \frac{2\pi p^4}{1-p^2}$ $(0 < p < 1)$,

(iv) $\displaystyle\int_0^{\pi} \frac{1}{p^2+\sin^2\theta} d\theta = \frac{\pi}{p(1+p^2)^{1/2}}$ $(p > 0)$.

5. Show that

(i) $\displaystyle\int_0^\infty \frac{1}{x^2+a^2} dx = \frac{\pi}{2a}$ $(a > 0)$,

(ii) $\displaystyle\int_0^\infty \frac{1}{x^4+a^4} dx = \frac{\pi}{2\sqrt{2}a^3}$ $(a > 0)$,

(iii) $\displaystyle\int_0^\infty \frac{\cos mx}{(x^2+a^2)^2} dx = \frac{\pi(1+ma)e^{-ma}}{4a^3}$ $(a > 0, m > 0)$,

(iv) $\displaystyle\int_0^\infty \frac{x^3\sin mx}{(x^2+a^2)^2} dx = \frac{\pi(2-ma)e^{-ma}}{4}$ $(a > 0, m > 0)$.

[The circuit introduced for Example 3.7.24 (i) may be used in each case.]

6. Show that

(i) $\displaystyle\int_0^\infty \frac{\log x}{x^2+a^2} dx = \frac{\pi}{2a}\log a$ $(a > 0)$,

(ii) $\displaystyle\int_0^\infty \frac{\log x}{(x^2+1)^2} dx = -\frac{\pi}{4}$,

(iii) $\displaystyle\int_0^\infty \frac{(\log x)^2}{x^2+1} dx = \frac{\pi^3}{8}$.

[The circuit introduced for Example 3.7.24 (iv) may be used in each case.]

7. Show that

$$\int_0^\infty \left(\frac{\sin x}{x}\right)^2 dx = \frac{\pi}{2}.$$

[Hint: consider the integral of $z^{-2}(1 - \exp(2iz))$ over the circuit introduced in Example 3.7.24 (iv).]

8. Prove that

$$\int_0^\infty \frac{x^{2\lambda-1}}{1+x^2}dx = \frac{\pi}{2}\operatorname{cosec}(\lambda\pi) \ (0 < \lambda < 1),$$

and deduce that

$$\int_0^\infty \frac{x^{\lambda-1}}{1+x}dx = \pi\operatorname{cosec}(\lambda\pi) \ (0 < \lambda < 1).$$

[Cf. Example 3.7.24 (v).]

9. Suppose that $1 < R < \infty$ and let $\gamma_k : [0, 1] \to \mathbf{C}$ $(k = 1, 2, 3)$ be arcs defined by

$$\gamma_1(s) = Rs, \ \gamma_2(s) = R\exp(2\pi si/5), \ \gamma_3(s) = R(1 - s)\exp(2\pi i/5).$$

By integrating $z/(1 + z^5)$ over a simple circuit formed from the γ_k, suitably reparametrised, show that

$$\int_0^\infty \frac{x}{1+x^5}dx = \frac{\pi}{5\sin(2\pi/5)}.$$

10. Suppose that $0 < R < \infty$ and $\lambda = \exp(i\pi/4)$. Let $\sigma_1, \ldots, \sigma_4$ be arcs defined by

$$\sigma_1(s) = s - \lambda R(-1/2 \leq s \leq 1/2), \sigma_2(s) = 1/2 + \lambda s(-R \leq s \leq R),$$
$$\sigma_3(s) = -s + \lambda R(-1/2 \leq s \leq 1/2), \sigma_4(s) = -1/2 - \lambda s(-R \leq s \leq R),$$

and let $f \in H(\mathbf{C}\backslash\mathbf{Z})$ be given by

$$f(z) = \exp(i\pi z^2)\operatorname{cosec}(\pi z).$$

Prove that if $k \in \{1, 3\}$ then

$$\left|\int_{\sigma_k} f\right| \leq \frac{\sqrt{2}}{\pi R}\exp(-\pi R^2),$$

while if $k \in \{2, 4\}$ then

$$\int_{\sigma_k} f = 2i\int_0^R \exp(-\pi s^2)ds.$$

Further, by integrating f over a simple circuit whose track is a parallelogram with vertices $\pm\frac{1}{2} \pm \lambda R$, deduce that

$$\int_0^\infty \exp(-s^2)ds = \frac{\sqrt{\pi}}{2}.$$

[Cf. Example 2.3.32.]

11. Suppose that $0 < R < \infty$, let $f(z) = \exp(iz^2)$ ($z \in \mathbf{C}$) and let $\sigma : [0, 3] \to \mathbf{C}$ be the simple circuit given by

$$\sigma(s) = \begin{cases} Rs, & 0 \le s \le 1, \\ R\exp(i\pi(s-1)/4), & 1 \le s \le 2, \\ R(3-s)\exp(i\pi/4), & 2 \le s \le 3. \end{cases}$$

Show that $\int_\sigma f = 0$ and, given that $\int_0^\infty \exp(-x^2)dx = \sqrt{\pi}/2$, deduce that

$$\int_0^\infty \cos(x^2)dx = \int_0^\infty \sin(x^2)dx = \frac{1}{2}\sqrt{\frac{\pi}{2}}.$$

12. Suppose that $0 < R < \infty$ and $0 < \phi \le \pi$. Let $\gamma_k : [0, 1] \to \mathbf{C}$ ($k = 1, 2, 3$) be arcs defined by

$$\gamma_1(s) = Rs, \ \gamma_2(s) = R\exp(i\phi s), \ \gamma_3(s) = R(1-s)\exp(i\phi).$$

Let $f(z) = (iz)^{-1}(\exp(iz) - 1)$ ($z \ne 0$), $f(0) = 1$. By integrating f over a simple circuit formed from the γ_k suitably reparametrised, show that

$$\int_0^\infty t^{-1}\exp(-t\sin\phi)\sin(t\cos\phi)dt = \frac{1}{2}\pi - \phi.$$

Deduce that if a and b are real and positive,

$$\int_0^\infty e^{-ax}\frac{\sin bx}{x}dx = \frac{1}{2}\pi - \tan^{-1}\left(\frac{a}{b}\right)$$

13. Let a, λ and R be positive real numbers. By integrating the function $e^{i\lambda z}/(\cosh z + \cosh a)$ over a circuit whose track is composed of the segments $[-R, R]$, $[R, R+2\pi i]$, $[R+2\pi i, -R+2\pi i]$ and $[-R+2\pi i, -R]$, show that

$$\int_0^\infty \frac{\cos\lambda t}{\cosh t + \cosh a}dt = \frac{\pi\sin\lambda a}{\sinh\pi\lambda\sinh a}.$$

14. Let $n \in \mathbf{N}$ and $0 < \alpha < \pi$. Prove that the residue of the function $(z^2 - 2z\cos\alpha + 1)^{-2}\exp(inz)$ at the pole which lies in the upper half plane is $-i\lambda\exp(in\cos\alpha)$, where

$$\lambda = \frac{(1 + n\sin\alpha)\exp(-n\sin\alpha)}{4\sin^3\alpha}.$$

Hence show that, if $\alpha \neq \pi/2$, then

$$\int_0^\infty \frac{x(x^2+1)\sin nx}{(x^4 - 2x^2 \cos 2\alpha + 1)^2} dx = \frac{\pi \lambda \sin(n \cos \alpha)}{4 \cos \alpha}.$$

Find the value of the integral in the excluded case.

15. Let $\alpha \in \mathbf{C}\backslash\{0\}$. Prove that

$$\exp(\alpha(z + z^{-1})) = a_0 + \sum_{n=1}^\infty a_n(z^n + z^{-n}) \ (z \in \mathbf{C}\backslash\{0\}),$$

where the a_n are given by

$$a_n = \frac{1}{\pi} \int_0^\pi \exp(2\alpha \cos \theta) \cos n\theta d\theta = \alpha^n \sum_{j=0}^\infty \frac{\alpha^{2j}}{(n+j)!j!}.$$

Show also that

$$\exp(\alpha(z - z^{-1})) = b_0 + \sum_{n=1}^\infty b_n(z^n + (-z)^{-n}) \ (z \in \mathbf{C}\backslash\{0\}),$$

where the b_n are given by

$$b_n = \frac{1}{\pi} \int_0^\pi \cos(n\theta - 2\alpha \sin \theta) d\theta = \alpha^n \sum_{j=0}^\infty \frac{(-1)^j \alpha^{2j}}{(n+j)!j!}.$$

16. Prove that all the solutions of $z^5 + (1+i)z - 16 = 0$ lie in the annulus $1 < |z| < 2$.
17. Use Rouché's theorem to show that, if $n \in \mathbf{N}$ and $|\alpha| > e$, then $\alpha z^n = e^z$ has, counted according to multiplicity, exactly n zeros in the disk $|z| < 1$.
18. Let $\lambda > 1$. Show that the equation $\lambda - z - e^{-z} = 0$ has exactly one solution in the right half-plane, and that this solution lies in $B(\lambda, 1)$ and is real.
19. Prove that

$$\pi \operatorname{cosec} \pi z = \frac{1}{z} + \sum_{k=1}^\infty (-1)^k \frac{2z}{z^2 - k^2} \ (z \in \mathbf{C}\backslash\mathbf{Z}),$$

and show also that

$$\frac{\pi^2}{12} = \sum_{k=1}^\infty \frac{(-1)^{k+1}}{k^2}.$$

3.8 Simply-Connected Regions: The Riemann Mapping Theorem

Throughout this section we shall assume that G is a region in $\mathbf{C}$. From Remark 3.6.33 (iii) and Corollary 3.6.36, it is known that, if G is simply-connected, then every $f \in H(G)$ has a primitive in G. The objective of this section is to establish the converse; namely, if every $f \in H(G)$ has a primitive in G, then G is simply-connected. In pursuit of that objective, celebrated results, notably Montel's theorem and the Riemann mapping theorem, are obtained as intermediate steps.

Definition 3.8.1 A family of functions $\mathscr{F} \subset H(G)$ is said to be **bounded in a subset** S of G if there exists a real number $M > 0$ such that, for all $f \in \mathscr{F}$,

$$|f|_S := \sup_{z \in S} |f(z)| \le M.$$

The family $\mathscr{F}$ is said to be **locally bounded in** G if, given any point in G, $\mathscr{F}$ is bounded in a neighbourhood of that point.

Remark 3.8.2 Let $\mathscr{F} \subset H(G)$. It is elementary, and left to the reader, to verify that the three statements which follow are equivalent:

(a) $\mathscr{F}$ is locally bounded in G.
(b) $\mathscr{F}$ is bounded in every compact subset of G.
(c) $\mathscr{F}$ is bounded in every open ball with closure contained in G.

A useful variant of (c) is based on the observation that, given any ball $B(a, r)$ such that $\overline{B(a, r)} \subset G$, there exists a ball $B(c, s)$ with centre c having rational real and imaginary parts and rational radius s, such that

$$B(a, r) \subset B(c, s) \subset \overline{B(c, s)} \subset G.$$

It follows that (a), (b) and (c) above are equivalent to
(d) $\mathscr{F}$ is bounded in every open ball $B(z, R)$ with $\overline{B(z, R)} \subset G$ and re(z), im(z) and R all rational.
  Evidently, the family of balls described in statement (d) is countable.

Lemma 3.8.3 *Let $\mathscr{F} \subset H(G)$ be locally bounded in G. Then $\mathscr{F}$ is equicontinuous at each point of G: given any $a \in G$ and any $\varepsilon > 0$, there exists $\delta > 0$ such that, for all $z \in B(a, \delta)$,*

$$\sup_{f \in \mathscr{F}} |f(z) - f(a)| < \varepsilon.$$

Proof Let $r > 0$ be chosen so that $\overline{B(a, 2r)} \subset G$, and let $\gamma(t) = a + 2re^{it}$ $(0 \le t \le 2\pi)$. Use of the Cauchy integral formula (Theorem 3.6.15) shows that, for all $f \in \mathscr{F}$ and all $z \in B(a, r)$,

$$f(z) - f(a) = \frac{1}{2\pi i} \int_\gamma \left\{ \frac{1}{\zeta - z} - \frac{1}{\zeta - a} \right\} f(\zeta) d\zeta = \frac{(z-a)}{2\pi i} \int_\gamma \frac{f(\zeta)}{(\zeta - z)(\zeta - a)} d\zeta.$$

Thus, appealing to Theorem 3.5.10, for all such f and z,

$$|f(z) - f(a)| \le \frac{|z-a|}{r} |f|_{\overline{B(a,2r)}},$$

since $|(\zeta - z)(\zeta - a)| \ge 2r^2$ whenever $\zeta \in \gamma^*$. As $\mathscr{F}$ is locally bounded,

$$M := 1 + \sup_{f \in \mathscr{F}} |f|_{\overline{B(a,2r)}} < \infty.$$

Hence, for all $z \in B(a, r)$,

$$\sup_{f \in \mathscr{F}} |f(z) - f(a)| \le M |z - a| / r,$$

from which the result follows immediately: put $\delta = r \min\{1, \varepsilon/M\}$. $\square$

Definition 3.8.4 A family $\mathscr{F} \subset H(G)$ is called **normal** if every sequence of functions in $\mathscr{F}$ has a subsequence which converges uniformly on each compact subset of G.

Note that it is **not** required that the limit function of the subsequence belongs to $\mathscr{F}$. Nevertheless, in view of Theorem 3.7.18, this limit function does belong to $H(G)$.

Theorem 3.8.5 *(Montel's theorem) A family $\mathscr{F} \subset H(G)$ is normal if and only if it is locally bounded.*

Proof To obtain a contradiction, suppose that $\mathscr{F} \subset H(G)$ is normal and not locally bounded. Then there is a compact set $K \subset G$ such that $\sup_{f \in \mathscr{F}} |f|_K = \infty$ and, in consequence, a sequence (f_n) in $\mathscr{F}$ such that

$$|f_n|_K \ge n (n \in \mathbf{N}).$$

Since $\mathscr{F}$ is normal, there exists $f \in H(G)$ and a subsequence $(f_{m(n)})$ of (f_n) which converges uniformly to f on K:

$$\lim_{n \to \infty} |f_{m(n)} - f|_K = 0.$$

But for all $n \in \mathbf{N}$,

$$n \le m(n) \le |f_{m(n)}|_K \le |f|_K + |f_{m(n)} - f|_K,$$

and so $|f|_K = \infty$, which is impossible.

It remains to show that if $\mathscr{F}$ is locally bounded, then it is normal. Let (f_n) be a sequence of functions in $\mathscr{F}$ and let $\{B_j : j \in \mathbf{N}\}$ be the family of all open balls with closures contained in G, with centres having rational real and imaginary parts, and with rational radii. By Lemma 3.8.3, $\mathscr{F}$ is equicontinuous at each point of G and is therefore equicontinuous on each $\overline{B_j}$. As (f_n) is uniformly bounded on $\overline{B_1}$, it follows from the Arzelà-Ascoli theorem (Theorem 2.3.22) that (f_n) has a subsequence that is uniformly convergent on $\overline{B_1}$. (We note that while Lemma 2.3.21 and Theorem 2.3.22 are stated and proved for real-valued functions only, extension to the complex case is elementary since the proofs are formally identical.) We now apply the same argument to this subsequence, considered this time on $\overline{B_2}$. Proceeding in this way, we see from the Cantor diagonalisation technique that (f_n) has a subsequence that is uniformly convergent on each B_j. Since every compact subset K of G can be covered by a finite collection of the B_j, it follows that this subsequence is uniformly convergent on each such K. Hence $\mathscr{F}$ is normal. □

Theorem 3.8.6 (Schwarz's lemma) *Let $B = B(0, 1)$ and $\lambda \in \mathbf{C}$. Let $f \in H(B)$ be such that $f(B) \subset B$, $f(0) = 0$ and $f'(0) = \lambda$. Then*

(i) $|\lambda| \leq 1$ *and*

$$|f(z)| \leq |z| \, (z \in B).$$

(ii) $|\lambda| = 1$ *if and only if there exists $z_0 \in B \backslash \{0\}$ such that $|f(z_0)| = |z_0|$; further, in either event,*

$$f(z) = \lambda z \, (z \in B).$$

Proof (i) Let $g : B \to \mathbf{C}$ be given by

$$g(z) = z^{-1} f(z) \text{ if } z \neq 0; \, g(0) = \lambda.$$

Evidently g is continuous; moreover, it is analytic in $B \backslash \{0\}$. Hence, using Theorem 3.6.9 and the remarks following it, $g \in H(B)$. Let $0 < r < 1$. Since $f(B) \subset B$, application of Theorem 3.6.28 (the maximum modulus theorem) shows that, for all $z \in B(0, r)$,

$$|g(z)| \leq \sup_{|w|=r} |g(w)| \leq r^{-1}.$$

Letting $r \to 1-$, it follows that

$$|g(z)| \leq 1 (z \in B).$$

Hence $|\lambda| \leq 1$ and

$$|f(z)| \leq |z| \, (z \in B).$$

(ii) Suppose that $|g(w)| = 1$ for some $w \in B$. Then $|g|$ has a local maximum in B and, since Theorem 3.6.28 applies, g is a constant mapping. Hence $|\lambda| = |g(0)| = 1$ implies that, for all $z \in B$, $|g(z)| = 1$ and therefore $|f(z)| = |z|$; alternatively,

$|f(z_0)| = |z_0|$ for some $z_0 \in B\setminus\{0\}$ implies that $1 = |g(z_0)| = |g(0)| = |\lambda|$. In either event, $g(z) = g(0) = \lambda$ $(z \in B)$ and so

$$f(z) = \lambda z \ (z \in B).$$

□

Definition 3.8.7 Let G be a region in **C**. An analytic bijective map $f : G \to G$ is called an **automorphism of** G.

Note that by Corollary 3.7.15, the inverse f^{-1} of such a map f belongs to $H(G)$.

Lemma 3.8.8 *Let $\alpha \in$ **C**, $|\alpha| < 1$, and let $\phi_\alpha : B = B(0, 1) \to B$ be given by*

$$\phi_\alpha(z) = (z - \alpha)(\overline{\alpha}z - 1)^{-1}.$$

Then ϕ_α is an automorphism of B.

Proof Evidently $\phi_\alpha \in H(B)$. If $z \in B$, then

$$|\overline{\alpha}z - 1|^2 - |z - \alpha|^2 = (1 - |\alpha|^2)(1 - |z|^2) > 0,$$

and so $|\phi_\alpha(z)| < 1$. Hence $\phi_\alpha(B) \subset B$. Routine calculation shows that $\phi_\alpha(\phi_\alpha(z)) = z$ $(z \in B)$, which means that ϕ_α is its own inverse.   □

We now show that up to rotations, the maps ϕ_α are the sole automorphisms of B: by a rotation is meant a map of the form $z \longmapsto e^{it}z$, where $t \in$ **R**.

Theorem 3.8.9 *Let $f : B \to B$ be an automorphism of B and let $\alpha = f^{-1}(0)$. Then there is a real number t such that*

$$f(z) = e^{it}\phi_\alpha(z) \ (z \in B),$$

where ϕ_α is defined as in the last lemma.

Proof Let $h := f \circ \phi_\alpha$; h is an automorphism of B and $h(0) = 0$. The first part of the Schwarz lemma applies to both h and h^{-1}. Thus, for all $z \in B$,

$$|h(z)| \leq |z| \quad \text{and} \quad \left|h^{-1}(z)\right| \leq |z|,$$

so that

$$|h(z)| = |z| \ (z \in B).$$

By the second part of Schwarz's lemma, there is a real number t such that $h(z) = e^{it}z$ $(z \in B)$. Hence

$$f(z) = h(\phi_\alpha(z)) = e^{it}\phi_\alpha(z) \ (z \in B).$$

□

Lemma 3.8.10 *Let $\alpha, \beta \in \mathbf{C}$, $|\alpha| < 1$, $|\beta| < 1$ and let ϕ_α, ϕ_β be automorphisms of B defined as in Lemma 3.8.8. Let $j : B \to B$ be given by $j(z) = z^2$, and let Ψ be defined on B by $\Psi = \phi_\alpha \circ j \circ \phi_\beta$. Then $|\Psi'(0)| < 1$.*

Proof Clearly $\Psi \in H(B)$ and $\Psi(B) \subset B$; also, since j is not injective, neither is Ψ. Put $\gamma = \Psi(0)$ and let ϕ_γ be the automorphism of B given by $\phi_\gamma(z) = (z - \gamma)(\overline{\gamma}z - 1)^{-1}$ $(z \in B)$. Set $f = \phi_\gamma \circ \Psi$. Then $f \in H(B)$, $f(B) \subset B$ and $f(0) = 0$. Application of the Schwarz lemma to f gives

$$\left| \phi_\gamma'(\gamma)\Psi'(0) \right| \leq 1,$$

and routine calculation shows that

$$\left| \phi_\gamma'(\gamma) \right| = (1 - |\gamma|^2)^{-1}.$$

Hence

$$\left| \Psi'(0) \right| \leq 1 - |\gamma|^2 \leq 1.$$

If $\left| \Psi'(0) \right| = 1$, then $\Psi(0) = \gamma = 0$ and, by the Schwarz lemma, Ψ has the form $\Psi(z) = \lambda z$, where $|\lambda| = 1$. But Ψ is not injective and so cannot have this form. Thus $\left| \Psi'(0) \right| < 1$. $\qquad \square$

Definition 3.8.11 A region G in $\mathbf{C}$ is said to have the square root property if each zero-free $f \in H(G)$ has a square root; that is, there exists $g \in H(G)$ such that $g^2 = f$.

Lemma 3.8.12 *Let G have the square root property and suppose that $G \neq \mathbf{C}$. Then there is an injective map $g \in H(G)$ such that $g(G) \subset B$.*

Proof Let $a \in \mathbf{C} \backslash G$. By hypothesis, there exists $h \in H(G)$ such that $h^2(z) = z - a$ $(z \in G)$; plainly, h is injective. By the open mapping theorem, $h(G)$ contains an open ball $B(c, r)$. Since h is zero-free, evidently $0 \notin B(c, r)$.

Let $w \in B(-c, r)$. We show that $w \notin h(G)$. To obtain a contradiction, suppose that $z \in G$ and $h(z) = w$. Since $-w \in B(c, r) \subset h(G)$, there exists $z_1 \in G$ such that $-w = h(z_1)$. The equality $h^2(z) = h^2(z_1)$ shows that $z = z_1$, that $w = h(z) = h(z_1) = -w$, and that $w = 0$. But $0 \notin B(c, r)$ and we have a contradiction. It follows that $B(-c, r) \cap h(G) = \emptyset$: for all $z \in G$, $|h(z) + c| \geq r$.

For $0 < |v| < 1$, define $g_v : G \to \mathbf{C}$ by

$$g_v(z) = vr(h(z) + c)^{-1}.$$

Then $g_v \in H(G)$, g_v is injective and $g_v(G) \subset B$. Evidently, any g_v has the properties desired of g. $\qquad \square$

Lemma 3.8.13 *Let G have the square root property and suppose that $G \neq \mathbf{C}$. Let $\mathscr{U} = \{ f \in H(G) : f \text{ is injective and } f(G) \subset B \}$. Suppose that $z_0 \in G$, $g \in \mathscr{U}$ and $g(G) \neq B$. Then there exists $f \in \mathscr{U}$ such that $|g'(z_0)| < |f'(z_0)|$.*

Proof Let $\alpha \in B \backslash g(G)$ and let ϕ_α be the automorphism of B given by

$$\phi_\alpha(z) = (\bar{\alpha}z - 1)^{-1}(z - \alpha) \ (z \in B).$$

Evidently $\phi_\alpha \circ g \in H(G)$ and, since $g(z) \neq \alpha \ (z \in G)$, $\phi_\alpha \circ g$ is zero-free. Hence there exists $h \in H(G)$ such that

$$h^2 = \phi_\alpha \circ g.$$

As both ϕ_α and g are injective, so also is h; further, $h(G) \subset B$, since $h^2(G) \subset B$. It follows that $h \in \mathcal{U}$.

Put $\beta = h(z_0)$. With ϕ_β defined in the obvious way, let $f = \phi_\beta \circ h$: note that $f \in \mathcal{U}$ and $f(z_0) = 0$. With $j : B \to B$ given by $j(z) = z^2 \ (z \in B)$,

$$g = \phi_\alpha \circ h^2 = \phi_\alpha \circ (\phi_\beta \circ f)^2 = (\phi_\alpha \circ j \circ \phi_\beta) \circ f = \Psi \circ f,$$

where $\Psi = \phi_\alpha \circ j \circ \phi_\beta$. It follows that

$$g'(z_0) = \Psi'(f(z_0)) f'(z_0) = \Psi'(0) f'(z_0).$$

Now, by Lemma 3.8.10, $\left| \Psi'(0) \right| < 1$. Thus, since g is injective and therefore, by Corollary 3.7.15, $g'(z_0) \neq 0$,

$$\left| g'(z_0) \right| < \left| f'(z_0) \right|. \qquad \qquad \qquad \square$$

Theorem 3.8.14 (The Riemann mapping theorem) *Let G have the square root property and suppose that $G \neq \mathbf{C}$. Then G is analytically isomorphic to B in the sense that there exists an analytic injective map g of G onto B (Corollary 3.7.15, the inverse function theorem, shows that $g^{-1} \in H(B)$).*

Proof Lemma 3.8.12 ensures the existence of an analytic injective map of G into B. Let v be such a map. Let $z_0 \in G$ and put $\alpha = \left| v'(z_0) \right|$; by Corollary 3.7.15, $\alpha > 0$. Let

$$\mathcal{V} = \{ f \in H(G) : f \text{ is injective, } f(G) \subset B \text{ and } \left| f'(z_0) \right| \geq \alpha \};$$

$\mathcal{V} \neq \emptyset$, since $v \in \mathcal{V}$. We establish below that there exists $g \in \mathcal{V}$ such that

$$\left| g'(z_0) \right| = \sup_{f \in \mathcal{V}} \left| f'(z_0) \right|. \qquad \qquad (3.8.1)$$

This fact accepted, then $g(G) = B$. For otherwise, by Lemma 3.8.13, there exists an analytic injective map h of G into B such that

$$\left| h'(z_0) \right| > \left| g'(z_0) \right| \geq \alpha.$$

But this implies that $h \in \mathcal{V}$, in which event

$$\left|g'(z_0)\right| \geq \left|h'(z_0)\right|,$$

a contradiction. It follows that G and B are analytically isomorphic, as required.

It remains to prove (3.8.1). Since the family $\mathcal{V}$ is bounded and therefore locally bounded in G, by Montel's theorem, $\mathcal{V}$ is normal. Hence the set $\{\left|f'(z_0)\right| : f \in \mathcal{V}\}$ is bounded: otherwise, there exists a sequence (f_n) in $\mathcal{V}$ such that $\left|f_n'(z_0)\right| \geq n$ $(n \in \mathbf{N})$ and so, using Theorem 3.7.18, no subsequence of either (f_n') or (f_n) converges uniformly on compact subsets of G. Let $\Lambda = \sup_{f \in \mathcal{V}} \left|f'(z_0)\right|$ and let (g_n) be a sequence in $\mathcal{V}$ such that

$$\left|g_n'(z_0)\right| \to \Lambda.$$

The sequence (g_n) has a subsequence, itself labelled (g_n) for convenience, which converges uniformly on compact subsets of G. Let g be its limit function. By Theorem 3.7.18, $g \in H(G)$ and (g_n') converges uniformly to g' on compact subsets of G. Hence

$$\left|g_n'(z_0)\right| \to \left|g'(z_0)\right|$$

and so $\Lambda = \left|g'(z_0)\right|$. Further, since $|g_n(z)| < 1$ $(z \in G, n \in \mathbf{N})$ and $\left|g_n'(z_0)\right| \geq \alpha$ $(n \in \mathbf{N})$, it follows that $g(G) \subset \bar{B}$ and $\left|g'(z_0)\right| \geq \alpha$. As $\left|g'(z_0)\right| \neq 0$, g is not constant. Hence, by Theorem 3.7.14 (the open mapping theorem), $g(G) \subset B$; also, by Theorem 3.7.21, g is injective. Summarising, $g \in \mathcal{V}$ and

$$\left|g'(z_0)\right| = \sup_{f \in \mathcal{V}} \left|f'(z_0)\right|. \qquad \square$$

Theorem 3.8.15 *Let G be a region in $\mathbf{C}$. The following statements are equivalent:*

(i) *G is simply-connected.*
(ii) *$\mathrm{ind}_\gamma(z) = 0$ for all $z \in \mathbf{C} \backslash G$ and all circuits γ in G;*
(iii) *$\int_\gamma f = 0$ for all $f \in H(G)$ and all circuits γ in G;*
(iv) *every $f \in H(G)$ has a primitive on G;*
(v) *every zero-free $f \in H(G)$ has an analytic logarithm on G;*
(vi) *every zero-free $f \in H(G)$ has an analytic square root on G;*
(vii) *either $G = \mathbf{C}$ or G is analytically isomorphic to B;*
(viii) *G is homeomorphic to B.*

Proof (i)$\Longrightarrow$(ii)

Given (i), since every closed path in G is null-homotopic, Proposition 3.4.24 and Theorem 3.4.28 show that, whenever $z \in \mathbf{C} \backslash G$ and γ is a closed path in G, $n(\gamma, z) = 0$; if in addition, γ is a circuit in G, Theorem 3.6.20 identifies $n(\gamma, z)$ and $\mathrm{ind}_\gamma(z)$. The validity of (ii) follows.

(ii)$\Longrightarrow$(iii)

This follows from Theorem 3.6.32, the global form of Cauchy's theorem.

(iii)$\Longrightarrow$(iv)

This is immediate from Theorem 3.6.2, the fundamental theorem of contour integration.

(iv)$\Longrightarrow$(v)

This is a consequence of Theorem 3.6.17.

(v)$\Longrightarrow$(vi)

Let $f \in H(G)$, $0 \notin f(G)$. Since (v) holds, there exists $g \in H(G)$ such that $\exp(g) = f$. Set $h = \exp(g/2)$. Then $h \in H(G)$ and $h^2 = f$: thus h is an analytic square root of f.

(vi)$\Longrightarrow$(vii)

This is the Riemann mapping theorem, Theorem 3.8.14.

(vii)$\Longrightarrow$(viii)

C and B are homeomorphic: an explicit homeomorphism is given by

$$z \longmapsto (1 + |z|)^{-1} z : \mathbf{C} \to B.$$

When $G \neq \mathbf{C}$, the result is plain.

(viii)$\Longrightarrow$(i)

By Theorem 2.4.23, G is path-connected. Let $f : G \to B$ be a homeomorphism, let $\gamma : [0, 1] \to G$ be a closed path in G and let $z_0 = f^{-1}(0)$. The map $f \circ \gamma$ is a path in B which is freely homotopic to the constant path $s \longmapsto 0 : [0, 1] \to B$ under $H : [0, 1] \times [0, 1] \to B$ given by $H(s, t) = (1 - t)f(\gamma(s))$. Let $\widetilde{H} = f^{-1} \circ H$. It is trivial to verify that $\widetilde{H}$ is a homotopy between γ and the constant path $s \longmapsto z_0 : [0, 1] \to G$. Thus G is simply-connected.   □

As a consequence of Theorem 3.8.15 it follows that every simply-connected region $G \neq \mathbf{C}$ in **C** is analytically isomorphic to the unit disc $B(0, 1)$. This assertion is often referred to in the literature as the Riemann mapping theorem. We refer to the books by Burckel [3] and Remmert [13] for historical discussion of this famous result. Theorem 3.8.15 is extraordinary because of the equivalence it establishes between different kinds of properties: analytic ((ii), (iii), (iv), (v), (vii)), topological ((i), (viii)) and algebraic (vi). It has exceptional aesthetic appeal.

3.9 The Jordan Curve Theorem

The famous Jordan curve theorem (briefly mentioned in Remark 3.4.31) states that if γ is a simple closed path in **C**, then $\mathbf{C} \backslash \gamma^*$ has exactly two connected components, $\mathscr{I}$ and $\mathscr{O}$; the first of these is bounded and is called the inside of γ, while $\mathscr{O}$ is unbounded and referred to as the outside of γ. Moreover, γ^* is the boundary of each of these components; the winding number $n(\gamma, z)$ is zero for all $z \in \mathscr{O}$; either $n(\gamma, z) = 1$ for all $z \in \mathscr{I}$ or $n(\gamma, z) = -1$ for all $z \in \mathscr{I}$. While this is immediate for such elementary paths as circles, nevertheless to establish it as stated is by no means a trivial matter, even though it may appear intuitively obvious. We refer to Burckel's book [3], and the references given there, for accounts of the history of this theorem. Over the years many proofs have been devised, largely falling into two categories: those designated as 'elementary' but often involving extraordinary ingenuity and complicated geometric constructions; and those that employ more advanced material. The lack of aesthetic appeal is a common feature of much of this work. Here we try to avoid the more obvious pitfalls in approaching this topic and, inspired by the approach of Burckel, attempt to give an account that grows in a natural way out of the material in the earlier part of the chapter.

3.9.1 Closed Paths and Continuous Maps on S^1

We remind the reader that the notation S^1 was introduced in Example 2.4.21 (iv): identifying $\mathbf{R}^2$ and $\mathbf{C}$ as metric spaces, $S^1 = \{z \in \mathbf{C} : |z| = 1\}$.

Theorem 3.9.1 *Let (X, d) be a metric space and denote by I the closed interval $[0, 1]$. Let $\Phi : I \times I \to X$ be continuous and suppose that, for all $t \in I$,*

$$\Phi(0, t) = \Phi(1, t).$$

Then there is a continuous $F : S^1 \times I \to X$ such that, for all $(s, t) \in I \times I$,

$$F(\exp(2\pi i s), t) = \Phi(s, t).$$

Proof The map $(s, t) \longmapsto (\exp(2\pi i s), t) : [0, 1) \times I \to S^1 \times I$ is bijective. Hence, if $(z, t) \in S^1 \times I$, then there exists a unique $(s, t) \in [0, 1) \times I$ such that

$$(\exp(2\pi i s), t) = (z, t).$$

Let $F(z, t) = \Phi(s, t)$. Since $\Phi(0, t) = \Phi(1, t)$ for all $t \in I$, it follows that for all $(s, t) \in I \times I$,

$$F(\exp(2\pi i s), t) = \Phi(s, t).$$

It remains to prove that F is continuous. Suppose that F is not continuous at $(z_0, t_0) \in S^1 \times I$. Then an $\varepsilon > 0$ and a sequence $((z_n, t_n))$ in $S^1 \times I$ exist such that $(z_n, t_n) \to (z_0, t_0)$ and

$$d(F(z_n, t_n), F(z_0, t_0)) \geq \varepsilon \ (n \in \mathbf{N}). \tag{3.9.1}$$

For each $n \in \mathbf{N}$ let s_n be that unique element of $[0, 1)$ such that $z_n = \exp(2\pi i s_n)$. Some subsequence of (s_n), $(s_{m(n)})$ say, is convergent: suppose that $s_{m(n)} \to s_0 \in I$. Then

$$(s_{m(n)}, t_{m(n)}) \to (s_0, t_0),$$

$$(z_{m(n)}, t_{m(n)}) = (\exp(2\pi i s_{m(n)}), t_{m(n)}) \to (\exp(2\pi i s_0), t_0) = (z_0, t_0)$$

and
$$d(F(z_{m(n)}, t_{m(n)}), F(z_0, t_0)) = d(\Phi(s_{m(n)}, t_{m(n)}), \Phi(s_0, t_0)) \to 0,$$

which contradicts (3.9.1).   □

Corollary 3.9.2 *Let X be a metric space and let $\phi : I \to X$ be a closed path in X. Then there exists a continuous map $f : S^1 \to X$ such that for all $s \in I$,*

$$f(\exp(2\pi i s)) = \phi(s).$$

Proof Let $\Phi : I \times I \to X$ be defined by $\Phi(s, t) = \phi(s)$. Then Φ is continuous and, for all $t \in I$, $\Phi(0, t) = \Phi(1, t)$. The theorem just proved shows that there is a continuous $F : S^1 \times I \to X$ such that, for all $(s, t) \in I \times I$,

$$F(\exp(2\pi i s), t) = \Phi(s, t) = \phi(s).$$

Let $f : S^1 \to X$ be defined by $f(z) = F(z, 0)$. Then f is continuous and $f(\exp(2\pi i s)) = \phi(s)$ for all $s \in I$.   □

Theorem 3.9.3 *Let X be a metric space. For $j \in \{0, 1\}$, let $f_j : S^1 \to X$ be continuous, $\phi_j : I \to X$ be a closed path in X, and suppose that*

$$f_j(\exp(2\pi i s)) = \phi_j(s) \ (s \in I).$$

Then f_0 and f_1 are homotopic if, and only if, ϕ_0 and ϕ_1 are freely homotopic.

Proof Suppose $f_0 \simeq f_1$, so that there is a continuous map $F : S^1 \times I \to X$ such that
$$F(z, 0) = f_0(z), \ F(z, 1) = f_1(z) \ (z \in S^1).$$

Define $\Phi : I \times I \to X$ by $\Phi(s, t) = F(\exp(2\pi i s), t)$. As it is a composition of continuous maps, Φ is continuous. Moreover,

$$\Phi(s, 0) = \phi_0(s), \ \Phi(s, 1) = \phi_1(s) \ (s \in I)$$

and
$$\Phi(0, t) = F(1, t) = \Phi(1, t) \ (t \in I).$$

Thus ϕ_0 and ϕ_1 are freely homotopic.

Conversely, suppose ϕ_0 and ϕ_1 are freely homotopic. Then there is a continuous map $\Phi : I \times I \to X$ such that

$$\Phi(s, 0) = \phi_0(s), \ \Phi(s, 1) = \phi_1(s) \ (s \in I)$$

and
$$\Phi(0, t) = \Phi(1, t) \ (t \in I).$$

By Theorem 3.9.1, there is a continuous map $F : S^1 \times I \to X$ such that, for all $(s, t) \in I \times I$,

$$F(\exp(2\pi i s), t) = \Phi(s, t).$$

Let $z \in S^1$. A unique $s \in [0, 1)$ exists such that $\exp(2\pi i s) = z$. Evidently

$$f_0(z) = \phi_0(s) = \Phi(s, 0) = F(z, 0)$$

and

$$f_1(z) = \phi_1(s) = \Phi(s, 1) = F(z, 1).$$

Hence F is a homotopy between f_0 and f_1.   □

3.9.2 Existence of Continuous Logarithms

Theorem 3.9.4 *Let $U \subset \mathbf{C}$ be open and $f : U \to \mathbf{C}\backslash\{0\}$ be continuous. Then the following statements are equivalent:*

(i) $n(f \circ \gamma, 0) = 0$ *for every closed path γ in U.*
(ii) *f has a continuous logarithm.*

Proof Suppose that (i) holds and that U is connected. It is enough to establish (ii) in the case that U is connected because of Theorem 2.4.27 : for, if this case is established, then, more generally, a continuous logarithm exists on each component of U and therefore on U itself.

Let $z_0 \in U$. By Theorem 3.4.6, since $f(z_0) \neq 0$, there exists $w_0 \in \mathbf{C}$ such that $\exp(w_0) = f(z_0)$. By Theorem 2.4.23, since U is open and connected, it is path-connected. Associate with each $z \in U$ a path $\gamma_z : [0, 1] \to U$ joining z_0 to z, so that $\gamma_z(0) = z_0, \gamma_z(1) = z$. We show that the family $\{\gamma_z : z \in U\}$ determines a continuous logarithm of f. Evidently, $f \circ \gamma_z$ is a path in $\mathbf{C}\backslash\{0\}$ and, by Theorem 3.4.19, it has a continuous logarithm, ϕ_z say, so that

$$\exp(\phi_z(t)) = f(\gamma_z(t)) \ (0 \le t \le 1).$$

Since $\exp(\phi_z(0)) = f(\gamma_z(0)) = f(z_0) = \exp(w_0)$, say, this logarithm may be adjusted so that $\phi_z(0) = w_0$. Put

$$\psi(z) = \phi_z(1) \ (z \in U).$$

Then

$$\exp(\psi(z)) = \exp(\phi_z(1)) = f(z) \ (z \in U),$$

and it remains to show that ψ is continuous.

Let $z_1 \in U$ and $r > 0$ be such that $B(z_1, r) \subset U$ and $f(B(z_1, r)) \subset B(f(z_1), |f(z_1)|)$. By Theorem 3.4.13, there is a branch L of the logarithm on $B(f(z_1), |f(z_1)|)$:

$$\exp(L(w)) = w \quad (w \in B(f(z_1), |f(z_1)|)).$$

It follows that, for all $z \in B(z_1, r)$,

$$\exp(L(f(z))) = f(z) = \exp(\phi_z(1)).$$

Since these equalities hold at the point z_1 we may and shall suppose L is so chosen that

$$L(f(z_1)) = \phi_{z_1}(1).$$

Let $z \in B(z_1, r)$ and let $v : [0, 1] \to \mathbf{C}$ be the path joining z_1 to z defined by $v(t) = z_1 + t(z - z_1)$. Define a closed path γ by

$$\gamma(t) = \begin{cases} \gamma_z(1 - t), & 0 \le t \le 1, \\ \gamma_{z_1}(t - 1), & 1 \le t \le 2, \\ v(t - 2), & 2 \le t \le 3, \end{cases}$$

and a function ϕ by

$$\phi(t) = \begin{cases} \phi_z(1 - t), & 0 \le t \le 1, \\ \phi_{z_1}(t - 1), & 1 \le t \le 2, \\ L(f(v(t - 2))), & 2 \le t \le 3. \end{cases}$$

The function ϕ is continuous on $[0, 3]$, being continuous on each of the subintervals $[0, 1]$, $[1, 2]$ and $[2, 3]$, and suitably defined at the points 1 and 2. Further, it is simple to check that it is a continuous logarithm of $f \circ \gamma$:

$$\exp(\phi(t)) = f(\gamma(t)) \, (0 \le t \le 3).$$

Since (i) holds,

$$0 = 2\pi i n(f \circ \gamma, 0) = \phi(3) - \phi(0) = L(f(z)) - \phi_z(1),$$

and therefore

$$\psi(z) = \phi_z(1) = L(f(z)).$$

It follows that $L \circ f$ and ψ coincide on $B(z_1, r)$; moreover, since $L \circ f$ is continuous on $B(z_1, r)$, so also is ψ. Since z_1 is an arbitrarily chosen point of U, ψ is continuous on U and (ii) holds.

Conversely, suppose (ii) holds. Let $g : U \to \mathbf{C}$ be a continuous logarithm of f, and let $\gamma : [0, 1] \to U$ be a closed path. Then

$$\exp(g(\gamma(t))) = f(\gamma(t))(0 \le t \le 1);$$

$g \circ \gamma$ is a continuous logarithm of $f \circ \gamma$; and

$$0 = g(\gamma(1)) - g(\gamma(0)) = 2\pi i n(f \circ \gamma, 0),$$

thus establishing (i). □

Theorem 3.9.5 *Let $U \subset \mathbf{C}$ be open and simply connected, and let $f : U \to \mathbf{C}\backslash\{0\}$ be continuous. Then f has a continuous logarithm.*

Proof Let γ be a closed path in U. Since U is simply connected, γ is freely homotopic to a constant path in U, and so $f \circ \gamma$ is freely homotopic to a constant path in $\mathbf{C}\backslash\{0\}$. Appeal to Theorem 3.4.28, which establishes the homotopy invariance of the winding number, and to Proposition 3.4.24, shows that $n(f \circ \gamma, 0) = 0$. The proof is now completed by use of Theorem 3.9.4. □

We consider next zero-free continuous functions on compact subsets of $\mathbf{C}$, with initial attention paid to functions on the compact sets $S^1 = \{z \in \mathbf{C} : |z| = 1\}$ and ∂Q, where $Q = [-1, 1]^2$.

Theorem 3.9.6 *Let $f : S^1 \to \mathbf{C}\backslash\{0\}$ be continuous and odd in the sense that $f(-z) = -f(z)$ for all $z \in S^1$. Then f does not have a continuous square root and so does not have a continuous logarithm.*

Proof To obtain a contradiction, suppose f has a continuous square root g, so that $g^2 = f$ and g is zero-free. Let $\phi : S^1 \to \mathbf{C}\backslash\{0\}$ be defined by

$$\phi(z) = g(-z)/g(z). \tag{3.9.2}$$

Then ϕ is continuous and $\phi^2 = -1$, so that ϕ is a continuous square root of the function -1. Since ϕ is continuous and S^1 is connected, either $\phi(S^1) = \{-i\}$ or $\phi(S^1) = \{i\}$. Whichever is the case,

$$-1 = \phi(z)\phi(-z) \text{ for all } z \in S^1;$$

but this contradicts the equality

$$\phi(z)\phi(-z) = 1 \text{ for all } z \in S^1$$

which is an immediate consequence of (3.9.2). Thus f does not have a continuous square root.

Lastly, f does not have a continuous logarithm: if it did have one, say h, then $\exp(h/2)$ would be a continuous square root. □

Note that this theorem applies to the case in which f is the identity map on S^1.

Theorem 3.9.7 (i) *Let* $f : S^1 \to \mathbf{C}\setminus\{0\}$ *be continuous. Then there is an integer m such that*

$$z \longmapsto z^m f(z) : S^1 \to \mathbf{C}\setminus\{0\}$$

has a continuous logarithm.

(ii) *Let* $g : S^1 \to \mathbf{C}\setminus\{0\}$ *be continuous, let p be an integer other than 0, and suppose that* g^p *has a continuous logarithm. Then g itself has a continuous logarithm.*

Proof (i) Let $\gamma : [0, 1] \to \mathbf{C}\setminus\{0\}$ be the closed path defined by $\gamma(t) = f(\exp(2\pi i t))$ and let $m = -n(\gamma, 0)$. By Theorem 3.4.19, γ has a continuous logarithm and so

$$\gamma(t) = \exp(2\pi i \psi(t)) \, (0 \le t \le 1),$$

where ψ is continuous and
$$-m = \psi(1) - \psi(0).$$

It follows that

$$\exp(2\pi i m t) f(\exp(2\pi i t)) = \exp(2\pi i (\psi(t) + m t)) \, (0 \le t \le 1). \tag{3.9.3}$$

The function $t \longmapsto 2\pi i (\psi(t) + m t) : [0, 1] \to \mathbf{C}$ has the same value at $t = 0$ as at $t = 1$. Thus, by Corollary 3.9.2, there exists a continuous $\phi : S^1 \to \mathbf{C}$ such that

$$\phi(\exp(2\pi i t)) = 2\pi i (\psi(t) + m t) \, (0 \le t \le 1).$$

Substitution of this equality in (3.9.3) shows that

$$\exp(2\pi m i t) f(\exp(2\pi i t)) = \exp(\phi(\exp(2\pi i t))) \, (0 \le t \le 1),$$

which translated to S^1 gives

$$z^m f(z) = \exp(\phi(z)) \, (z \in S^1).$$

Hence ϕ is a continuous logarithm of the map $z \longmapsto z^m f(z)$.

(ii) We may and shall assume that $p > 0$. Let $\phi : S^1 \to \mathbf{C}$ be a continuous logarithm of $g^p : \exp(\phi) = g^p$. The set of roots of $z^p = 1$ is

$$\{\exp(2\pi k i/p) : k = 1, 2, \ldots, p\}.$$

For $k \in \{1, 2, \ldots, p\}$ let

$$A_k := \{z \in S^1 : g(z) = \exp((\phi(z) + 2k\pi i)/p)\}.$$

Each of the A_k is closed; they are pairwise disjoint and have union S^1. Since S^1 is connected, exactly one of the $A_k = S^1$ and the rest are empty. □

Theorem 3.9.7 (i) has an analogue for squares.

Theorem 3.9.8 *Let $Q = [-1, 1]^2$ and suppose that $f : \partial Q \to \mathbf{C} \backslash \{0\}$ is continuous. Then there exists an integer m such that*

$$z \longmapsto z^m f(z) : \partial Q \to \mathbf{C} \backslash \{0\}$$

has a continuous logarithm.

Proof The map $\sigma : \partial Q \to S^1$ given by $\sigma(z) = z/ |z|$ is a homeomorphism of ∂Q onto S^1 (see Example 2.1.39 (iii)). Applying part (i) of the preceding theorem to $f \circ \sigma^{-1}$ we see that there exists an integer m and a continuous map $\psi : S^1 \to \mathbf{C}$ such that

$$w^m f(\sigma^{-1}(w)) = \exp(\psi(w)) \ (w \in S^1).$$

Thus for all $z \in \partial Q$,

$$(\sigma(z))^m f(z) = \exp(\psi(\sigma(z)))$$

and therefore

$$z^m f(z) = |z|^m \exp(\psi(\sigma(z))) = \exp(\psi(\sigma(z)) + m \log |z|) = \exp(\phi(z)),$$

where ϕ has the obvious definition and is a continuous logarithm of the map $z \longmapsto z^m f(z)$. $\qquad \Box$

Theorem 3.9.9 *Let $K \subset \mathbf{C}$ be one or other of the compact sets $[-1, 1]^2$ and $\overline{B(0, 1)}$, and suppose that $f : K \to \mathbf{C} \backslash \{0\}$ is continuous. Then f has a continuous logarithm.*

Proof By the Tietze extension theorem (extended in an obvious manner to complex-valued functions), there exists a continuous function $F : \mathbf{C} \to \mathbf{C}$ such that $F \mid_K = f$. Let $V = F^{-1}(\mathbf{C} \backslash \{0\})$. Then V is open, $\text{dist}(K, {}^c V) > 0$ and, where U is the interior of K, there exists $\lambda > 1$ such that $K \subset \lambda U \subset V$. By Theorem 3.9.5, $F \mid_{\lambda U}$ has a continuous logarithm and thus so does f. $\qquad \Box$

We have seen via Theorems 3.9.5 and 3.9.6 that whether or not a zero-free continuous map has a continuous logarithm is influenced by its domain. If an extension to a disk or to $\mathbf{C}$ is possible, then the existence of continuous logarithms is assured. Extension theorems come into their own in the context of compact domains of definition. The theorem that follows preserves not only the property of being continuous but also that of being zero-free.

Theorem 3.9.10 (The homotopy extension theorem) *Suppose $A \subset B \subset \mathbf{R}^n$, with A compact and B closed. Let the functions $f : A \to \mathbf{C} \backslash \{0\}$ and $g : B \to \mathbf{C} \backslash \{0\}$ be continuous, and suppose that f and the restriction $g \mid_A$ of g to A are homotopic, so that there exists a continuous map $h : A \times [0, 1] \to \mathbf{C} \backslash \{0\}$ such that*

$$h(x, 0) = f(x), h(x, 1) = g(x) \ (x \in A).$$

Then there exists a continuous function $F : B \to \mathbf{C}\backslash\{0\}$ such that $F \mid_A = f$.

Proof Let $T = (A \times [0, 1]) \cup (B \times \{1\})$ and extend the domain of h to T by defining

$$h(x, 1) = g(x)\, (x \in B) :$$

the glueing lemma shows that this extension is continuous. Applied to the real and imaginary parts of h, the Tietze extension theorem establishes that there is a continuous map $H : \mathbf{R}^{n+1} \to \mathbf{C}$ whose restriction to T is h. Evidently $T \subset H^{-1}(\mathbf{C}\backslash\{0\})$, since $H(T) = h(T) \subset \mathbf{C}\backslash\{0\}$. For each $k \in \mathbf{N}$ put

$$A_k = \{x \in \mathbf{R}^n : d(x, A) \leq k^{-1}\};$$

each A_k is compact and

$$\cap_{k=1}^{\infty} A_k \times [0, 1] = A \times [0, 1] \subset T \subset H^{-1}(\mathbf{C}\backslash\{0\}).$$

Since $^c\left(H^{-1}(\mathbf{C}\backslash\{0\})\right)$ is closed and its distance from the compact set $A \times [0, 1]$ is positive, there exists $k \in \mathbf{N}$ such that

$$A_k \times [0, 1] \subset H^{-1}(\mathbf{C}\backslash\{0\}).$$

Let $x \in B$. Then

$$(x, \min\{1, kd(x, A)\}) \in (A_k \times [0, 1]) \cup (B \times \{1\})$$

and, if we put

$$F(x) = H\,(x, \min\{1, kd(x, A)\})\,,$$

then a continuous zero-free extension of f to B is obtained.   □

Theorem 3.9.11 *Let $c \in \mathbf{C}\backslash\{0\}$, K be a compact subset of $\mathbf{C}$ and $f : K \to \mathbf{C}\backslash\{0\}$ be continuous. With the understanding that 'the constant map c' refers to the map $z \longmapsto c : K \to \mathbf{C}\backslash\{0\}$, the following statements are equivalent:*

(i) *f and the constant map c are homotopic.*
(ii) *f has a continuous, zero-free extension to $\mathbf{C}$; that is, there is a continuous function $F : \mathbf{C} \to \mathbf{C}\backslash\{0\}$ whose restriction to K equals f.*
(iii) *f has a continuous logarithm.*

Proof The results that (i) implies (ii) and (ii) implies (iii) are immediate consequences of Theorems 3.9.10 and 3.9.5, respectively. To show that (iii) implies (i), let $\phi : K \to \mathbf{C}$ be a continuous logarithm of f, so that

$$f(z) = \exp(\phi(z))\, (z \in K),$$

and let $w \in \mathbf{C}$ be such that $c = \exp(w)$. Such a w exists in view of Theorem 3.4.6. Define $h : K \times [0, 1] \to \mathbf{C} \backslash \{0\}$ by

$$h(z, t) = \exp((1 - t)w + t\phi(z)).$$

Evidently h is a homotopy between f and the constant map c. $\qquad\square$

Corollary 3.9.12 *Let K be a compact subset of $\mathbf{C}$ and let functions $f_0, f_1 : K \to \mathbf{C} \backslash \{0\}$ be continuous and homotopic. Then f_0 has a continuous logarithm if and only if f_1 has.*

Proof Suppose that f_0 has a continuous logarithm and that $c \in \mathbf{C} \backslash \{0\}$. By Theorem 3.9.11, f_0 and the constant map c are homotopic. Hence, by Theorem 2.5.5, f_1 and the constant map c are also homotopic, and so a further appeal to Theorem 3.9.11 shows that f_1 has a continuous logarithm. $\qquad\square$

Theorem 3.9.13 *Let K be a compact subset of $\mathbf{C}$, $f : K \to \mathbf{C} \backslash \{0\}$ be continuous and suppose 0 lies in the unbounded component of $\mathbf{C} \backslash f(K)$. Then f has a continuous logarithm.*

Proof Let $r > \sup |f(K)|$ and let $f + r$ denote the map $z \longmapsto f(z) + r : K \to \mathbf{C} \backslash \{0\}$. Let U be the unbounded component of $\mathbf{C} \backslash f(K)$. Since U is open and connected, it is path-connected. Let $\gamma : [0, 1] \to U$ be a path joining 0 to $r : \gamma(0) = 0$, $\gamma(1) = r$. Define maps $g, h : K \times [0, 1] \to \mathbf{C} \backslash \{0\}$ as follows:

$$g(z, t) = f(z) + \gamma(t), h(z, t) = (1 - t)f(z) + r.$$

Plainly, g is a homotopy between f and $f + r$, and h is a homotopy between $f + r$ and the constant map r. By Theorem 2.5.5 it follows that f and the constant map r are homotopic, and application of Theorem 3.9.11 now gives the result. $\qquad\square$

Theorem 3.9.14 *Let K be a compact subset of $\mathbf{C}$ and let $f : K \to \mathbf{C} \backslash \{0\}$ be continuous. Then there are finitely many points $p_1, \ldots, p_N \in \mathbf{C} \backslash K$ and integers $n_1, \ldots, n_N$ such that the map $F : K \to \mathbf{C} \backslash \{0\}$ defined by*

$$F(z) = f(z) \prod_{j=1}^{N} (z - p_j)^{n_j}$$

has a continuous logarithm.

Proof Let $\rho > \sup |K|$. Since $z \longmapsto (2\rho)^{-1}(z + \rho(1 + i))$ is a homeomorphism from $[-\rho, \rho]^2$ onto $Q := [0, 1]^2$, we may and shall assume that $K \subset (0, 1)^2$. By the Tietze theorem for complex-valued functions, a continuous $f_0 : Q \to \mathbf{C}$ exists such that $f_0 \mid_K = f$. Let $L = f_0^{-1}(\{0\})$. If $L = \emptyset$, then the result is immediate: after an obvious transformation Theorem 3.9.9 may be applied and so f_0 has a continuous logarithm, as does f.

Suppose that $L \neq \emptyset$. Since L is closed in Q and $L \cap K = \emptyset$, $\mathrm{dist}(K, L) > 0$ and there exists $\delta > 0$ such that

$$|z - w| > \delta \quad (z \in K, w \in L).$$

Let $m \in \mathbf{N}$, $m > \sqrt{2}/\delta$ and, for $j, k = 1, 2, \ldots, m$, consider the cells

$$Q_{jk} = [m^{-1}(j - 1), m^{-1}j] \times [m^{-1}(k - 1), m^{-1}k]$$

with centres

$$p_{jk} = m^{-1}((j - 1/2) + i(k - 1/2))$$

and vertices $v_{j-1,k-1}, v_{j-1,k}, v_{j,k-1}, v_{j,k}$, where $v_{j,k} = m^{-1}(j + ik)$. Put

$$\mathscr{K} = \{(j, k) : 1 \leq j, k \leq m, Q_{jk} \cap K \neq \emptyset\}$$

and

$$\mathscr{L} = \{(j, k) : 1 \leq j, k \leq m, Q_{jk} \cap K = \emptyset\}.$$

Let

$$K_1 = \cup_{(j,k) \in \mathscr{K}} Q_{jk}.$$

It is plain that K_1 is closed, $K \subset K_1 \subset Q \backslash L$ and $f_1 := f_0 |_{K_1}$ is a continuous zero-free extension of f to K_1. Let

$$K_2 = K_1 \cup \left(\cup_{j,k=1}^{m} \partial Q_{jk} \right).$$

We next seek a continuous zero-free function f_2 on the closed set K_2 whose restriction to K_1 is f_1. Let $f_2(z) = f_1(z)$ for all $z \in K_1$, and let $f_2(v_{j,k}) = 1$ whenever $v_{j,k} \notin K_1$. To complete the definition of f_2 on the ∂Q_{jk} it will be convenient to use the symbol $\sigma(a, b)$ to denote the line segment with complex endpoints a and b :

$$\sigma(a, b) = \{(1 - t)a + tb : 0 \leq t \leq 1\}.$$

The intersection of each 'horizontal' line segment $\sigma(v_{j-1,k}, v_{j,k})$ $(1 \leq j \leq m, 0 \leq k \leq m)$ and each 'vertical' line segment $\sigma(v_{j,k-1}, v_{j,k})$ with K_1 is either the segment itself or is contained in the set of endpoints of the segment. Let $\sigma(a, b)$ be a segment of the type $\sigma(a, b) \cap K_1 \subset \{a, b\}$. Then $f_2(a)$ and $f_2(b)$ are non-zero complex numbers and f_2 is not so far defined on $\sigma(a, b) \backslash \{a, b\}$. Since $\mathbf{C} \backslash \{0\}$ is path-connected, there is a path $v : [0, 1] \to \mathbf{C} \backslash \{0\}$ joining $f_2(a)$ to $f_2(b)$. For $z \in \sigma(a, b)$, define $f_2(z) = v((z - a)/(b - a))$. Thus a map $f_2 : K_2 \to \mathbf{C} \backslash \{0\}$ has been defined, it is continuous and its restriction to K_1 is f_1.

Finally, let $(j, k) \in \mathscr{L}$. Then $\partial Q_{jk} = K_2 \cap Q_{jk}$ and Theorems 3.9.8 and 3.9.11 show, through use of the map

$$w \longmapsto p_{jk} + (2m)^{-1}w : [-1, 1]^2 \to Q_{jk},$$

that an integer n_{jk} exists such that

$$z \longmapsto (z - p_{jk})^{n_{jk}} f_2(z) : \partial Q_{jk} \to \mathbf{C}\backslash\{0\}$$

has a continuous zero-free extension $F_{jk} : Q_{jk} \to \mathbf{C}\backslash\{0\}$, say. Consider the map $G_{jk} : Q_{jk} \to \mathbf{C}\backslash\{0\}$ defined by

$$G_{jk}(z) = F_{jk}(z) \prod_{(r,s)\in\mathscr{L}\backslash\{(j,k)\}} (z - p_{rs})^{n_{rs}};$$

if $z \in \partial Q_{jk}$, then

$$G_{jk}(z) = f_2(z) \prod_{(r,s)\in\mathscr{L}} (z - p_{rs})^{n_{rs}}.$$

It follows that a function $F_0 : Q \to \mathbf{C}\backslash\{0\}$ is well-defined by

$$F_0(z) = \begin{cases} f_2(z) \prod_{(r,s)\in\mathscr{L}}(z - p_{rs})^{n_{rs}} & (z \in K_2), \\ G_{jk}(z) & (z \in Q_{jk}, (j, k) \in \mathscr{L}); \end{cases}$$

the glueing lemma shows it to be continuous; and an application of Theorem 3.9.9 extended in the obvious way, shows it to have a continuous logarithm. Let F be the restriction of F_0 to K. Then F has a continuous logarithm and is of the form required. $\square$

The next result strengthens Theorem 3.9.14, casting out those p_j which lie in the unbounded component of $\mathbf{C}\backslash K$ and replacing by a single representative those which lie in the same bounded component.

Theorem 3.9.15 *Let $K \subset \mathbf{C}$ be compact, let $\mathscr{C}$ be the set of all **bounded** components of $\mathbf{C}\backslash K$ and associate with each $C \in \mathscr{C}$ a point $p_C \in C$. Then, given any continuous function $f : K \to \mathbf{C}\backslash\{0\}$, there exist components $C_1, \ldots, C_M \in \mathscr{C}$ and integers $m_1, \ldots, m_M$ such that the function $F : K \to \mathbf{C}\backslash\{0\}$ defined by*

$$F(z) = f(z) \prod_{k=1}^{M} (z - p_{C_k})^{m_k}$$

has a continuous logarithm. (If $\mathscr{C}$ is void, then the product is to be interpreted as 1.)

Proof Let $p_1, \ldots, p_N$ and $n_1, \ldots, n_N$ be as in Theorem 3.9.14, so that $\widetilde{F} : K \to \mathbf{C}\backslash\{0\}$ defined by

$$\widetilde{F}(z) = f(z) \prod_{j=1}^{N} (z - p_j)^{n_j}$$

has a continuous logarithm. By Theorem 3.9.13, if some p_j exists which lies in the unbounded component of $\mathbf{C}\backslash K$, then

$$z \longmapsto z - p_j : K \to \mathbf{C}\backslash\{0\}$$

has a continuous logarithm and, therefore, so also does

$$z \longmapsto \widetilde{F}(z)(z - p_j)^{-n_j} : K \to \mathbf{C}\backslash\{0\}.$$

It follows henceforth that we may assume that each p_j lies in a bounded component of $\mathbf{C}\backslash K$.

Let C_k $(1 \le k \le M)$ be those elements of $\mathscr{C}$ which contain at least one of the points p_j and, for each j $(1 \le j \le N)$ let $g_j : K \to \mathbf{C}\backslash\{0\}$ be defined by

$$g_j(z) = (z - p_j)^{-1}(z - p_{C_k}),$$

where $p_j \in C_k$. Since C_k is open, connected and thus path-connected, there is a path $v : [0, 1] \to C_k$ such that $v(0) = p_{C_k}, v(1) = p_j$. The map $H : K \times [0, 1] \to \mathbf{C}\backslash\{0\}$ defined by

$$H(z, t) = (z - p_j)^{-1}(z - v(t))$$

is evidently a homotopy between g_j and the constant map 1, and so, by Theorem 3.9.11, g_j has a continuous logarithm. Setting $m_k = \sum_{\{j:p_j \in C_k\}} n_j$, it follows that $F : K \to \mathbf{C}\backslash\{0\}$ given by

$$F(z) = \widetilde{F}(z) \prod_{j=1}^{N} (g_j(z))^{n_j} = f(z) \prod_{k=1}^{M} (z - p_{C_k})^{m_k}$$

has a continuous logarithm, as required. $\square$

An immediate consequence of the last theorem is that the connectedness of $\mathbf{C}\backslash K$ (which implies that $\mathbf{C}\backslash K$ has no bounded component) is a sufficient condition to ensure that every continuous $f : K \to \mathbf{C}\backslash\{0\}$ has a continuous logarithm. This condition turns out to be also necessary: the following result is an aid in establishing this.

Theorem 3.9.16 *Let $K \subset \mathbf{C}$ be compact, C be a bounded component of $\mathbf{C}\backslash K$, $p \in C$, and define $f : K \to \mathbf{C}\backslash\{0\}$ by $f(z) = z - p$. Then, for all non-zero integers n, f^n does not have a continuous zero-free extension to $K \cup C$.*

Proof We may plainly suppose that $p = 0$. To obtain a contradiction, let n be a non-zero integer and suppose that $F : K \cup C \to \mathbf{C}\backslash\{0\}$ is a continuous map that extends f^n. Note that since $\overline{C}$ has empty intersection with any component of $\mathbf{C}\backslash K$ other than C, $\overline{C} \subset K \cup C$. Let $r > \sup |C|$ and, for $z \in \overline{B(0, r)}$, define

$$g(z) = \begin{cases} z^n & \text{if } z \in \overline{B(0,r)}\backslash C, \\ F(z) & \text{if } z \in \overline{C}. \end{cases}$$

Since $\overline{C} \cap \left(\overline{B(0,r)}\backslash C\right) = \overline{C}\backslash C \subset K$ and $F(z) = z^n$ whenever $z \in K$, there is consistency of definition. Because F is zero-free and $0 \in C$, the map g defined on $\overline{B(0,r)}$ is also zero-free; further, by the glueing lemma it is continuous. Appeal to Theorem 3.9.9 shows that g has a continuous logarithm and so a continuous $\phi : \overline{B(0,r)} \to \mathbf{C}$ exists such that $g(z) = \exp(\phi(z))$. In particular, whenever $|z| = r$, $z^n = \exp(\phi(z))$. Thus $z \longmapsto z^n$ has a continuous logarithm on $|z| = r$ as, in view of Theorem 3.9.7 (ii), does the identity map $z \longmapsto z$. But this contradicts Theorem 3.9.6. $\qquad\square$

Theorem 3.9.17 (Borsuk) *Let $K \subset \mathbf{C}$ be compact. Then the following conditions are equivalent:*

(i) $\mathbf{C}\backslash K$ *is connected.*
(ii) *Each continuous function $f : K \to \mathbf{C}\backslash\{0\}$ has a continuous logarithm.*
(iii) *Each continuous function $f : K \to \mathbf{C}\backslash\{0\}$ has a continuous, zero-free extension to $\mathbf{C}$.*

Proof That (i) implies (ii) is immediate from Theorem 3.9.15; further, (ii) implies (iii) by Theorem 3.9.11.

Suppose that (iii) holds. If $\mathbf{C}\backslash K$ had a bounded component then, by Theorem 3.9.16, a continuous $f : K \to \mathbf{C}\backslash\{0\}$ exists which does not have a continuous zero-free extension to $\mathbf{C}$. Hence $\mathbf{C}\backslash K$ has no bounded component and (i) holds. $\quad\square$

3.9.3 Properties of Jordan Curves

Definition 3.9.18 A set $J \subset \mathbf{C}$ is said to be a **Jordan curve** if it is a homeomorphic image of the unit circle $S^1 = \{z \in \mathbf{C} : |z| = 1\}$.

Theorem 3.9.19 *Let J be a Jordan curve. Then $\mathbf{C}\backslash J$ has exactly one bounded component.*

Proof To begin with, suppose that $\mathbf{C}\backslash J$ has no bounded component and therefore is connected. Let $\phi : S^1 \to J$ be a homeomorphism and suppose that $f : S^1 \to \mathbf{C}\backslash\{0\}$ is continuous. Then $f \circ \phi^{-1} : J \to \mathbf{C}\backslash\{0\}$ is continuous and, by Theorem 3.9.17, a continuous map $\psi : J \to \mathbf{C}$ exists such that

$$f(\phi^{-1}(z)) = \exp(\psi(z)) \ (z \in J).$$

This equality shows that

$$f(w) = \exp(\psi(\phi(w))) \ (w \in S^1)$$

and so f has a continuous logarithm. But Theorem 3.9.6 contradicts this conclusion: not every such f has a continuous logarithm. It follows that $\mathbf{C}\backslash J$ has at least one bounded component.

Next, suppose that $\mathbf{C}\backslash J$ has two distinct bounded components C_1 and C_2 and, as earlier, let $\phi : S^1 \to J$ be a homeomorphism. Choose $p_j \in C_j$ and put $f_j(z) = z - p_j$ ($j = 1, 2; z \in J$). For $j = 1, 2$ the map $f_j \circ \phi$ is continuous and zero-free on S^1 and so, by Theorem 3.9.7 (i), there are an integer n_j and a continuous function ψ_j on S^1 such that

$$\phi(w) - p_j = w^{n_j} \exp(\psi_j(w)) \ (w \in S^1).$$

It follows that

$$(\phi(w) - p_1)^{n_2} (\phi(w) - p_2)^{-n_1} = \exp(n_2 \psi_1(w) - n_1 \psi_2(w)) \ (w \in S^1),$$

and therefore

$$(z - p_1)^{n_2} (z - p_2)^{-n_1} = \exp(\Psi(z)) \ (z \in J),$$

where $\Psi = n_2 \psi_1 \circ \phi^{-1} - n_1 \psi_2 \circ \phi^{-1}$.

Suppose $n_2 \neq 0$. Then

$$(z - p_1)^{n_2} = (z - p_2)^{n_1} \exp(\Psi(z)) \ (z \in J).$$

Evidently $z \longmapsto (z - p_2)^{n_1}$ has a continuous zero-free extension to $J \cup C_1$; also, because of the Tietze theorem, Ψ has a continuous extension to $\mathbf{C}$. Hence $z \longmapsto (z - p_1)^{n_2}$ has a continuous zero-free extension to $J \cup C_1$, an outcome which contradicts Theorem 3.9.16. Thus $n_2 = 0$ and, similarly, $n_1 = 0$. Hence

$$\phi(w) - p_1 = \exp(\psi_1(w)) \ (w \in S^1)$$

and

$$f_1(z) = z - p_1 = \exp(\psi_1(\phi^{-1}(z))) \ (z \in J).$$

Another appeal to the Tietze extension theorem shows that $\psi_1 \circ \phi^{-1}$ has a continuous extension to $\mathbf{C}$ and therefore f_1 has a continuous zero-free extension to $J \cup C_1$. But, by Theorem 3.9.16, no such extension exists. Hence $\mathbf{C}\backslash J$ has at most one bounded component. $\square$

Lemma 3.9.20 *Let K be a proper closed subset of a Jordan curve J. Then $\mathbf{C}\backslash K$ is connected.*

Proof Initially, suppose $J = S^1$. Then since $\mathbf{C}\backslash K$ is open, it is enough to observe that it is path-connected: if $z_0 \in S^1\backslash K$, then it is clear that each $z \in \mathbf{C}\backslash K$ can be joined to z_0 by a path in $\mathbf{C}\backslash K$.

Next, suppose $J = \phi(S^1)$ for some homeomorphism $\phi : S^1 \to J$. Let $K = \phi(T)$, where T is a proper subset of S^1, and let $f : K \to \mathbf{C}\backslash\{0\}$ be continuous. Then

$f \circ \phi : T \to \mathbf{C} \backslash \{0\}$ is continuous and so, by Theorem 3.9.17, there exists a continuous $\psi : T \to \mathbf{C}$ such that

$$f(\phi(w)) = \exp(\psi(w)) \; (w \in T).$$

Hence

$$f(z) = \exp(\psi(\phi^{-1}(z))) \; (z \in K)$$

and so, using Theorem 3.9.17 again, we see that $\mathbf{C} \backslash K$ is connected. $\quad\square$

Theorem 3.9.21 *Let J be a Jordan curve and let C be a component of $\mathbf{C} \backslash J$. Then $\partial C = J$.*

Proof By Theorem 3.9.19, $\mathbf{C} \backslash J$ has exactly one bounded component. Let U be that component and let V be the unbounded component. The sets U, J and V are pairwise disjoint and their union is $\mathbf{C}$; further, U and V are open and J is compact. Since U is contained in the closed set $^c V$, we must have $\overline{U} \subset {}^c V$; similarly, $\overline{V} \subset {}^c U$. Hence $\partial U = \overline{U} \backslash U \subset {}^c V \cap {}^c U = {}^c(U \cup V) = J$ and, likewise, $\partial V \subset J$.

To obtain a contradiction, suppose that $\partial U \neq J$. Then, by Lemma 3.9.20, $\mathbf{C} \backslash \partial U$ is connected. However,

$$\mathbf{C} \backslash \partial U = {}^c(\overline{U} \cap {}^c U) = U \cup {}^c(\overline{U}),$$
$$U \cap {}^c(\overline{U}) \subset (U \cap {}^c U) = \emptyset, U \neq \emptyset \text{ and } {}^c(\overline{U}) \supset V \neq \emptyset,$$

from which it follows that $\mathbf{C} \backslash \partial U$ is disconnected. Thus $\partial U = J$. Analogous reasoning shows that $\partial V = J$. $\quad\square$

Theorem 3.9.22 *Let J be a Jordan curve and C be the bounded component of $\mathbf{C} \backslash J$. Let $\phi : S^1 \to J$ be be a homeomorphism and let $\gamma : [0, 1] \to J$ be the closed path defined by*

$$\gamma(t) = \phi(\exp(2\pi i t)) \; (0 \leq t \leq 1).$$

Then either $n(\gamma, z) = 1$ for all $z \in C$ or $n(\gamma, z) = -1$ for all $z \in C$.

Proof Applied to the map $\phi^{-1} : J \to S^1 \subset \mathbf{C} \backslash \{0\}$, Theorem 3.9.15 establishes the existence of a point $p \in C$, an integer m and a continuous map $\psi : J \to \mathbf{C}$ such that

$$(z - p)^{-m} \phi^{-1}(z) = \exp(\psi(z)) \; (z \in J).$$

Hence

$$w = (\phi(w) - p)^m \exp(\psi(\phi(w))) \; (w \in S^1)$$

and therefore

$$v = (\gamma - p)^m \exp(\psi \circ \gamma), \text{ where } v(t) = \exp(2\pi i t) \; (0 \leq t \leq 1).$$

By Proposition 3.4.25, relative to $0 \in \mathbf{C}$, the winding numbers of the closed paths ν, $\gamma - p$ and $\exp(\psi \circ \gamma)$ satisfy the equality

$$n(\nu, 0) = mn(\gamma - p, 0) + n(\exp(\psi \circ \gamma), 0).$$

By Example 3.4.22 (i), $n(\nu, 0) = 1$; also, since $\psi \circ \gamma$ is a continuous logarithm of $\exp(\psi \circ \gamma)$, $n(\exp(\psi \circ \gamma), 0) = 0$. Thus $mn(\gamma - p, 0) = mn(\gamma, p) = 1$ and so $n(\gamma, p) = \pm 1$. Lastly, use of Theorem 3.4.27 shows that

$$n(\gamma, z) = n(\gamma, p) \ (z \in \mathbf{C}),$$

and the proof is complete.

$\square$

For the convenience of the reader we summarise the preceding results in the following form, often called the Jordan curve theorem.

Theorem 3.9.23 *Let J be a Jordan curve. Then $\mathbf{C} \backslash J$ has exactly one bounded component $\mathscr{I}(J)$ (called the **inside** of J), and exactly one unbounded component $\mathscr{O}(J)$ (called the **outside** of J); $\partial \mathscr{I}(J) = \partial \mathscr{O}(J) = J$. Let $\phi : S^1 \to J$ be be a homeomorphism and let $\gamma : [0, 1] \to J$ be the closed path defined by*

$$\gamma(t) = \phi(\exp(2\pi i t)) \ (0 \le t \le 1).$$

Then either $n(\gamma, z) = 1$ for all $z \in \mathscr{I}(J)$ or $n(\gamma, z) = -1$ for all $z \in \mathscr{I}(J)$; $n(\gamma, z) = 0$ for all $z \in \mathscr{O}(J)$.

Note that the the claim $n(\gamma, z) = 0$ for all $z \in \mathscr{O}(J)$ follows from Theorem 3.4.27. The 'inside, outside' terminology is in accordance with Definition 3.4.30.

Identification of Jordan curves is made simpler by the next result.

Theorem 3.9.24 *A set $J \subset \mathbf{C}$ is a Jordan curve if and only if it is the track of a simple closed path in $\mathbf{C}$.*

Proof First suppose that J is a Jordan curve. Then there is a homeomorphism $f : S^1 \to J$. Define a continuous map γ of $[0, 1]$ onto S^1 by $\gamma(s) = \exp(2\pi i s)$. Then $J = (f \circ \gamma)^*$ and $f \circ \gamma$ is a simple closed path in $\mathbf{C}$.

Conversely, if $\gamma : [0, 1] \to \mathbf{C}$ is a simple closed path, then by Corollary 3.9.2 there is a continuous map $f : S^1 \to \mathbf{C}$ such that $f(\exp(2\pi i s)) = \gamma(s)$. Since γ is simple, f is injective, and so, by Theorem 2.3.24 (ii), it is a homeomorphism onto $f(S^1)$. Thus $\gamma^* = f(S^1)$ is a homeomorphic image of S^1 and so is a Jordan curve.

$\square$

Appendix A
Sets and Functions

A.1 Sets

For a systematic development of set theory we refer to [14]. Here we restrict ourselves to terminology. We accept as undefined basic concepts those of 'set' and 'membership of a set'. To express this last we use the symbol $\in$ and write $x \in A$ to indicate that x is a member (an element) of A. If x is not a member of A we write $x \notin A$.

The set which has no elements is called the *empty* set (*void* set) and is written $\emptyset$.

If A and B are sets and every element of A is an element of B, we say that A is a *subset* of B and write $A \subset B$ or $B \supset A$. If, in addition, there is an element of B which is not in A, then A is said to be a *proper* subset of B. Note that, for each set A, $\emptyset \subset A$ and $A \subset A$.

If $A \subset B$ and $B \subset A$, we write $A = B$; otherwise $A \neq B$.

Common ways of specifying a subset of a given set A are as follows:

(i) If $a, b, c, \ldots$ are members of A, then the subset of consisting of precisely those members is written

$$\{a, b, c, \ldots\}.$$

(ii) If, for each $x \in A$, $S(x)$ is a statement which is either true or false, then

$$\{x \in A : S(x)\}$$

is the set of those $x \in A$ such that $S(x)$ is true.

To illustrate, when $A = \mathbf{N}$,

$$\{3, 7\} = \{n \in \mathbf{N} : n^2 - 10n + 21 = 0\}.$$

Note that sets may themselves be elements of other sets. For example,

$$\{\{1, 3, 5, 7\}, \{2, 4, 6\}\}$$

R. H. Dyer and D. E. Edmunds, *From Real to Complex Analysis*,
Springer Undergraduate Mathematics Series, DOI: 10.1007/978-3-319-06209-9,
© Springer International Publishing Switzerland 2014

is a set with two elements, namely the set of all positive odd integers <8 and the set of all positive even integers <8.

If A is any set, then the collection of its subsets constitutes a set with these subsets as its elements. This collection is denoted by $P(A)$, or 2^A, and is termed the *power set* of A. Thus if $A = \{1, 2\}$, then $P(A) = \{\emptyset, \{1\}, \{2\}, \{1, 2\}\}$. In general, if A has n elements, then $P(A)$ has 2^n elements.

A set whose elements are sets is often referred to as a *family* of sets.

Let A and B be sets. The set $A \times B$, called the *Cartesian product* of A and B, is the totality of all ordered pairs (x, y) whose first coordinate $x \in A$ and whose second $y \in B$. Here the ordering is essential: $(x, y) = (u, v)$ if and only if $x = u$ and $y = v$. Note that $(x, y) \neq (y, x)$ if $x \neq y$. When $A = B$ we sometimes write A^2 for $A \times A$; for example, $\mathbf{R}^2$ in place of $\mathbf{R} \times \mathbf{R}$.

Similarly, we may write $A \times B \times C$ as a set of ordered triples, etc.

A.2 Set Operations

Fix a set E and let $P(E)$ be the family of all subsets of E. Let $A, B \in P(E)$. Their *union*

$$A \cup B := \{x \in E : x \in A \text{ or } x \in B\};$$

their *intersection*

$$A \cap B := \{x \in E : x \in A \text{ and } x \in B\}.$$

When $A \cap B = \emptyset$ the sets A and B are said to be *disjoint*.

If A, B and C are subsets of E, then

$$A \cup (B \cup C) = (A \cup B) \cup C,$$

and either side is written $A \cup B \cup C$. A similar remark holds for $A \cap B \cap C$. Further, operations with $\cup$ and $\cap$ obey certain distributive laws:

$$A \cup (B \cap C) = (A \cup B) \cap (A \cup C),$$

$$A \cap (B \cup C) = (A \cap B) \cup (A \cap C).$$

Proof of these assertions is straightforward. To illustrate:

$$x \in A \cup (B \cap C) \Longleftrightarrow x \in A \text{ or } x \in B \cap C \Longleftrightarrow x \in A \cup B \text{ and } x \in A \cup C$$
$$\Longleftrightarrow x \in (A \cup B) \cap (A \cup C).$$

For $A \in P(E)$, the *complement of A relative to E* is the set

$$^c A := \{x \in E : x \notin A\};$$

note that $^c(^cA) = A$. The De Morgan laws relate complements to intersections and unions: for all $A, B \in P(E)$,

$$^c(A \cup B) = {}^cA \cap {}^cB, \, ^c(A \cap B) = {}^cA \cup {}^cB.$$

To prove the first:

$$x \in {}^c(A \cup B) \Longleftrightarrow x \in E \text{ and } x \notin (A \cup B) \Longleftrightarrow x \in E, x \notin A \text{ and } x \notin B$$
$$\Longleftrightarrow x \in {}^cA \text{ and } x \in {}^cB \Longleftrightarrow x \in {}^cA \cap {}^cB.$$

Let $A, B \in P(E)$. We define the *difference $B \backslash A$* (the *relative complement* of A in B) by

$$B \backslash A = \{x \in E : x \in B \text{ and } x \notin A\};$$

plainly $B \backslash A = B \cap {}^cA$.

The process of taking the union (intersection) of two sets can be extended to larger families. For $\mathscr{F} \subset P(E)$,

$$\bigcup_{F \in \mathscr{F}} F := \{x \in E : \text{for some } F \in \mathscr{F}, x \in F\}$$

and

$$\bigcap_{F \in \mathscr{F}} F := \{x \in E : \text{for all } F \in \mathscr{F}, x \in F\}.$$

The union of the sets in $\mathscr{F}$ may be denoted by $\cup \mathscr{F}$; their intersection by $\cap \mathscr{F}$.

Sometimes a family of subsets of E may be given in indexed form and then we use a different notation for the union (intersection). Let I be a non-void set and suppose that to each $i \in I$ corresponds a set $A_i \in P(E)$. The set $\{A \in P(E) : \text{for some } i \in I, \, A = A_i\}$, commonly written $\{A_i : i \in I\}$ or $\{A_i\}_{i \in I}$, is called an *indexed family of sets* and I is called an *indexing set*. The union of this indexed family

$$\bigcup_{i \in I} A_i := \{x \in E : \text{ for some } i \in I, x \in A_i\};$$

and the intersection

$$\bigcap_{i \in I} A_i := \{x \in E : \text{ for all } i \in I, x \in A_i\}.$$

A.3 Functions

Let A and B be non-empty sets. A *function* f from A to B associates with each $x \in A$ exactly one element $f(x) \in B$. The set A is the *set of definition* (the *domain*) of f, B is the *target set* (the *codomain*) of f. To display the three components of a function (rule of correspondence, domain, codomain), we write

$$f : A \to B \text{ or } x \mapsto f(x) : A \to B.$$

As an example, if $E \subset A$, then the characteristic function of E with domain A and codomain $\mathbf{R}$, denoted by $\chi_E : A \to \mathbf{R}$, is defined by

$$\chi_E(x) = \begin{cases} 1 & \text{if } x \in E, \\ 0 & \text{if } x \in A \backslash E. \end{cases}$$

As synonyms for the word function we may use *map* or *mapping*. The *graph* of f is the set

$$G(f) := \{(x, f(x)) \in A \times B : x \in A\}.$$

A set $H \subset A \times B$ is the graph of a function from A to B if, and only if,

(i) for all $x \in A$, there exists $y \in B$ such that $(x, y) \in H$;
(ii) $(x, y) \in H, (x, y') \in H \Longrightarrow y = y'$.

A map $f : A \to B$ generates further maps between the power sets of A and of B:

$$f : P(A) \to P(B)$$

is defined by

$$f(X) = \{f(x) \in B : x \in X\};$$

while

$$f^{-1} : P(B) \to P(A)$$

is defined by

$$f^{-1}(Y) = \{x \in A : f(x) \in Y\}.$$

The set $f(X)$ is called the *image* (under f) of $X \subset A$; $f(A)$ is called the *range* of f; $f^{-1}(Y)$ is called the *inverse image* (under f) of $Y \subset B$.

Example A.3.1 Let f be the map $x \mapsto x^2 : \mathbf{R} \to \mathbf{R}$. Then

$$f([0, 1]) = [0, 1] = f([-1, 0]), \ f([1, 2]) = [1, 4], \ f(\{1, 2, 3, 4\})$$
$$= \{1, 4, 9, 16\};$$
$$f^{-1}([0, 1]) = [-1, 1], \ f^{-1}([1, 4]) = [-2, -1] \cup [1, 2], \ f^{-1}([-1, 0]) = \{0\},$$
$$f^{-1}([-4, -1]) = \emptyset.$$

The *composition* of mappings $f : A \to B$ and $g : B \to C$ is the map $g \circ f$ from A to C which associates with each $x \in A$ the element $(g \circ f)(x) := g(f(x))$ of C. For any map $h : C \to D$ it is easy to show that $h \circ (g \circ f) = (h \circ g) \circ f$.

A function $f : A \to B$ is said to be *injective (one-to-one)* if $a, a' \in A$, $a \neq a'$ $\implies f(a) \neq f(a')$; it is called *surjective (onto)* if $f(A) = B$; if it is both injective and surjective, then it is termed *bijective*. As an example of a bijective map we cite the *identity* map i_A from A onto itself defined by $i_A(x) = x$.

Bijective maps prove to be of considerable importance. Suppose that $f : A \to B$ is bijective. Then, for each $b \in B$, there is one and only one $a \in A$ such that $f(a) = b$. Hence a function $g : B \to A$ may be defined by the rule which assigns to each $b \in B$ that unique element $g(b) \in A$ such that $f(g(b)) = b$. The function g is injective since $f \circ g = i_B$:

$$b, b' \in B, g(b) = g(b') \implies b = f(g(b)) = f(g(b')) = b'.$$

Moreover, since $g \circ f = i_A$, g is surjective:

$$a \in A \implies f(a) \in B \implies f(g(f(a))) = f(a) \implies g(f(a)) = a.$$

Thus, corresponding to a bijective map $f : A \to B$ there is a bijective map $g : B \to A$ such that

$$f \circ g = i_B \text{ and } g \circ f = i_A.$$

It is elementary to show that g is the unique map with these properties; it is called the *inverse* of f. Since for each $b \in B$,

$$f^{-1}(\{b\}) = \{a \in A : f(a) = b\} = \{g(b)\},$$

it is customary to denote the map g by f^{-1}. One has to understand by context the sense in which the symbol f^{-1} is used.

Particular terminology is needed for maps $f : I \to \mathbf{R}$, where I is an interval in the real line. Such a map is said to be *increasing (strictly increasing)* if $f(y) \geq f(x)$ $(f(y) > f(x))$ whenever $x, y \in I$ and $y > x$; it is said to be *decreasing (strictly decreasing)* if $f(y) \leq f(x)$ $(f(y) < f(x))$ whenever $x, y \in I$ and $y > x$.

A.4 The Real Number System

The real number system consists of a set $\mathbf{R}$, a subset $\mathbf{P}$ of $\mathbf{R}$, and two maps $(x, y) \longmapsto x + y$ and $(x, y) \longmapsto x \cdot y$ (commonly written xy) of $\mathbf{R} \times \mathbf{R}$ to $\mathbf{R}$ such that the following three axioms are satisfied.

Axiom A.4.1 $(\mathbf{R}, +, \cdot)$ is a field: $(\mathbf{R}, +)$ and $(\mathbf{R}\backslash\{0\}, \cdot)$ are abelian groups and, for all $x, y, z \in \mathbf{R}$, $x(y + z) = xy + xz$.

Axiom A.4.2 The sets $\mathbf{P}$, $\{0\}$, $-\mathbf{P} := \{-x : x \in \mathbf{P}\}$ are pairwise disjoint and their union is $\mathbf{R}$; further, $x + y$ and xy belong to $\mathbf{P}$ whenever $x, y \in \mathbf{P}$.

The elements of $\mathbf{P}$ are called positive numbers and those of $-\mathbf{P}$ negative numbers. We write $x < y$ or $y > x$ to mean that $y - x \in \mathbf{P}$. We also write $x \leq y$ or $y \geq x$ to mean that $y - x \in \mathbf{P} \cup \{0\}$.

Axiom A.4.3 (*Dedekind completeness*) Let A, B be non-empty subsets of $\mathbf{R}$ with union $\mathbf{R}$, and suppose that $a < b$ whenever $a \in A$ and $b \in B$. Then there exists a unique $c \in \mathbf{R}$ such that

(i) $x \in \mathbf{R}, x < c \Longrightarrow x \in A$,
(ii) $x \in \mathbf{R}, c < x \Longrightarrow x \in B$.

The existence of an object $\mathbf{R}$ satisfying the above axioms is assumed. The next two results flow from Axioms 1 and 2.

Theorem A.4.4 *Let $x, y, z \in \mathbf{R}$. Then*

(i) $x < y, y < z \Longrightarrow x < z$;
(ii) *exactly one of $x < y, x = y, x > y$ holds;*
(iii) $x < y, z \in \mathbf{R} \Longrightarrow x + z < y + z$;
(iv) $x < y, z > 0 \Longrightarrow xz < yz$;
(v) $x < y, z < 0 \Longrightarrow xz > yz$;
(vi) $1 > 0$ *and* $-1 < 0$;
(vii) $z > 0 \Longrightarrow z^{-1} \left(= \frac{1}{z}\right) > 0$;
(viii) $0 < x < y \Longrightarrow 0 < y^{-1} < x^{-1}$.

A similar theorem holds with '$\leq$' in place of '$<$' in some instances.

Theorem A.4.5 *For each $x \in \mathbf{R}$, let*

$$|x| := \begin{cases} x & \text{if } x \geq 0, \\ -x & \text{if } x < 0. \end{cases}$$

Then, for all $x, y \in \mathbf{R}$,
(i) $|xy| = |x| |y|$, (ii) $|x + y| \leq |x| + |y|$, (iii) $||x| - |y|| \leq |x - y|$.

Definition A.4.6 Let E be a non-empty subset of $\mathbf{R}$. If $b \in \mathbf{R}$ and, for every $x \in E$, $x \leq b$, then b is termed an upper bound for E and the set E is said to be bounded above. If $c \in \mathbf{R}$ is such that

(i) $x \leq c$ for all $x \in E$, and
(ii) $x \leq b$ for all $x \in E \Longrightarrow c \leq b$,

then c is called the least upper bound (supremum) of E and is written as $\sup E$. [The terms lower bound and greatest lower bound (infimum) are defined similarly.]

Theorem A.4.7 (The axiom of order completeness; the supremum principle) *Every non-empty set of real numbers which is bounded above has a least upper bound.*

Proof Let E be a non-empty subset of $\mathbf{R}$ which is bounded above. Let $B = \{b \in \mathbf{R} : x \leq b \text{ for all } x \in E\}$ and $A = \mathbf{R} \backslash B$. Plainly $B \neq \emptyset$; also, if $x \in E$, then $x - 1 \in A$ and so $A \neq \emptyset$. Given $a \in A$ and $b \in B$, there exists $x \in E$ such that $a < x \leq b$. Thus $a \in A, b \in B \implies a < b$. Appealing now to Axiom A.4.3, there exists a unique $c \in \mathbf{R}$ such that $(-\infty, c) \subset A$ and $(c, \infty) \subset B$. Suppose $c \in A$. Then $c < y$ for some $y \in E$ and, with $2 := 1 + 1$, $c < \frac{1}{2}(c + y) < y$. But $c < \frac{1}{2}(c + y) \implies \frac{1}{2}(c + y) \in B$, while $\frac{1}{2}(c + y) < y \implies \frac{1}{2}(c + y) \in A$. Hence $c \notin A$, $c \in B$ and $c = \sup E$. $\qquad\square$

Corollary A.4.8 *Every non-empty set of real numbers which is bounded below has an infimum.*

The natural numbers
A set $I \subset \mathbf{R}$ is said to be *inductive* if
(i) $1 \in I$, and (ii) $x \in I \implies x + 1 \in I$.

Let $\mathscr{I}$ be the class of all inductive sets in $\mathbf{R}$. Note that $\mathscr{I} \neq \emptyset$; for example, the sets $\mathbf{R}, \mathbf{P}$ and $\{x \in \mathbf{R} : x \geq 1\}$ all belong to $\mathscr{I}$. The set of *natural numbers (positive integers)*

$$\mathbf{N} := \cap_{I \in \mathscr{I}} I.$$

Clearly $\mathbf{N} \subset I$ for every $I \in \mathscr{I}$ and, for all $n \in \mathbf{N}$, $n \geq 1$. Further, $\mathbf{N}$ is inductive: $1 \in \mathbf{N}$ (since $1 \in I$ for all $I \in \mathscr{I}$); $x \in \mathbf{N} \implies x \in I$ $(I \in \mathscr{I}) \implies x + 1 \in I$ $(I \in \mathscr{I}) \implies x + 1 \in \mathbf{N}$.

The symbol $2 := 1 + 1, 3 := 2 + 1, 4 := 3 + 1$, etc.

Theorem A.4.9 (The finite induction principle) *Let $S \subset \mathbf{N}$ be such that* (i) $1 \in S$, *and* (ii) $x \in S \implies x + 1 \in S$. *Then $S = \mathbf{N}$.*

Proof By hypothesis, S is inductive and so $\mathbf{N} \subset S$. But also $S \subset \mathbf{N}$. Hence $S = \mathbf{N}$. $\square$

This theorem may be used to establish the following results, the proofs of which are left to the reader:

(i) If $n \in \mathbf{N}$ and $n > 1$, then $n - 1 \in \mathbf{N}$.
(ii) If $m, n \in \mathbf{N}$, then $m + n, mn \in \mathbf{N}$.
(iii) If $n \in \mathbf{N}, x \in \mathbf{R}$ and $n < x < n + 1$, then $x \notin \mathbf{N}$.

Corollary A.4.10 (The well-ordering principle) *If A is a non-empty subset of $\mathbf{N}$, then it has a smallest element.*

Theorem A.4.11 (The Archimedean order property) *Let $a, b \in \mathbf{R}$ and $a > 0$. Then there is an $n \in \mathbf{N}$ such that $na > b$. In particular, $\mathbf{N}$ is not bounded above in $\mathbf{R}$.*

Proof Suppose there exist $a, b \in \mathbf{R}$ with $a > 0$ for which the result is false. Let $A = \{na : n \in \mathbf{N}\}$. Then b is an upper bound for A and so, by Theorem A.4.7, A has a supremum. Choose $k \in \mathbf{N}$ such that $ka > \sup A - a$. Then $(k + 1)a > \sup A$. But this is not possible since $na \le \sup A$ for all $n \in \mathbf{N}$.

To show that $\mathbf{N}$ is not bounded above take $a = 1$. $\square$

It is assumed that the reader is familiar with the extension of $\mathbf{N}$ to $\mathbf{Z}$, the set of all integers:

$$\mathbf{Z} = -\mathbf{N} \cup \{0\} \cup \mathbf{N}.$$

Sequences in R

Let X be a non-empty set. A map $f : \mathbf{N} \to X$ is called a *sequence* in X. Writing $x_n = f(n)$, x_n is the nth *term* of the sequence. It is customary to write $(x_n)_{n \in \mathbf{N}}$, or simply (x_n), in place of $f : \mathbf{N} \to X$.

A sequence (y_n) in X is said to be a *subsequence* of (x_n) if there exists a map $m : \mathbf{N} \to \mathbf{N}$ such that, for all $n \in \mathbf{N}$,

(i) $m(n) < m(n + 1)$,
(ii) $y_n = x_{m(n)}$.

A sequence (x_n) in $\mathbf{R}$ is said to be *monotone* if one of the following holds:

(i) $x_n < x_{n+1}$ ($n \in \mathbf{N}$), strictly increasing;
(ii) $x_n \le x_{n+1}$ ($n \in \mathbf{N}$), increasing;
(iii) $x_n > x_{n+1}$ ($n \in \mathbf{N}$), strictly decreasing;
(iv) $x_n \ge x_{n+1}$ ($n \in \mathbf{N}$), decreasing.

A sequence (x_n) in $\mathbf{R}$ is said to *converge* if there exists $x \in \mathbf{R}$ such that, given any $\varepsilon > 0$, there exists $N = N(\varepsilon) \in \mathbf{N}$ such that

$$n \ge N \implies |x_n - x| < \varepsilon.$$

In this event we may write $\lim_{n \to \infty} x_n = x$, $\lim x_n = x$ or $x_n \to x$.

Theorem A.4.12 *Every bounded monotone sequence in* $\mathbf{R}$ *converges.*

Proof Let (x_n) be an increasing sequence in $\mathbf{R}$ such that $x_n \le K$ ($n \in \mathbf{N}$), where $K \in \mathbf{R}$. Let $E = \{x_n : n \in \mathbf{N}\}$ and $x = \sup E$ (Theorem A.4.7 ensures existence). Let $\varepsilon > 0$. Since $x - \varepsilon$ is not an upper bound for E, there exists $N \in \mathbf{N}$ such that $x - \varepsilon < x_N$. It follows that, for all $n \ge N$, $x - \varepsilon < x_n \le x$. Hence $n \ge N \implies |x - x_n| < \varepsilon$, and $x_n \to x$.

In the case that (x_n) is decreasing, consider $(-x_n)$. $\square$

Theorem A.4.13 (Bolzano-Weierstrass) *Every bounded sequence in* $\mathbf{R}$ *has a convergent subsequence.*

Proof Let (x_n) be a bounded sequence in $\mathbf{R}$; let $K \in \mathbf{R}$ be such that $|x_n| \le K$ ($n \in \mathbf{N}$). For each $n \in \mathbf{N}$ let $u_n := \sup\{x_k : k \ge n\}$. The sequence (u_n) is decreasing. Let $u := \lim u_n = \inf\{u_n : n \ge 1\} \ge -K$. Define $(x_{m(n)})$ inductively as follows. Choose $m(1)$ to be the least $k \in \mathbf{N}$ such that $x_k > u - 1$. Having chosen $m(1) < m(2) < \ldots < m(n)$ for some $n \geq 1$, select $m(n + 1)$ to be the least integer $k > m(n)$ such that $x_k > u - (n + 1)^{-1}$. Then

$$u_n \geq u_{m(n)} \geq x_{m(n)} > u - n^{-1} \ (n \in \mathbf{N}),$$

and so, by a sandwich argument, $\lim x_{m(n)} = u$. $\qquad\square$

A sequence (x_n) in $\mathbf{R}$ is said to be a *Cauchy sequence* if, given any $\varepsilon > 0$, there exists $N = N(\varepsilon) \in \mathbf{N}$ such that

$$m, n \geq N \implies |x_m - x_n| < \varepsilon.$$

Theorem A.4.14 *The real number system is Cauchy complete; that is, every Cauchy sequence in $\mathbf{R}$ converges.*

Proof Let (x_n) be a Cauchy sequence in $\mathbf{R}$. The sequence is bounded, since there exists an $N \in \mathbf{N}$ such that $|x_n - x_N| < 1$ whenever $n \geq N$ and therefore $|x_n| \leq K := 1 + \max_{1 \leq k \leq N} |x_k| \ (n \in \mathbf{N})$. By Theorem A.4.13, it follows that (x_n) has a convergent subsequence, $(x_{m(n)})$ say. Suppose $\lim x_{m(n)} = x$. We show that $\lim x_n = x$.

Let $\varepsilon > 0$. There exists $N_1 \in \mathbf{N}$ such that

$$n \geq N_1 \implies \left| x_{m(n)} - x \right| < \varepsilon/2.$$

Also, there exists $N_2 \in \mathbf{N}$ such that

$$p, q \geq N_2 \implies \left| x_p - x_q \right| < \varepsilon/2.$$

Hence

$$|x_n - x| \leq \left| x_n - x_{m(n)} \right| + \left| x_{m(n)} - x \right| < \varepsilon$$

whenever $n > \max(N_1, N_2)$. Thus $\lim x_n = x$. $\qquad\square$

Theorem A.4.15 (The Cantor nested intervals principle) *Let $\{I_n\}$ be a family of non-empty, bounded, closed intervals such that*

(i) $I_{n+1} \subset I_n \ (n \in \mathbf{N})$, *and*
(ii) $\lim l(I_n) = 0$.

Then there exists $x \in \mathbf{R}$ such that $\{x\} = \cap_{n=1}^{\infty} I_n$.
    [Here $l(I_n)$, the length of I_n, is defined by $l(I_n) = \sup I_n - \inf I_n$.]

Proof Let (x_n) be a sequence in $\mathbf{R}$ such that $x_n \in I_n \ (n \in \mathbf{N})$. By (i), if $n \geq m$, then $x_n \in I_m$ and $|x_n - x_m| \leq l(I_m)$. Let $\varepsilon > 0$. By (ii), there exists $N \in \mathbf{N}$ such that $l(I_N) < \varepsilon$. Hence $m, n \geq N \implies |x_n - x_m| \leq \max\{l(I_n), l(I_m)\} \leq l(I_N) < \varepsilon$, and so (x_n) is a Cauchy sequence in $\mathbf{R}$. Let $x = \lim x_n$. Since I_n is a closed interval,

$x \in I_n$ ($n \in \mathbf{N}$). Hence $x \in \cap_{n=1}^{\infty} I_n$. If $y \in \cap_{n=1}^{\infty} I_n$, then $x = y$, since $|x - y| \le l(I_n) \to 0$. $\qquad\square$

For further details of the development and properties of the real number system, see [17], Chap. 1.

A.5 The Axiom of Choice

First we recall some basic concepts. A set S is said to be **finite** if either it is empty or there exists $n \in \mathbf{N}$ such that S can be mapped bijectively onto $\{k \in \mathbf{N} : 1 \le k \le n\}$. Sets that are not finite are said to be **infinite**. A set E is called **countable** if it can be mapped bijectively onto $\mathbf{N}$; any infinite set that is not countable is said to be **uncountable**.

The axiom of choice has various equivalent forms (see, for example, [10]). We give the following:

Axiom A.5.1 (*The axiom of choice*) For every non-empty family $\mathscr{F}$ of non-empty sets, there is a function $f : \mathscr{F} \to \cup_{S \in \mathscr{F}} S$ such that for all $S \in \mathscr{F}$, $f(S) \in S$.

The function f is said to be a **choice function** on $\mathscr{F}$. The only form of this axiom that will be used in this book is the following weaker version.

Axiom A.5.2 (*The countable axiom of choice*) Every countable family of non-empty sets has a choice function.

Proposition A.5.3 *The countable axiom of choice implies that every infinite set has a countable subset.*

Proof Let S be an infinite set and let $\mathscr{F}$ be the family of all finite sequences of distinct elements of S: that is, $\mathscr{F} = \{A_k : k \in \mathbf{N}\}$, where

$$A_k = \{(a_0, a_1, \ldots, a_k) : a_0, \ldots, a_k \text{ distinct elements of } S\}.$$

Then by the countable axiom of choice, $\mathscr{F}$ has a choice function f: $f(A_k) \in A_k$ for all k. The union of all the chosen finite sequences is countable. $\qquad\square$

We refer to [10] for a thorough discussion of the axiom of choice and its variants, together with an analysis of its place in contemporary mathematics.

Notes on the Exercises

Sketch solutions or hints are provided for selected exercises.

Exercise 1.1.10

2. By question 1 above,

$$\mathrm{osc}(f; A) = \sup\{|f(x) - f(y)| : x, y \in A\} = \sup\{f(x) - f(y) : x, y \in A\}$$
$$= \sup\{f(x) : x \in A\} + \sup\{-f(x) : x \in A\}$$
$$= \sup\{f(x) : x \in A\} - \inf\{f(x) : x \in A\}.$$

5. Suppose the claim in the hint is false. There is a nested sequence $([a_n, b_n])$ of closed subintervals of $[a, b]$ such that $0 < b_n - a_n < 1/n$ and $\sup\{f(t) : a_n \le t \le b_n\} < 1/n$. By Theorem A.4.15, the $[a_n, b_n]$ have intersection $\{u\}$ for some $u \in [a, b]$. Thus $f(u) < 1/n$ for all n, contradicting the hypothesis that $f(u) > 0$. Now let $P, Q \in \mathscr{P}[a, b]$ and $Q = \{a, c, d, b\}$. Then $0 < \varepsilon(d - c) \le U(P \cup Q, f) \le U(P, f)$: thus $\int_a^b f > 0$.

6. Suppose $f \in \mathscr{R}[a, b]$ and $\varepsilon > 0$. There exist $P_1, P_2 \in \mathscr{P}[a, b]$ such that

$$\int_a^b f - \varepsilon/2 < L(P_1, f), \quad U(P_2, f) < \int_a^b f + \varepsilon/2.$$

Let $P = P_1 \cup P_2$. Since $L(P_1, f) \le L(P, f) \le U(P, f) \le U(P_2, f)$, it follows that $U(P, f) - L(P, f) < \varepsilon$.

For the converse, as $L(P, f) \le \underline{\int_a^b} f \le \overline{\int_a^b} f \le U(P, f)$ whenever $P \in \mathscr{P}[a, b]$, we see that $0 \le \overline{\int_a^b} f - \underline{\int_a^b} f < \varepsilon$ whenever $\varepsilon > 0$.

Exercise 1.2.14

2. For the lack of uniform continuity, consider points $1/(2n + 1)$, $1/(2n)$.
3. Since f is uniformly continuous on $(0, 1]$, given $\varepsilon > 0$, there exists $\delta > 0$ such that $|f(x) - f(y)| < \varepsilon$ if $|x - y| < \delta$ ($x, y \in (0, 1]$). Suppose f is unbounded:

R. H. Dyer and D. E. Edmunds, *From Real to Complex Analysis*, Springer Undergraduate Mathematics Series, DOI: 10.1007/978-3-319-06209-9, © Springer International Publishing Switzerland 2014

there exists $(x_n) \subset (0, 1]$ such that $|f(x_n)| \to \infty$. But the bounded sequence (x_n) must have a convergent subsequence, denoted again by (x_n), and so there exists $N \in \mathbf{N}$ such that $|x_n - x_m| < \delta$ if $m, n > N$: thus $|f(x_n) - f(x_m)| < \varepsilon$ if $m, n > N$, contradicting $|f(x_n)| \to \infty$.

7. If f' is bounded, then by the mean-value theorem,

$$\left| \frac{f(x) - f(y)}{x - y} \right| = |f'(z)| \text{ for some } z \text{ between the distinct points } x, y.$$

Thus f is Lipschitz. The converse is obvious.

8. (a) $f(x) = |x - 1/2|$ is not differentiable at $1/2$.
   (b) Use question 7 above plus the fact that f' is not bounded.

9. Consider $f : [0, 1] \to \mathscr{R}$ given by $f(t) = |t - 1/2|$ if t is rational, $f(t) = 0$ otherwise.

10. Let $x \in [a, b)$ and $u_n = x + (b - x)/n$ $(n \in \mathbf{N})$. By Lemma 1.2.6 and Theorem 1.2.8, $f \in \mathbf{R}[u_{n+1}, u_n]$ and there exists $x_n \in (u_{n+1}, u_n)$ at which f is continuous. The case in which $x \in (a, b]$ is similar.

Exercise 1.3.10

1. By Example 1.1.4 (i) and Theorem 1.3.1 (v), $f \in \mathscr{R}[c_{j-1}, c_j]$ and

$$\int_{c_{j-1}}^{c_j} f = \alpha_j (c_j - c_{j-1}) \ (1 \le j \le k);$$

now use Corollary 1.3.5.

2. f is bounded on $[0, 1]$ and continuous on $(0, 1)$: use Example 1.3.9 (ii).

3. Use Theorem 1.2.13 to show that $|f|$, f^2 and $1/f^2 \in \mathscr{R}[a, b]$. Now appeal to Theorem 1.3.2 (a): $1/f = f \cdot (1/f^2) \in \mathscr{R}[a, b]$. Alternatively, observe that for any $P = \{x_0, x_1, \ldots, x_n\} \in \mathscr{P}[a, b]$,

$$U(P, 1/f) - L(P, 1/f) = \sum_{i=1}^{n} \text{osc} \, (1/f; [x_{i-1}, x_i]) \, \Delta x_i$$
$$\le \varepsilon^{-2} \sum_{i=1}^{n} \text{osc} \, (f; [x_{i-1}, x_i]) \, \Delta x_i$$
$$= \varepsilon^{-2} \{ U(P, f) - L(P, f) \},$$

and either apply Theorem 1.1.7 or Exercise 1.1.10/6.

6. For $n \in \mathbf{N}$, let $T_n = \{t \in [0, 1] : f(t) > 1/n\}$: T_n is finite.
   (a) Continuity: suppose $u \in [0, 1] \cap \mathbf{Q}$ and $f(u) = 1/q$. For all $\delta > 0$ there is an irrational number $v \in [0, 1]$ such that $|u - v| < \delta$ and $|f(u) - f(v)| = 1/q$. Hence f is not continuous at u. Now suppose $u \in [0, 1] \backslash \mathbf{Q}$, $\varepsilon > 0$ and $n \in \mathbf{N}$ is such that $1/n < \varepsilon$. Let $\delta = \min\{|u - t| : t \in T_n\} : \delta > 0$. Since $|f(u) - f(v)| = |f(v)| \le 1/n < \varepsilon$ if $v \in [0, 1]$ and $|v - u| < \delta$, f is continuous at u.
   (b) Integrability: for each $n \in \mathbf{N}$, let $f_n : [0, 1] \to \mathbf{R}$ be defined by

$$f_n(t) = \begin{cases} 0, & t \in T_n, \\ f(t), & t \in [0,1] \backslash T_n. \end{cases}$$

Since $f_n(t) \leq 1/n$ $(t \in [0,1])$ and $f \geq 0$, appeal to Theorem 1.3.1 (v) and Corollary 1.1.8 shows that

$$0 \leq \underline{\int_0^1} f \leq \overline{\int_0^1} f = \overline{\int_0^1} f_n \leq 1/n.$$

Exercise 1.4.15

1. Let $P_n \in \mathscr{P}[a,b]$ have as its points of partition $a + r(b-a)/n$ $(r = 0,1,2,\ldots,n)$; $w(P_n) = (b-a)/n \to 0$. Then

$$L(P_n, f) \leq \sum_{r=1}^n f\left(\xi_r^{(n)}\right)(b-a)/n \leq U(P_n, f).$$

Now use Theorem 1.1.7.

3. Use question 1 above to show that

$$\pi/4 = \int_0^1 (1+t^2)^{-1} dt = \lim \sum_{r=1}^n n^{-1} \left(1 + \frac{r^2}{n^2}\right)^{-1}.$$

4. Use question 1 above to show that

$$(b-a)^{-1} \int_a^b f = \lim \frac{1}{n} \sum_{r=0}^{n-1} f(x_{2r}) = \lim \frac{1}{n} \sum_{r=1}^n f(x_{2r-1})$$

$$= \lim \frac{1}{n} \sum_{r=1}^n f(x_{2r});$$

combine these.

5. Suppose $|f(t)|, |f'(t)| \leq K$ $(t \in [a,b])$. For $\lambda > 0$, integration by parts gives

$$\left| \int_a^b f(t) \cos \lambda t \, dt \right| = \left| \left[\frac{\sin \lambda t}{\lambda} f(t) \right]_a^b - \int_a^b f'(t) \left(\frac{\sin \lambda t}{\lambda} \right) dt \right|$$

$$\leq K(2 + b - a)/\lambda.$$

7. Since $t \longmapsto \mathrm{cosec}(t/2) : (0, 2\pi) \to \mathbf{R}$ is continuously differentiable, use of exercise 5 above shows that

$$\lim_{\lambda \to \infty} I_\lambda(x) = 0 \quad (0 < x < 2\pi). \tag{$*$}$$

For $n \in \mathbf{N}$ and $0 < x < 2\pi$,

$$I_{n+1/2}(x) - I_{n-1/2}(x) = \int_x^\pi \cos ntdt = -\frac{\sin nx}{n}$$

and hence

$$I_{k+1/2}(x) - I_{1/2}(x) = -\sum_{n=1}^k \frac{\sin nx}{n}.$$

Thus

$$\left|I_{k+1/2}(x)\right| = \left|\frac{1}{2}(\pi - x) - \sum_{n=1}^k \frac{\sin nx}{n}\right|.$$

Now use (*).

10. Use of Theorems 1.4.4 and 1.4.6 shows that $J(0) = \pi/2$, $J(1) = 1$ and $(n + 2)J(n + 2) = (n + 1)J(n)$ $(n \in \mathbf{N_0})$. Induction gives the formulae for $J(2n)$ and $J(2n + 1)$. Since $0 < 1 - \sin\theta < 1$ if $0 < \theta < \pi/2$, use of Theorem 1.3.1 (vi) shows that $J(2n + 1) < J(2n) < J(2n - 1)$. Now routine manipulation gives *Wallis's inequality*; the *product* follows trivially.

11. As $g'(x) = f'(x) - \gamma$, it follows that $g'(a) < 0 < g'(b)$. The continuity of g implies that g has a minimum on $[a, b]$, at $c \in [a, b]$, say. Since $0 > g'(a) = \lim_{y\to a} \frac{g(y)-g(a)}{y-a}$, $g(y) < g(a)$ when y is close enough to a: thus $c \neq a$; similarly $c \neq b$. Hence $c \in (a, b)$ and so $g'(c) = 0$.

12. Take $f = F'$, where $F(x) = x^2 \sin(1/x)$ $(0 < x \le 1)$, $F(0) = 0$.

15. On making the suggested substitution,

$$I(a) = \int_0^{\pi/2} \log\left\{(1 - 2a\cos x + a^2)(1 + 2a\cos x + a^2)\right\}$$

$$= \int_0^{\pi/2} \log(1 + a^4 - 2a^2\cos 2x)dx = I(a^2)/2.$$

Thus for all $n \in \mathbf{N}$, $I(a) = 2^{-n}I(a^{2^n})$: if $|a| < 1$, then

$$|I(a)| \le \left|\log(1 - |a|)^2\right|\pi/2^n,$$

and so $I(a) = 0$. If $|a| > 1$, then $I(a) = 2\pi\log|a| + I(1/a) = 2\pi\log|a|$.

Exercise 1.5.7

1. Both ϕ and $\log\phi$ belong to $\mathcal{R}[a, b]$. For each $n \in \mathbf{N}$, write $x_r^{(n)} = a + r(b-a)/n$ $(r = 0, 1, \ldots, n)$, $A_n = n^{-1}\sum_{r=1}^n \log\left(\phi\left(x_r^{(n)}\right)\right)$, $B_n = n^{-1}\sum_{r=1}^n \phi\left(x_r^{(n)}\right)$. Then $A_n = \log\left\{\prod_{r=1}^n \phi\left(x_r^{(n)}\right)\right\}^{1/n}$, and appeal to the arithmetic-geometric inequality and Exercise 1.4.15/1 shows that

$$A_n \le \log B_n, \quad A_n \to (b-a)^{-1} \int_a^b \log(\phi(t))dt, \quad B_n \to (b-a)^{-1} \int_a^b \phi(t)dt > 0.$$

2. Use the integral test.

4. Suppose that $\alpha < \beta$; when $\beta < \alpha$ the proof is similar. Define $R_n(\beta) = \frac{1}{(n-1)!} \int_\alpha^\beta (\beta - t)^{n-1} f^{(n)}(t)dt$. Integrate by parts to obtain $R_k(\beta) - R_{k+1}(\beta) = (\beta - \alpha)^k f^{(k)}(\alpha)/k! \; (k = 1, \ldots, n-1)$; by the fundamental theorem of calculus, $R_1(\beta) = f(\beta) - f(\alpha)$. Sum over k:

$$R_n(\beta) = f(\beta) - \sum_{k=0}^{n-1} \frac{f^{(k)}(\alpha)}{k!} (\beta - \alpha)^k.$$

To derive the result involving γ, let m, M be the inf and sup respectively of $f^{(n)}$ on $[\alpha, \beta]$. If $m = M$, use the fundamental theorem of calculus. Otherwise, by the continuity of $f^{(n)}$, there exists $t \in (\alpha, \beta)$ such that $m < f^{(n)}(t) < M$, which leads to $m(\beta - \alpha)^n/n! < R_n(\beta) < M(\beta - \alpha)^n/n!$. Now use the intermediate-value theorem. For the final part, proceed as above, this time by consideration of the inf and sup of $(\beta - t)^{n-1} f^{(n)}(t)$.

5. By Theorem 1.4.4, f_1 is differentiable, $f_1' = f$ and $f_1(a) = 0$. Use of induction together with Theorem 1.4.9 shows that each f_n is n—times differentiable and $f_n^{(j)} = f_{n-j}$ $(j = 0, 1, \ldots, n)$, setting $f_0 = f$. Since $f_j(a) = 0$ $(j = 1, \ldots, n)$, the result follows from Taylor's theorem.

7. (i) Plainly L is linear; since $L(1) = L(u_1) = L(u_2) = L(u_3) = 0$, where $u_j(t) = t^j$, it follows that $L(p) = 0$ if p is a polynomial of degree at most 3. Use Taylor's theorem with integral form of the remainder to give

$$L(f) = \frac{1}{3!} \left\{ \int_0^h \left(\int_0^x (x-t)^3 f^{(4)}(t)dt \right) dx - \frac{h}{3} \int_0^h (h-x)^3 f^{(4)}(x)dx \right\}$$

$$+ \frac{1}{3!} \left\{ \int_{-h}^0 \left(\int_0^x (x-t)^3 f^{(4)}(t)dt \right) dx - \frac{h}{3} \int_{-h}^0 (h+x)^3 f^{(4)}(x)dx \right\}$$

$$= -\frac{1}{72} \left\{ \int_0^h (h-x)^3(h+3x)f^{(4)}(x)dx \right.$$

$$\left. + \int_{-h}^0 (h+x)^3(h-3x)f^{(4)}(x)dx \right\},$$

from which the required inequality follows.

(ii) Let $M = \sup_{a \le t \le b} \left| f^{(4)}(t) \right|$, $h = (b-a)/n$, $g_r(t) = f(t + x_{2r-1})$ $(1 \le r \le n)$. Then g_r is four times differentiable, $g_r^{(4)} \in \mathscr{R}[-h, h]$ and

$$\left| \int_{-h}^h g_r - \frac{h}{3} \{g_r(-h) + 4g_r(0) + g_r(h)\} \right| \le \frac{Mh^5}{90}.$$

Thus

$$\left| \int_{x_{2r-2}}^{x_{2r}} f - \frac{(b-a)}{6n} \{ f(x_{2r-2}) + 4f(x_{2r-1}) + f(x_{2r}) \} \right| \leq \frac{M(b-a)^5}{2880n^5}.$$

Exercise 1.6.15

2. Use Theorems 1.6.11 and 1.6.12.
3. The integrand is locally integrable. For $0 < u < 1 < v < \infty$, let $a(u) = \int_u^1 x^{p-1} e^{-x} dx$, $b(v) = \int_1^v x^{p-1} e^{-x} dx$. Then $a(u) \leq \int_u^1 x^{p-1} dx \leq 1/p$ and $a(u)$ increases as u decreases: hence $\lim_{u \to 0} a(u)$ exists in $\mathbf{R}$. Moreover, if $n \in \mathbf{N}$, $n > p+1$, then $b(v) \leq n! \int_1^v x^{-(n+1-p)} dx \leq n! \int_1^v x^{-2} dx \leq n!$; since $b(v)$ increases with v, $\lim_{v \to \infty} b(v)$ exists in $\mathbf{R}$. Thus $\Gamma(p) = \lim_{u \to 0} a(u) + \lim_{v \to \infty} b(v)$. For the rest, use integration by parts.
4. (a) To deal with the lower limit of integration, $\alpha > 0$; for the upper limit we need $\alpha < 1$: for both, $\alpha \in (0, 1)$.
   (b) $\alpha < 1$.
   (c) Integrand is $x^{-\alpha+1} \frac{\sin x}{x}$; need $\alpha < 2$.
   (d) $\alpha > -1$.
6. For $n \in \mathbf{N}$, define $f_n, g_n : [0, \pi/2] \to \mathbf{R}$ by $f_n(x) = \cot x \sin 2nx$ ($x \neq 0$), $f_n(0) = 2n$; $g_n(x) = x^{-1} \sin 2nx$ ($x \neq 0$), $g_n(0) = 2n$. The functions f_n and g_n are continuous; moreover,

$$\int_0^{\pi/2} f_n(x) dx = \int_0^{\pi/2} \left\{ 1 + 2 \sum_{r=1}^{n-1} \cos 2rx + \cos 2nx \right\} dx = \pi/2$$

and, using a change of variable and Remark 1.6.13,

$$\lim_{n \to \infty} \int_0^{\pi/2} g_n(x) dx = \lim_{n \to \infty} \int_0^{n\pi} x^{-1} \sin x dx = \int_0^\infty x^{-1} \sin x dx.$$

Define $h : [0, \pi/2] \to \mathbf{R}$ by $h(x) = \cot x - x^{-1}$ ($x \neq 0$), $h(0) = 0$; h is continuously differentiable. Use Exercise 1.4.15/5 to show that

$$0 = \lim_{n \to \infty} \int_0^{\pi/2} h(x) \sin 2nx dx = \lim_{n \to \infty} \int_0^{\pi/2} (f_n(x) - g_n(x)) dx$$

$$= \frac{\pi}{2} - \int_0^\infty x^{-1} \sin x dx.$$

7. $\int_\varepsilon^{1/e} x^{-1} (\log(1/x))^{-\theta} dx = \left[(\log(1/x))^{-\theta+1} \right]_\varepsilon^{1/e} /(\theta - 1)$.
8. Use integration by parts to evaluate $\int_0^v e^{-x} \sin(\lambda x) dx$ and then let $v \to \infty$.

9. $\int_2^v \frac{\cos x}{\log x} dx = \left[\frac{\sin x}{\log x}\right]_2^v + \int_2^v \frac{\sin x}{x(\log x)^2} dx$; note that the integral on the right is bounded above by $\int_2^v \frac{1}{x(\log x)^2} dx$. The other part of the question is similar but easier.

10. $\int_u^v \sin(e^x) dx = \int_{e^u}^{e^v} \frac{\sin y}{y} dy$.

Exercise 1.7.17

1. (a) $\lim_{n\to\infty} f_n(x) = 0$ but $f_n(1/n) = n^2/(n^2 + 1) \geq 1/2$: not uniformly convergent.

   (b) $\lim_{n\to\infty} f_n(x) = 0$ but $f_n\left(\sqrt[n]{1/2}\right) = 1/4$: not uniformly convergent.

   (c) $\lim_{n\to\infty} f_n(x) = 0$ $(0 \leq x < 1)$, $1/2$ $(x = 1)$. Each f_n is continuous but the limit function is not: not uniformly convergent.

   (d) $|f_n(x)| \leq 1/n$: uniformly convergent.

2. Since $0 \leq f_n(x) \leq n^p(1 - x)^n$ it follows that $\lim_{n\to\infty} f_n(x) = 0$ $(0 < x \leq 1)$; also plainly $\lim_{n\to\infty} f_n(0) = 0$. As $f_n'(x) = n^p(1 - x)^{n-1}\{1 - (n + 1)x\} = 0$ iff $x = 1$ or $1/(n+1)$, we see that f_n has a global maximum at $1/(n+1)$. Thus

$$0 \leq f_n(x) \leq f_n\left(1/(n + 1)\right) = n^{p-1}/\left(1 + \frac{1}{n}\right)^{n+1},$$

which shows that (f_n) is uniformly convergent to 0 on $[0, 1]$ iff $p < 1$. Since $\int_0^1 f_n = \frac{n^p}{(n+1)(n+2)}$ the rest follows.

3. For the first part use the Weierstrass M−test. The series does not converge when $x = 1$ as the nth term does not tend to zero.

4. If $0 < |t| < 1$, then $t^{-1}\log(1 - t^2) = -\sum_{r=0}^\infty t^{2r+1}/(r+1)$. Use the Weierstrass M−test to show that $\sum_{r=0}^\infty \left(\frac{1}{2}\sin\theta\right)^{2r+1}/(r+1)$ converges uniformly on $[0, 1]$, justifying term by term integration.

8. Since f_n is continuous and $|f_n(x)| \leq 2.4^{-n}$ $(x \in \mathbf{R}, n \in \mathbf{N})$, Theorems 1.7.5 and 1.7.7 show that $\sum_{n=1}^\infty f_n$ converges uniformly on $\mathbf{R}$ and that f is continuous. Fix $u \in \mathbf{R}$. Note that if $r \in \mathbf{Z}$ and $2r \leq x, y \leq 2(r + 1)$, then

$$|g(x) - g(y)| = |x - y|. \tag{*}$$

Given any $k \in \mathbf{N}$, there is a unique $r_k \in \mathbf{Z}$ such that $2r_k \leq 4^k u < 2(r_k + 1)$. Let $I_k = [2r_k, 2(r_k + 1)]$; each I_k has length 2 and $I_{k+1} \subset 4I_k$ $(k \in \mathbf{N})$. Set $v_k = 1$ if $4^k u + 1 \in I_k$, $v_k = -1$ otherwise. Then $4^k u + v_k \in I_k$ $(k \in \mathbf{N})$, where $v_k \in \{-1, 1\}$; moreover, $4^j u + 4^{j-k} v_k \in I_j$ if $1 \leq j \leq k$, since $I_k \subset 4I_{k-1} \subset 4^2 I_{k-2} \subset \ldots \subset 4^{k-1} I_1$. For each $n \in \mathbf{N}$,

$$\left|f_n\left(u + 4^{-k}v_k\right) - f_n(u)\right| = 4^{-n}\left|g\left(4^n u + 4^{n-k}v_k\right) - g\left(4^n u\right)\right|;$$

further, since g is periodic with period 4 and $(*)$ holds, it follows that

$$\left| f_n\left(u + 4^{-k}v_k\right) - f_n(u) \right| = \begin{cases} 4^{-k}, & 1 \le n \le k, \\ 0, & n > k. \end{cases}$$

Let $h_k = 4^{-k}v_k$. Then

$$h_k^{-1}\left(f(u + h_k) - f(u)\right) = \sum_{n=1}^{k} h_k^{-1}\left(f_n(u + h_k) - f_n(u)\right) = \sum_{n=1}^{k} v_k,$$

an integer which is even or odd according as k is even or odd. Since $h_k \to 0$ and $\lim_{k \to \infty} h_k^{-1}\left(f(u + h_k) - f(u)\right)$ does not exist, f is not differentiable at u.

9. The maximum value of $x^n(1 - x)$ on $[0, 1]$ is $\left(\frac{n}{n+1}\right)^n / (n + 1)$; by the M–test, the series is uniformly convergent on $[0, 1]$.

10. For all $x \in \mathbf{R}$, $\left|(2n)! \sin^3(x/n!)\right| \le (2n)!/(n!)^3 \le 2^{2n}/n! := a_n$. Since $\lim a_{n+1}/a_n = 0$, the M–test show that the series is uniformly convergent on $\mathbf{R}$; continuity follows.

11. If $0 \le x < 1$, then $(1 - x)/(1 - x^3) = (1 - x)(1 + x^3 + \ldots + x^{3(n-1)}) + x^{3n}/(1 + x + x^2)$, so that

$$\left| \int_0^1 \frac{1 - x}{1 - x^3}dx - \sum_{r=0}^{n-1} \int_0^1 (1 - x)x^{3r}dx \right| = \int_0^1 \frac{x^{3n}}{1 + x + x^2}dx \le \int_0^1 x^{3n}dx$$

$$= \frac{1}{3n + 1}.$$

Hence

$$\sum_{r=0}^{\infty} \frac{1}{(3r + 1)(3r + 2)} = \int_0^1 \frac{1 - x}{1 - x^3}dx = \int_0^1 (1+x+x^2)^{-1}dx = \pi/(3\sqrt{3}).$$

Apply this technique to the given integral.

13. Expand $\log(1 - x)$.

14. Use of the inequalities

$$\int_0^1 (1 - t^2)^n dt \ge \int_0^{1/n} (1 - t^2)^n dt \ge \int_0^{1/n} (1 - nt^2)dt > \frac{1}{2n}$$

shows that $c_n < n$ for all $n \in \mathbf{N}$. Let $0 < \delta \le 1$. The uniform convergence of (q_n) on $[-1, -\delta] \cup [\delta, 1]$ is clear: $\sup_{\delta \le |t| \le 1} q_n(t) \le n(1 - \delta^2)^n \to 0$. Each $p_n(s)$ is a polynomial. Since $f(t) = 0$ if $|t| \ge 1$, use of Exercise 1.3.10/5 shows that, for all $n \in \mathbf{N}$ and $s \in [0, 1]$,

$$p_n(s) = \int_{-s}^{1-s} f(u + s)q_n(u)du = \int_{-1}^{1} f(u + s)q_n(u)du$$

and hence

$$p_n(s) - f(s) = \int_{-1}^{1} (f(u+s) - f(s)) q_n(u) du.$$

Let $K = 1 + \sup_{t \in \mathbf{R}} |f(t)|$ and $\varepsilon > 0$. As f is uniformly continuous, there exists $\delta \in (0, 1)$ such that for all $x \in \mathbf{R}$, $|f(x+h) - f(x)| < \varepsilon$ if $|h| < \delta$. Further, there exists $N \in \mathbf{N}$ such that $\sup_{\delta \le |u| \le 1} q_n(u) < \varepsilon$ if $n \ge N$. Thus if $s \in [0, 1]$ and $n \ge N$, then

$$\int_{-\delta}^{\delta} |f(u+s) - f(s)| q_n(u) du < \varepsilon \text{ and } \int_{-1}^{-\delta} |f(u+s) - f(s)| q_n(u) du$$
$$< 2K\varepsilon;$$

similarly, $\int_{\delta}^{1} |f(u+s) - f(s)| q_n(u) du < 2K\varepsilon$. Hence, for all $n \ge N$,

$$\sup_{s \in [0,1]} |p_n(s) - f(s)| \le \int_{-1}^{1} |f(u+s) - f(s)| q_n(u) du < (4K+1)\varepsilon,$$

from which the desired uniform convergence follows. The rest is a matter of routine manipulation.

18. Let $g_n = f_n - f$ for each n and suppose that (g_n) is decreasing and converges pointwise but not uniformly to 0. For some $\varepsilon > 0$ there is a sequence (x_n) in $[a, b]$ such that $g_n(x_n) \ge \varepsilon$ $(n \in \mathbf{N})$. By Theorem A.4.13, a point $x \in [a, b]$ and a subsequence $(x_{m(n)})$ of (x_n) exist such that $x_{m(n)} \to x$. Let $k \in \mathbf{N}$. For $n \ge k$, $g_k (x_{m(n)}) \ge g_{m(n)} (x_{m(n)}) \ge \varepsilon$; further, since g_k is continuous, $\lim_{n \to \infty} g_k (x_{m(n)}) = g_k(x) \ge \varepsilon$. Thus $0 = \lim_{k \to \infty} g_k(x) \ge \varepsilon$: contradiction.

Exercise 2.1.45

1. Since $f'(t) = t^{p-1} - 1$, the minimum of f is attained only at 1 and the minimum value is 0. Hence $t^p/p + 1/p' - t \ge 0$ for all $t \ge 0$. Now take $t = ab^{-1/(p-1)}$ (assuming $b \ne 0$; the result is obvious if $b = 0$). Hölder's inequality is clear if either $\sum |x_k|^p$ or $\sum |y_k|^p$ is zero. Otherwise, put $a = |x_j| / (\sum |x_k|^p)^{1/p}$, $b = |y_j| / (\sum |y_k|^{p'})^{1/p'}$ in the first part to obtain

$$\frac{|x_j y_j|}{(\sum |x_k|^p)^{1/p} (\sum |y_k|^{p'})^{1/p'}} \le \frac{|x_j|^p}{p \sum |x_k|^p} + \frac{|y_j|^{p'}}{p' \sum |y_k|^{p'}};$$

now sum over j. For Minkowski's inequality, note that

$$\sum_1^n |x_k + y_k|^p \le \sum_1^n |x_k + y_k|^{p-1} |x_k| + \sum_1^n |x_k + y_k|^{p-1} |y_k|$$

$$\le \left(\sum_1^n |x_k + y_k|^p\right)^{1/p'} \left\{ \left(\sum_1^n |x_k|^p\right)^{1/p} + \left(\sum_1^n |y_k|^{p'}\right)^{1/p'} \right\}.$$

Verification of the metric space axioms is routine.

2. Only the triangle inequality presents any problems: proceed as in question 1 above to obtain the integral version of Minkowski's inequality.

3. As above, the triangle inequality holds for sums $\sum_1^N$; now let $N \to \infty$.

4. The triangle inequality follows from the hint and the fact that $|x_n - y_n| \le |x_n - z_n| + |z_n - y_n|$.

6. (i) Closed, not open, (ii) neither, (iii) neither, (iv) neither.

7. Last part: no.

8. Let S be any proper non-empty subset. If $x \in S$, then $B(x, 1) = \{x\} \subset S$ and so S is open. Now apply the same argument to the complement of S.

9. Let $x, y, z \in S$ and suppose that $d(x, z) = 1/k, d(z, y) = 1/l$. Then $x_r = z_r$ $(r < k), x_k \ne z_k; z_r = y_r$ $(r < l), z_l \ne y_l$. Hence either $x = y$, in which case $d(x, y) = 0$, or $d(x, y) = 1/t$ for some $t \ge \min(k, l)$.

11. Take $x = y$ in (ii): $d(y, z) \le d(z, y)$. Interchange of z and y gives $d(y, z) = d(z, y)$.

12. Let U be open in (X, d_1). Given $x \in U$, there exists $\varepsilon > 0$ such that $\{y \in X : d_1(x, y) < \varepsilon\} \subset U$: thus $\{y \in X : d_2(x, y) < \alpha\varepsilon\} \subset U$ and so U is open in (X, d_2). Interchange the rôles of d_1 and d_2.

13. If $x \in \overset{o}{A} \cap \overset{o}{B}$ then $B(x, \varepsilon_1) \subset A$ and $B(x, \varepsilon_2) \subset B$ for some $\varepsilon_1, \varepsilon_2 > 0$: thus $B(x, \varepsilon) \subset A \cap B, \varepsilon = \min(\varepsilon_1, \varepsilon_2)$, giving $\overset{o}{A} \cap \overset{o}{B} \subset \overset{\frown}{A \cap B}$. The reverse inclusion is obvious. Also, plainly $A \cup B$ is contained in the closed set $\overline{A} \cup \overline{B}$, so that $\overline{A \cup B} \subset \overline{A} \cup \overline{B}$; the reverse inclusion is obvious. Take $X = \mathbf{R}, A = \mathbf{Q}, B = {}^c\mathbf{Q}$: $\overset{o}{A} \cup \overset{o}{B} = \emptyset, \overset{\frown}{A \cup B} = \mathbf{R}, A \cap B = \mathbf{R}, A \cap B = \emptyset$.

D is closed; for its interior, replace $\ge$ by $>$.

14. (i) Interior $\{(x, y) : 0 < x < y < 1\}$, closure $\{(x, y) : 0 \le x \le y \le 1\}$; (ii) interior empty, closure $\{(x, 0) : 0 \le x \le 1\}$; (iii) interior empty, closure $\mathbf{R}^2$.

15. Let $y \in [0, 1]\backslash S$. Uniqueness of decimal representation is obtained by ruling out recurring 9's. There is a first digit in the decimal representation of y which is not 0 or 1: suppose it is the nth digit. Then the distance of y from S is greater than 10^{-n-1} and so $B(y, 10^{-n-1}) \cap S = \emptyset$: thus S is closed.

16. Let (E, d) be discrete and let x, y be distinct points of E. Then $B(x, 1) = \{x\} = \overline{\{x\}}$, while $\{z \in E : d(x, z) \le 1\} = E$.

17. (i) A is a closed subset of Y iff $Y\backslash A$ is open in Y, which by Lemma 2.1.5 (iii) is equivalent to the existence of an open set U in X such that $Y\backslash A = Y \cap U$, which is equivalent to $A = Y \cap (X\backslash U)$.

(ii) By (i), $cl_Y(S) = B \cap Y$ for some closed subset B of X: hence $cl_Y(S)$ is closed in X and so $cl_Y(S) \supset cl_X(S)$. Given $x \in cl_Y(S)$, there exists $(x_n) \subset S$ such that $x_n \to x$: thus $cl_Y(S) \subset cl_X(S)$.

18. Continuity at $(x, y) \neq (0, 0)$: $(x_n, y_n) \to (x, y)$ implies that $x_n \to x$ and $y_n \to y$, so that $f(x_n, y_n) \to xy/(x^2 + y^2)$. Discontinuity at $(0, 0)$: $f(1/n, 1/n) = 1/2 \not\to 0$.

20. f is continuous everywhere except at $(0, 0)$: $f(2y, y) = (3/5, 9y^2/5) \not\to (0, 0)$ as $y \to 0$.

21. The map $f : \mathbf{R}^2 \to \mathbf{R}$ given by $f(x, y) = x^2 - y^2 + 2xy$ is continuous: $S = f^{-1}((-\infty, 0))$ is open.

23. There is a sequence (a_n) of elements of A that converges to a: thus $a \in \overline{A}$.

24. Take $U = f^{-1}((1/2, 3/2))$, $V = f^{-1}((-3/2, -1/2))$, where f is as in Urysohn's lemma.

Exercise 2.2.29

2. (i) If (x_n) is a Cauchy sequence in F, then $x_n \to x \in X : x \in F$ as F is closed, and so F is complete.

(ii) Let $x \in \overline{F}$, and let (x_n) be a sequence in F that converges to x. Then (x_n) is a Cauchy sequence in F; (x_n) converges to a point of F; thus $x \in F$ and F is closed.

3. For all $f, g \in C(I)$ and $x \in I$,

$$|(Tf)(x) - (Tg)(x)| \leq d_\infty(f, g) \int_0^x (x - t)dt \leq (1/2)d_\infty(f, g).$$

Thus $d_\infty(Tf, Tg) \leq d_\infty(f, g)/2$. Hence T has a unique fixed point; routine verification or use of the iterative process shows that it is given by $\sinh x$.

4. Define $T : C([0, k]) \to C([0, k])$ by $(Tf)(x) = 1 + \int_0^x f(t^2)dt$. Since

$$|(Tf)(x) - (Tg)(x)| \leq \int_0^x \left| f(t^2) - g(t^2) \right| dt \leq x d_\infty(f, g),$$

it follows that $d_\infty(Tf, Tg) \leq k d_\infty(f, g)$. By the contraction mapping theorem the integral equation has exactly one solution f_k, when $k \in (0, 1)$.

To deal with the case $k = 1$, define f by $f(x) = f_l(x)$ whenever $x \leq l < 1$. Since the sequence $(T^n g)_{n \in \mathbf{N}}$ converges to f, direct calculation of $T^n g$ shows that f is bounded above on $(0, 1)$. Thus f satisfies the integral equation on $(0, 1)$ and may be extended by continuity to $[0, 1]$.

5. (i) $T : (0, 1) \to (0, 1)$, $Tx = x^2/4$.

(ii) $T : \mathbf{R} \to \mathbf{R}$, $Tx = \frac{\pi}{2} + x - \tan^{-1} x$. Given $x, y \in \mathbf{R}$, $x \neq y$, there exists z strictly between x and y such that $|Tx - Ty| = |x - y| z^2/(1 + z^2) < |x - y|$; but T has no fixed point.

(iii) $T : \mathscr{B}([0, 1]) \to \mathscr{B}([0, 1])$,

$$(Tf)(x) = \begin{cases} f\left(\frac{1}{2}+x\right), & 0 \le x < \frac{1}{2}, \\ 0, & \frac{1}{2} \le x \le 1. \end{cases}$$

Note that $T^2 = 0$.

6. Define $\phi : (0, \infty) \to (0, \infty)$ by $\phi(t) = 1/t : \phi$ coincides with its inverse and is continuous when $(0, \infty)$ is given the natural metric inherited from $\mathbf{R}$. Thus $|x - x_n| \to 0 \implies |\phi(x) - \phi(x_n)| \to 0 \implies |\phi(\phi(x)) - \phi(\phi(x_n))| \to 0 :$ $d_1(x, x_n) \to 0 \implies d_2(x, x_n) \to 0 \implies d_1(x, x_n) \to 0$. If (x_n) is a Cauchy sequence in (X, d_2), then there exists $y \in \mathbf{R}$, $y \ge 1$, such that $\phi(x_n) \to y$, so that $d_2(x_n, 1/y) \to 0$: thus $(\mathbf{R}, d_2)$ is complete. However, $(\mathbf{R}, d_1)$ is not complete: consider $(1/n)$.

Exercise 2.3.38

2. Suppose the result false. Then for each $n \in \mathbf{N}$, there exists $x_n \in X$ such that $d(x_n, F_i) < 1/n$ for all i: as X is compact, (x_n) must have a convergent subsequence, with limit $x \in X$; and as each F_i is closed, $x \in F_i$ for all i, giving a contradiction.

4. As X is compact, there are convergent subsequences $(T^{n_j}(a))_{j \in \mathbf{N}}$ and $(T^{n_j}(b))_{j \in \mathbf{N}}$: thus given $\varepsilon > 0$, there exists $J \in \mathbf{N}$ such that

$$d\left(T^{n_j}(a), T^{n_{j+l}}(a)\right) < \varepsilon \text{ and } d\left(T^{n_j}(b), T^{n_{j+l}}(b)\right) < \varepsilon$$

whenever $j \ge J$ and $l \in \mathbf{N}$. Thus $d\left(a, T^k(a)\right) < \varepsilon$ and $d\left(b, T^k(b)\right) < \varepsilon$, where $k = n_{J+1} - n_J$. Hence

$$d(T(a), T(b)) \le d(T^k(a), T^k(b)) \le d(T^k(a), a) + d(a, b) + d(b, T^k(b))$$
$$< d(a, b) + 2\varepsilon.$$

Thus $d(T(a), T(b)) = d(a, b)$. Density is clear from $d\left(a, T^k(a)\right) < \varepsilon$. Given $a \in X$, by the density of $T(X)$ and the compactness of X, there is a sequence (x_n) such that $x_n \to c$ and $Tx_n \to a$: thus

$$d(a, Tc) \le d(a, Tx_n) + d(Tx_n, Tc) = d(a, Tx_n) + d(x_n, c),$$

which implies that $a = Tc$.

5. Uniqueness: if $Tx = x$, $Ty = y$ and $x \ne y$, then $d(x, y) = d(Tx, Ty) < d(x, y)$: contradiction.

Existence: suppose there is no fixed point. The map $x \longmapsto d(x, Tx)$ is continuous on the compact set X and so attains its minimum, at x_0, say. But $d(Tx_0, T^2x_0) < d(x_0, Tx_0)$: contradiction.

6. Suppose (f_n) is decreasing, put $g_n = f_n - f : g_n \ge g_{n+1} \ge 0$. Put $M_n = \sup\{g_n(x) : x \in X\} = d(g_n, 0)$: we must show that $M_n \to 0$. Given $\varepsilon > 0$, put $U_n = g_n^{-1}((-\infty, \varepsilon))$. Then U_n is open, increases with n and for all $x \in X$, $g_n(x) \to 0 : x \in U_n$ for some n. Hence the U_n cover the compact X, which is

thus covered by a finite collection of the U_n: as the U_n are nested, $U_N = X$ for some N, which means that $g_N(x) < \varepsilon$ for all $x \in X$, and so $M_N \leq \varepsilon$. Since $M_n \downarrow$ as $n \uparrow$, and $M_n \geq 0$, $\lim M_n = 0$.

10. Boundedness is clear; for equicontinuity and pointwise convergence use the mean-value theorem. If the set were relatively compact, there would be a convergent sequence $\left(f_{n_k}\right)$, with limit the zero function. But $f_{n_k}\left((4n_k + 1)\pi^2\right) = -1$ for all k: contradiction.

11. Use the mean-value theorem.

12. Suppose that f is not lower semi-continuous at x but $f(x) \leq \liminf f(x_n)$ whenever $x_n \to x$. Then there exist $\varepsilon > 0$ and a sequence (y_n) such that $d(x, y_n) < 1/n$ and $f(y_n) \leq f(x) - \varepsilon$: thus $y_n \to x$ and $\liminf f(y_n) < f(x)$, and we have a contradiction. Conversely, suppose f is lower semi-continuous at x, $x_n \to x$ and $\varepsilon > 0$. Then there exists $\delta > 0$ such that $f(y) > f(x) - \varepsilon$ if $d(x, y) < \delta$. Thus for some $n \in \mathbf{N}$, $\inf\{f(x_k) : k \geq n\} \geq f(x) - \varepsilon$, which implies that $\liminf f(x_n) \geq f(x) - \varepsilon$. This holds for all $\varepsilon > 0$.

15. Suppose the result false. Then for each $n \in \mathbf{N}$, there exists $A_n \subset X$ not contained in any U and with diam $A_n < 1/n$. Let $x_n \in A_n$; the sequence (x_n) has a convergent subsequence with limit x; $x \in U$ for some U; $B(x, \varepsilon) \subset U$ for some $\varepsilon > 0$. There exists $n > 2/\varepsilon$ such that $x_n \in B(x, \varepsilon/2)$. If $a \in A_n$, then

$$d(a, x) \leq d(a, x_n) + d(x_n, x) < \text{diam } A_n + \varepsilon/2 < \varepsilon.$$

Thus $A_n \subset B(x, \varepsilon) \subset U$: contradiction.

Exercise 2.4.33

1. Suppose B is not connected. Then there are open sets U, V such that $U \cap B \neq \emptyset$, $V \cap B \neq \emptyset$, $U \cap V \cap B = \emptyset$ and $B \subset U \cup V$. Then $U \cap V \cap A = \emptyset$ and $A \subset U \cup V$. If $U \cap A = \emptyset$, then $A \subset {}^cU$, which is closed: hence $\overline{A} \subset {}^cU$, so that $U \cap B = \emptyset$: contradiction. Thus $U \cap A \neq \emptyset$; similarly $V \cap A \neq \emptyset$. This means that A is not connected: contradiction.

2. $\mathbf{R}^2$ is path-connected and thus connected. Since $S \cap (\mathbf{C}\backslash S) = \emptyset$, $S \cup (\mathbf{C}\backslash S) = \mathbf{C}$ and both S and $\mathbf{C}\backslash S$ are open, either S or $\mathbf{C}\backslash S$ is empty.

3. E is path-connected and thus connected. F is not connected: its components are $\left\{(x, y) : x^2 + y^2 < 1\right\}$ and the one-point sets $\{(1 + 1/n, 0)\}$ $(n \in \mathbf{N})$.

4. Define $f : GL(n, \mathbf{R}) \to \mathbf{R}\backslash\{0\}$ by $f(a) = \det a$. Since f is continuous and surjective, and $\mathbf{R}\backslash\{0\}$ is not connected, $GL(n, \mathbf{R})$ is not connected.

5. Let $c \in A \cap B$ and suppose $a, b \in A \cup B$. There are paths in $A \cup B$ joining a to c and c to b: combine them.

6. Let $c, d \in f(E)$, $c = f(a)$, $d = f(b)$. There is a path γ in E joining a and b: consider $f \circ \gamma$.

Exercise 2.5.30

1. Define $g : S^n \times I \to \mathbf{R}^{n+1}$ by $g(x, t) = (1 - t)f(x) + tx$; both g and $(x, t) \longmapsto \|g(x, t)\|$ are continuous. The hypothesis guarantees that for all $(x, t) \in S^n \times I, g(x, t) \neq 0$, for the only zero of g could be when $t = 1/2$ and $f(x) = -x$, which is impossible. Hence H is continuous. Since $H(x, 0) = f(x)$ and $H(x, 1) = x$ $(x \in S^n)$, $f \simeq \mathrm{id}_{S^n}$.

2. Suppose $\mu \sim \nu$. Then $\mu * \widehat{\nu} \sim \nu * \widehat{\nu}$ (Theorem 2.5.9) $\sim e_x$ (Theorem 2.5.13). By Theorem 2.5.5, $\mu * \widehat{\nu} \sim e_x$. Conversely, suppose $\mu * \widehat{\nu} \sim e_x$. Then $\mu \sim \mu * e_y$ (Theorem 2.5.12) $\sim \mu * (\widehat{\nu} * \nu)$ (Theorems 2.5.9 and 2.5.13) $\sim (\mu * \widehat{\nu}) * \nu$ (Theorem 2.5.11) $\sim e_x * \nu$ (Theorem 2.5.9) $\sim \nu$ (Theorem 2.5.12).

3. Consider the maps $f, g : I \to \mathbf{R}^2$ defined by $f(t) = (\cos 2\pi t, \sin 2\pi t)$, $g(t) = (1, 0)$.

4. To avoid triviality, suppose $0 < a < 1$. Let $K_1 = \{(s, t) \in I \times I : s + at \leq 1\}$, $K_2 = \{(s, t) \in I \times I : s + at \geq 1\}$; both these sets are closed. The maps $(s, t) \longmapsto s + at \longmapsto f(s + at) : K_1 \to X$, $(s, t) \longmapsto s + at - 1 \longmapsto f(s + at - 1) : K_2 \to X$ are continuous and, since $f(1) = f(0)$, their values coincide at points $(s, t) \in K_1 \cap K_2$. Hence H is consistently defined and the glueing lemma shows it to be continuous. Plainly $H(s, 0) = f(s)$, $H(s, 1) = g(s)$ $(s \in I)$ and $H(0, t) = f(ta) = H(1, t)$ $(t \in I)$. The rest is left to the reader.

5. Let $F : X \times I \to Y$ be defined by $F(x, t) = (1 - t)f_0(x) + tf_1(x)$. Then $F(x, 0) = f_0(x)$, $F(x, 1) = f_1(x)$ and it remains to prove that F is continuous. Since

$$d_2(F(x_0, t_0), F(x_1, t_1)) \leq |t_0 - t_1| \, d_2(f_0(x_1), f_1(x_1))$$
$$+ (1 - t_0) d_2(f_0(x_1), f_0(x_0)) + t_0 d_2(f_1(x_1), f_1(x_0)),$$

routine procedures give the result.

7. (i) X and Y are not homeomorphic as the removal of $(1, 0)$ from Y disconnects Y but the removal of any point from X leaves X connected.

Let $T = \{(x, 0) : 1 \leq x \leq 2\}$, $p = (1, 0)$; let d be the standard metric on $\mathbf{R}^2$. Since $d(f(w), f(x)) = d(w, x)$ $(w, x \in X)$, f is continuous. Further, g is continuous: if $y \in Y \setminus \{p\}$, $r = d(y, p)$ and $u \in B(y, r)$, then

$$d(g(u), g(y)) = \begin{cases} d(u, y) & \text{if } y \in S^1, \\ 0 & \text{if } y \in T; \end{cases}$$

if $y = p$, then

$$d(g(u), g(y)) = \begin{cases} d(u, y) & \text{if } u \in S^1, \\ 0 & \text{if } u \in T. \end{cases}$$

Evidently $g \circ f = \mathrm{id}_X$. The map $f \circ g : Y \to Y$ is defined by

$$(f \circ g)(y) = \begin{cases} y & \text{if } y \in S^1, \\ p & \text{if } y \in T. \end{cases}$$

Consider $H : Y \times I \to Y$ defined by $H(y, t) = (1 - t)f(g(y)) + ty$. This map is continuous: $(y_n, t_n) \to (y, t) \implies H(y_n, t_n) \to H(y, t)$. Moreover, $H(y, 0) = f(g(y))$, $H(y, 1) = y$ $(y \in Y)$. Hence X and Y are homotopy-equivalent.

(ii) Referring to (i), f is continuous; the continuity of g follows from that of $y \longmapsto |y|^{-1} : Y \to \mathbf{R}$ and id_Y. Plainly $g \circ f \simeq id_X$. The map $f \circ g : Y \to Y$ is defined by $(f \circ g)(y) = |y|^{-1} y$. Let $H : Y \times I \to Y$ be defined by $H(y, t) = (1 - t)f(g(y)) + t\, id_Y(y) = (1 - t + t|y|)|y|^{-1} y$. Then H is continuous, $H(y, 0) = (f \circ g)(y)$ and $H(y, 1) = id_Y(y)$ $(y \in Y)$. Hence $f \circ g \simeq id_Y$: thus X and Y are homotopy-equivalent. They are not homeomorphic: $S^1 \setminus \{(-1, 0), (1, 0)\}$ is not connected.

8. Suppose that X is contractible. Let $a \in X$ and assume that $id_X \simeq c$, where $c : X \to X$ is the constant map $c(x) = a$ $(x \in X)$. Let $g : \{a\} \to X$ be defined by $g(a) = a$. Then $g \circ c = c$, $c \circ g = id_{\{a\}}$, $g \circ c \simeq id_X$ and $c \circ g \simeq id_{\{a\}}$. Thus X and $\{a\}$ are homotopy-equivalent. Conversely, let $a \in X$ and suppose that X and $\{a\}$ are homotopy-equivalent. There are continuous maps $f : X \to \{a\}$ and $g : \{a\} \to X$ such that $g \circ f \simeq id_X$ and $f \circ g \simeq id_{\{a\}}$. Since $g \circ f$ is the constant map $x \longmapsto g(a)$, X is contractible.

Let $K \subset \mathbf{R}^n$ be convex, $a \in K$ and suppose c is the constant map $c(x) = a$ $(x \in K)$. The map $H : K \times I \to K$ given by $H(x, t) = (1 - t)x + ta$ is continuous, $H(\cdot, 0) = id_K$ and $H(\cdot, 1) = c$. Hence K is contractible.

Exercise 3.1.24

1. If $\mathbf{C}$ were an ordered field, then either $i > 0$ or $i < 0$. Suppose $i > 0$. Then $-1 = i^2 > 0$, so that $0 = 1 + (-1) > 1$; but $1 = (-1)(-1) > 0$: contradiction. Proceed similarly if it is supposed that $i < 0$.

2. $|z_1| = |z_1 - z_2 + z_2| \le |z_1 - z_2| + |z_2|$, so that $|z_1| - |z_2| \le |z_1 - z_2|$. Now interchange z_1 and z_2.

3. First part: induction. If equality holds, then

$$|z_1| + \cdots + |z_n| = |(z_1 + z_2) + \cdots + z_n| \le |z_1 + z_2| + \cdots + |z_n|,$$

so that $|z_1| + |z_2| = |z_1 + z_2|$, which gives $z_1 \bar{z}_2 = |z_1| |z_2|$. As the numbering of terms is arbitrary, necessity follows. Sufficiency: clear if all z_j are zero. Suppose some $z_j \ne 0$. Then

$$|z_1 + z_2 + \cdots + z_n| = \frac{1}{|\bar{z}_j|} |z_1 \bar{z}_j + \cdots + z_n \bar{z}_j|$$

$$= \frac{1}{|z_j|} (|z_1| |z_j| + \cdots + |z_n| |z_j|).$$

4. If $\mathrm{im}(a\bar{b}) = 0$, then $a\bar{b} = \pm |ab|$ and so $a|b|^2 = \pm |ab|\, b$. For the converse, note that

$$ab = \begin{cases} \mu\lambda^{-1}|b|^2, & \lambda \neq 0, \\ \lambda\mu^{-1}|a|^2, & \mu \neq 0. \end{cases}$$

5. For all $\lambda \in \mathbf{C}$,

$$0 \leq \sum |a_k - \lambda b_k|^2 = A + |\lambda|^2 B - 2\,\mathrm{re}(\bar{\lambda} C).$$

If $B = 0$ the result is obvious; if $B \neq 0$, choose $\lambda = C/B$.

6. For all $z \in \mathbf{C}$,

$$|(1 - z)p(z)| \geq a_0 - \left\{ \sum_{k=1}^{n} (a_{k-1} - a_k)|z|^k + a_n |z|^{n+1} \right\};$$

by exercise 3 above, the inequality is strict if and only if $z \neq |z|$. If $z \neq |z|$ and $|z| \leq 1$, then

$$|(1 - z)p(z)| > a_0 - \left\{ \sum_{k=1}^{n} (a_{k-1} - a_k)|z|^k + a_n |z|^{n+1} \right\} \geq 0.$$

Plainly $p(z) \neq 0$ if $z = |z|$.

Exercise 3.2.12

1. For all rational r, $f(r) = rf(1)$ and $f(ir) = rf(i)$; since the rationals are dense in $\mathbf{R}$, $f(x) = xf(1)$ and $f(ix) = xf(i)$ for all real x. For all $z = x + iy \in \mathbf{C}$,

$$f(z) = xf(1) + yf(i) = \frac{1}{2}(z + \bar{z})f(1) + \frac{1}{2i}(z - \bar{z})f(i).$$

2. f is not continuous at 0: if $x \neq 0$, x real, then $f(x) = 1 \nrightarrow f(0)$ as $x \to 0$. Since $|g(z)| \leq |z|^2 \to 0$ as $z \to 0$, g is continuous at 0.

3. Since

$$\frac{f(r\cos\alpha + ir\sin\alpha)}{r} = \frac{r^4\cos^3\alpha\sin\alpha(\sin\alpha - i\cos\alpha)}{r^6\cos^6\alpha + r^2\sin^2\alpha},$$

this expression is zero if either $\cos\alpha$ or $\sin\alpha$ is zero, and otherwise has modulus bounded above by $r^4/(r^6\cos^6\alpha + r^2\sin^2\alpha) \to 0$ as $r \to 0$. If $z = x + imx^3$ (x, m real, $x \neq 0$), $f(z)/z = -im/(1 + m^2)$.

4. (i) $||z| - |a|| \leq |z - a| \to 0$ as $z \to a$: continuity follows. The function is not differentiable anywhere as the Cauchy-Riemann equations are not satisfied.
   (ii) As $\lim_{z\to 0}|z|^2/z = 0$ the function is differentiable at 0. Since the Cauchy-Riemann equations do not hold at any point other than 0, the function is differentiable only at 0.

5. Put $f = u + iv$. If u is constant, then $u_1 = v_2 = 0$ and $u_2 = -v_1 = 0$: thus $f' = 0$ in G and f is constant (Theorem 3.2.11). The same holds when v is constant. If $u^2 + v^2$ is constant, then $uu_1 + vv_1 = 0$ and $uu_2 + vv_2 = 0$; use of

the Cauchy-Riemann equations gives $(u^2 + v^2)u_1 = 0$ and $(u^2 + v^2)u_2 = 0$. If $u^2 + v^2 = 0$ the result is clear; if $u^2 + v^2 \neq 0$, use the first part.

6. Since $u_1(x, y) = e^x \cos y + y$ and $u_2(x, y) = -e^x \sin y + x$, use of the Cauchy-Riemann equations leads us to consider v of the forms $v(x, y) = e^x \sin y + \frac{1}{2}y^2 + g(x)$ and $v(x, y) = e^x \sin y - \frac{1}{2}x^2 + h(y)$: thus a desired f is given by $f(z) = e^z - iz^2/2$.

7. Take $f_0(z) = z^{-2}e^{-iz}$. Let $f \in \mathscr{F}$ and define $g = f/f_0 : g \in H(G)$ and $|g(z)| = 1$ in G. Write $g = u + iv$: then $u^2 + v^2 = 1$ and by question 5 above, g is constant, $g = \lambda, |\lambda| = 1$.

Exercise 3.3.14

1. Put $b = \lim \sup b_n$ and suppose that b is finite. There is a subsequence (b_{n_k}) that converges to b; hence $a_{n_k}b_{n_k} \to ab$ and so $ab \leq \lim \sup a_n b_n$. We may suppose that for all n, $a_n \neq 0$. Apply what has been proved with a_n replaced by $1/a_n$ and b_n replaced by $a_n b_n$: thus $\lim \sup b_n = \lim \sup(1/a_n)a_n b_n \geq (1/a) \lim \sup a_n b_n$, and the result follows. The case $b = \infty$ is simpler.

2. $1, 0, \infty, 1$.

3. (i) $\lim \sup |n^3 a_n|^{1/n} = \lim \sup |a_n|^{1/n}$: radius of convergence is R;
   (ii) $R^{1/3}$;   (iii) R^3.

4. For all $n \in \mathbf{N}$, $\sum_{r=0}^{n} |c_r| \leq (\sum_{r=0}^{n} |a_r|) (\sum_{r=0}^{n} |b_r|) \leq (\sum_{r=0}^{\infty} |a_r|) (\sum_{r=0}^{\infty} |b_r|)$; $(\sum_{r=0}^{n} |c_r|)$ is monotonic increasing and bounded above; $\sum c_r$ is absolutely convergent. Put $A_n = \sum_{r=0}^{n} a_r$, $B_n = \sum_{r=0}^{n} b_r$, $C_n = \sum_{r=0}^{n} c_r$, $E_n = \sum_{r=0}^{n} |a_r|$, $F_n = \sum_{r=0}^{n} |b_r|$. Then

$$A_n \to A, \quad B_n \to B, \quad |C_{2n} - AB| \leq |C_{2n} - A_n B_n| + |A_n B_n - AB|$$
$$\leq |E_{2n} F_{2n} - E_n F_n| + |A_n B_n - AB| \to 0.$$

Since (C_n) is convergent and $C_{2n} \to AB$, we see that $C_n \to AB$.

5. $\sum a_n z^n$ is uniformly convergent on the closed unit disc D and is therefore continuous on it, by the Weierstrass theorem. Suppose $z, \zeta \in D, z \neq \zeta$. Then

$$\frac{f(z) - f(\zeta)}{z - \zeta} = 1 + \sum_{n=2}^{\infty} a_n \left(z^{n-1} + z^{n-2}\zeta + \ldots + \zeta^{n-1}\right),$$

and so

$$\left| \frac{f(z) - f(\zeta)}{z - \zeta} \right| \geq 1 - \sum_{n=2}^{\infty} n a_n > 0.$$

Exercise 3.4.36

1. Application of the ratio test shows that the radius of convergence is 1. If $|z| < 1$,

$$(1 + z)f'(z) = \alpha + \sum_{n=1}^{\infty} \left\{ \frac{\alpha(\alpha - 1) \ldots (\alpha - n)}{n!} + \frac{\alpha(\alpha - 1) \ldots (\alpha - n + 1)}{(n - 1)!} \right\} z^n$$
$$= \alpha f(z).$$

Put $D = \mathbf{C}\backslash\{z : \operatorname{re} z \le -1, \operatorname{im} z = 0\}$. Then $z \longmapsto (1+z)^{-\alpha} \in H(D)$ and has derivative $z \longmapsto -\alpha(1+z)^{-1-\alpha}$. For $|z| < 1$, $\phi'(z) = (1+z)^{-\alpha} f'(z) - \alpha(1+z)^{-1-\alpha} f(z) = 0$. Hence ϕ is constant; since $\phi(0) = f(0) = 1$, the result follows.

3. $\arg(-1-i) = -3\pi/4$, $\quad \log(-1-i) = \log\sqrt{2} - 3\pi i/4$,

$$(-1-i)^i = E\,(i\log(-1-i)) = E(3\pi/4)E(i\log\sqrt{2})$$
$$= E(3\pi/4)\left\{\cos(\log\sqrt{2}) + i\sin(\log\sqrt{2})\right\}.$$

$(3\pi/2 - \arg)(-1-i) = 5\pi/4$, $(3\pi/2 - \log)(-1-i) = \log\sqrt{2} + 5\pi i/4$,

$$\left((-1-i)^i\right)_{3\pi/2} = E\,(i(3\pi/2 - \log)(-1-i))$$
$$= E(-5\pi/4)\left\{\cos(\log\sqrt{2}) + i\sin(\log\sqrt{2})\right\}.$$

4. (i) $\{m + n\alpha : m, n \in \mathbf{Z}\} = \bigcup_{n\in\mathbf{Z}}\{m + n\alpha : m \in \mathbf{Z}\}$: countability follows. For density, enough to prove that given $\theta \in \mathbf{R}$, $\varepsilon > 0$, there exist $m, n \in \mathbf{Z}$ such that

$$|m + n\alpha - \theta| < \varepsilon. \tag{1}$$

Let $r \in \mathbf{N}, r > \varepsilon^{-1}$. By the Archimedean order property (Theorem A.4.11) and the well-ordering principle (Corollary A.4.10), for each $j \in \mathbf{N}$, there is a unique $s_j \in \mathbf{Z}$ such that $0 \le s_j + j\alpha < 1$. Among the $(r+1)$ numbers $s_k + k\alpha$ $(k = 1, \dots, r+1)$ there are at least two, indexed by k_1, k_2, such that

$$s_{k_1} + k_1\alpha < s_{k_2} + k_2\alpha, 0 < s_{k_2} - s_{k_1} + (k_2 - k_1)\alpha < r^{-1} < \varepsilon.$$

Let $p = s_{k_2} - s_{k_1}, q = k_2 - k_1, \lambda = p + q\alpha > 0$: then $0 < \lambda < \varepsilon$. Arguing as before, there is a least $l \in \mathbf{Z}$ such that $\lambda^{-1}(\theta - \varepsilon) < l$. Thus

$$\lambda^{-1}(\theta - \varepsilon) < l < \lambda^{-1}(\theta - \varepsilon) + 2 < \lambda^{-1}(\theta + \varepsilon).$$

Hence $|l\lambda - \theta| < \varepsilon$. Now take $m = lp, n = lq$ to obtain (1).

(ii) Let $w \in \mathbf{C}$, $|w| = 1$, $w = e^{i\theta}$. By (i), there are sequences $(l_n), (k_n) \subset \mathbf{Z}$ such that $l_n + k_n\alpha \to \theta/(2\pi)$. Let $w_n = \exp\left(2\pi\,(l_n + k_n\alpha)\,i\right) = \exp\left(2\pi k_n\alpha i\right)$; $w_n \to e^{i\theta} = w$ and each w_n belongs to the αth power of 1.

5. (i) Let $A_n = \{z \in \mathbf{C} : z^n = 1\}$, $A = \bigcup_1^\infty A_n$: A is countable but the unit circle is not. Alternatively, note that $\alpha \in \mathbf{R}\backslash\mathbf{Q}$, $\lambda = \exp(i\pi\alpha) \Longrightarrow$ for all $n \in \mathbf{N}$, $\lambda^n \ne 1$.

(ii) (a) Since $\sum a_n z^n$ is absolutely convergent when $|z| < R$,

$$A_n(z) = \frac{1}{n+1} \sum_{k=0}^{n} \sum_{p=0}^{\infty} a_p \lambda^{kp} z^p = \frac{1}{n+1} \sum_{p=0}^{\infty} \left\{ \sum_{k=0}^{n} \lambda^{kp} \right\} a_p z^p$$

$$= a_0 + \frac{1}{n+1} \sum_{p=1}^{\infty} \left(\frac{1 - \lambda^{(n+1)p}}{1 - \lambda^p} \right) a_p z^p \quad (|z| < R, n \in \mathbf{N_0}).$$

Given $\varepsilon > 0$, fix z $(0 < |z| < R)$, choose q such that $\sum_{p=q+1}^{\infty} |a_p z^p| < \varepsilon/2$: for all large enough n,

$$|A_n(z) - a_0| \leq \frac{1}{n+1} \sum_{p=1}^{q} \left| \frac{1 - \lambda^{(n+1)p}}{1 - \lambda^p} \right| |a_p z^p| + \sum_{p=q+1}^{\infty} |a_p z^p| < \varepsilon.$$

(b) If $|z| = r$, $|f(z)| \leq M(r)$, $|A_n(z)| \leq M(r)$, $\lim |A_n(z)| = |a_0| \leq M(r)$.
(iii) Proceed as in (ii), working with g, where $g(z) = f(z)/z^j$.
(iv) $|a_j| \leq Mr^{-j}$ $(r > 0, j \in \mathbf{N})$. Let $r \to \infty$.
7. Let $\theta : [a, b] \to \mathbf{R}$ be a continuous argument of $\gamma : 2\pi n(\gamma, 0) = \theta(b) - \theta(a)$. The map $\theta \circ \psi : [c, d] \to \mathbf{R}$ is a continuous argument of $\gamma \circ \psi$ and $2\pi n (\gamma \circ \psi, 0) = \theta(\psi(d)) - \theta(\psi(c)) = \theta(b)) - \theta(a)$.

Exercise 3.5.14

1. Values of integrals: $i, 2i, 2i$.
2. (i) $\int_\nu \mathrm{re}\, z dz = \int_0^\pi a \cos s(-a \sin s + ib \cos s)ds = \pi i ab/2$.
   (ii) $\int_\nu \bar{z} dz = \int_0^\pi (a \cos s - ib \sin s)(-a \sin s + ib \cos s)ds = \pi i ab$.
   The integrals involving μ are handled in a similar way.

Exercise 3.6.37

1. (i) $\int_\mu \cos z dz = \sin(-\pi/2) - \sin(\pi/2) = -2$.
   (ii) $v'(t) = (3t^2 + 2\pi t i) \exp(-2\pi i/t)$ $(t \neq 0)$, $v'(0) = 0 : v'$ is continuous on $[0, 1]$ and so v is a contour. $\int_\nu z^2 e^z dz = e - 2$.
2. γ is continuous, $\lim_{t \to 0} \gamma(t) = \lim_{t \to 1} \gamma(t) = 0$: γ is a closed path. γ' exists and is continuous in $[0, 1/2]$, $[1/2, 1]$: γ is a circuit.
   Both $z \longmapsto \cos^3(z^2)$ and $z \longmapsto \log(1 + z)$ are analytic in the unit disc: by Cauchy's theorem in a convex set, both integrals are 0.
4. $z^2 - 2az + 1 = (z - \alpha)(z - \beta)$, $\alpha = a + \sqrt{a^2 - 1}$, $\beta = a - \sqrt{a^2 - 1}$; $(z - \alpha)^{-1}$ is analytic in the disc $|z| < \alpha$; by Cauchy's integral formula,

$$\mathrm{ind}_\gamma(\beta) \cdot \frac{1}{\beta - \alpha} = \frac{1}{2\pi i} \int_\gamma \frac{1}{(z - \alpha)(z - \beta)} dz;$$

by Lemma 3.6.11, $\mathrm{ind}_\gamma(\beta) = 1$. Thus $\int_\gamma \frac{1}{z^2 - 2az + 1} dz = -\pi i/\sqrt{a^2 - 1}$. Now observe that $\int_\gamma \frac{1}{z^2 - 2az + 1} dz = -\frac{i}{2} \int_0^{2\pi} \frac{dt}{a - \cos t}$.

7. Let V be a convex open subset of G. By Theorem 3.6.6, there exists $F \in H(V)$ such that $F'(z) = f(z)$ $(z \in V)$; thus f is analytic on V, and as this holds for all such V, $f \in H(G)$.

10. Radii of convergence are (a) 1, (b) $\pi/2$.

11. 2, π, 4.

12. Radii of convergence are (a) 1, (b) $1 - e^{-1}$.

13. Suppose $f \neq 0$. There exist $a \in G$ and $r > 0$ such that $f(z) \neq 0$ if $z \in B(a, r)$. Hence $g(z) = 0$ for all $z \in B(a, r)$: by Theorem 3.6.27, $g = 0$.

15. To obtain a contradiction, suppose $w \notin S$ and that $(w - z_0)/(w - z_1) \notin D(\pi)$. For some $r > 0$, $w - z_0 = re^{i\pi}(w - z_1)$, so that $w = (1 - t')z_0 + t'z_1 \in S$, where $t' = r(1+r)^{-1}$, a contradiction. Since $w \longmapsto \log\left(\frac{w-z_1}{w-z_0}\right)$ is a primitive of $w \longmapsto (w-z_1)^{-1} - (w-z_0)^{-1}$ on $\mathbf{C} \backslash S$, by Theorem 3.6.2, $\mathrm{ind}_\gamma(z_1) = \mathrm{ind}_\gamma(z_0)$. It follows that $\mathrm{ind}_\gamma(\cdot)$ is constant on each component of $\mathbf{C} \backslash \gamma^*$, each being open, connected and polygonally connected (Theorem 2.4.23). Finally, suppose that $\sup\{|w| : w \in \gamma^*\} < r < |z|$, where z lies in the unbounded component of $\mathbf{C} \backslash \gamma^*$. Since $w \longmapsto (w - z)^{-1}$ is analytic on the convex set $B(0, r)$, by Theorem 3.6.9,

$$\mathrm{ind}_\gamma(z) = \frac{1}{2\pi i} \int_\gamma (w - z)^{-1} dw = 0.$$

Exercise 3.7.25

1. (a) Pole of order 1 (use Lemma 3.7.7), (b) removable singularity (use Lemma 3.7.6), (c) essential singularity (use a contradiction argument).

2. (a) $-(1 + i/\sqrt{3})/4$, (b) $-1/3$, (c) $(2ei)^{-1}$, (d) $1/4!$.

3. (a) $(1 + z^2)^{-2} = \sum_{n=0}^{\infty}(-1)^n(n + 1)z^{2n}$.
   (b) $(1 + z^2)^{-2} = z^{-4}\sum_{n=0}^{\infty}(-1)^n(n + 1)z^{-2n} = \sum_{n=2}^{\infty}(-1)^n(n - 1)z^{-2n}$.
   (c)

$$(1 + z^2)^{-2} = -\frac{1}{4(z - i)^2}\left(1 + \frac{z - i}{2i}\right)^{-2}$$

$$= -\frac{1}{4(z - i)^2} + \frac{1}{4i(z - i)} + \frac{1}{16}\sum_{p=0}^{\infty}(-1)^p(p + 3)\left(\frac{z - i}{2i}\right)^p.$$

f has a pole of order 2 at i and $\mathrm{res}(f; i) = \frac{1}{4i}$.
Use the standard semicircular contour γ: by the residue theorem,

$$2\pi i \cdot \frac{1}{4i} = \int_\gamma (1 + z^2)^{-2} dz$$

and so

$$\left| \frac{\pi}{4} - \int_0^R (1 + t^2)^{-2} dt \right| \leq \frac{R}{2}\left| \int_0^\pi \frac{\exp(i\theta)}{(1 + R^2\exp(2i\theta))^2} d\theta \right| \leq \pi R(R^2 - 1)^{-2}/2.$$

Thus $\int_0^\infty (1+t^2)^{-2} dt = \pi/4$.

4. Integrate appropriate functions over the unit circle.

9. The function $z \longmapsto z/(1+z^5)$ has simple poles at $z_n = \exp((2n+1)i\pi/5)$ with corresponding residues $-z_n^2/5$ $(n = 0, \pm 1, \pm 2)$.

17. The functions f and g given by $f(z) = \alpha z^n$, $g(z) = \alpha z^n - \exp z$ are entire. Since f has n zeros in the open unit disc and $|f(z) - g(z)| < |f(z)|$ if $|z| = 1$, the result follows from Rouché's theorem.

19. Consider the function g defined by

$$g(z) = \{\pi z \csc (\pi z) - 1\}/z^2 \ (z \neq 0), \ g(0) = \pi^2/6,$$

observe that g is meromorphic in $\mathbf{C}$ with simple poles at $\pm n (n \in \mathbf{N})$, proceed as in 3.7.1 to show that $g(z) = \sum_{k=1}^{\infty} \frac{(-1)^k}{z^2 - k^2}$, then put $z = 0$.

References

1. Arzelà, C.: Sulla integrazione per serie. Atti Acc. Lincei Rend. Roma 1(4), 523–537, 596–599 (1885)
2. Bourbaki, N.: Topologie Générale, Chap. 9 (Utilisation des nombres réels en topologie générale), Hermann, Paris (1958)
3. Burckel, R.B.: An Introduction to Classical Complex Analysis. Birkhäuser, Basel-Stuttgart (1979)
4. Davis, P.J.: Interpolation and Approximation, Blaisdell, New York (1962)
5. Dieudonné, J.: Foundations of Modern Analysis. Academic Press, New York-San Francisco-London (1969)
6. Dixon, J.D.: A brief proof of Cauchy's integral theorem. Proc. Amer. Math. Soc. 29, 625–626 (1971)
7. Gelbaum, B.R., Olmsted, J.M.H.: Counterexamples in Analysis. Holden-Day, San Francisco (1964)
8. Goffman, C.: Real Functions. Prindle, Weber and Schmidt, Boston (1953)
9. Jahnke, H.N. (ed.): A History of Analysis, History of Mathematics, vol. 24. American Mathematical Society, Providence, R. I. (2003)
10. Jech, T.J.: The Axiom of Choice. Dover, New York (2008)
11. Luxemburg, W.A.J.: Arzelà's dominated convergence theorem for the Riemann integral. Amer. Math. Monthly 78, 670–679 (1971)
12. Rana, I.K.: An Introduction to Measure and Integration, Graduate Studies in Mathematics, vol. 45, 2nd edn. American Mathematical Society, Providence, R.I. (2002)
13. Remmert, R.: Classical Topics in Complex Function Theory. Springer-Verlag, Berlin-Heidelberg-New York (1998)
14. Rotman, B., Kneebone, G.T.: Theory of Sets and Transfinite Numbers. Max Parrish, London (1966)
15. Rudin, W.: Real and Complex Analysis. McGraw-Hill, New York (1966)
16. Saks, S., Zygmund, A.: Analytic Functions, 2nd edn. Polish Scientific Publishers, Warsaw (1965)
17. Stromberg, K.R.: Introduction to Classical Real Analysis. Belmont, Wadsworth (1981)
18. Volterra, V.: Sui principii del calcolo integrale. Giornale Mat. 19, 333–372 (1881)
19. Wilder, R.L.: Introduction to the Foundations of Mathematics. Wiley, New York (1965)

R. H. Dyer and D. E. Edmunds, *From Real to Complex Analysis*,
Springer Undergraduate Mathematics Series, DOI: 10.1007/978-3-319-06209-9,
© Springer International Publishing Switzerland 2014

Index